Lecture Notes in Mathematics 2111

For further volumes:
http://www.springer.com/series/304

FONDAZIONE CIME
ROBERTO CONTI
CENTRO INTERNAZIONALE MATEMATICO ESTIVO
INTERNATIONAL MATHEMATICAL SUMMER CENTER

Fondazione C.I.M.E., Firenze

C.I.M.E. stands for *Centro Internazionale Matematico Estivo*, that is, International Mathematical Summer Centre. Conceived in the early fifties, it was born in 1954 in Florence, Italy, and welcomed by the world mathematical community: it continues successfully, year for year, to this day.

Many mathematicians from all over the world have been involved in a way or another in C.I.M.E.'s activities over the years. The main purpose and mode of functioning of the Centre may be summarised as follows: every year, during the summer, sessions on different themes from pure and applied mathematics are offered by application to mathematicians from all countries. A Session is generally based on three or four main courses given by specialists of international renown, plus a certain number of seminars, and is held in an attractive rural location in Italy.

The aim of a C.I.M.E. session is to bring to the attention of younger researchers the origins, development, and perspectives of some very active branch of mathematical research. The topics of the courses are generally of international resonance. The full immersion atmosphere of the courses and the daily exchange among participants are thus an initiation to international collaboration in mathematical research.

C.I.M.E. Director
Pietro ZECCA
Dipartimento di Energetica "S. Stecco"
Università di Firenze
Via S. Marta, 3
50139 Florence
Italy
e-mail: zecca@unifi.it

C.I.M.E. Secretary
Elvira MASCOLO
Dipartimento di Matematica "U. Dini"
Università di Firenze
viale G.B. Morgagni 67/A
50134 Florence
Italy
e-mail: mascolo@math.unifi.it

For more information see CIME's homepage: http://www.cime.unifi.it

Lou van den Dries • Jochen Koenigsmann •
H. Dugald Macpherson • Anand Pillay •
Carlo Toffalori • Alex J. Wilkie

Model Theory in Algebra, Analysis and Arithmetic

Cetraro, Italy 2012

Editors: H. Dugald Macpherson,
Carlo Toffalori

Springer

FONDAZIONE
CIME
ROBERTO CONTI
CENTRO INTERNAZIONALE MATEMATICO ESTIVO
INTERNATIONAL MATHEMATICAL SUMMER CENTER

Lou van den Dries
Department of Mathematics
University of Illinois
Urbana, IL, USA

Jochen Koenigsmann
Mathematical Institute
University of Oxford
Oxford, United Kingdom

H. Dugald Macpherson
University of Leeds School of Mathematics
Leeds, United Kingdom

Anand Pillay
Department of Mathematics
University of Notre Dame
Hurley, IN, USA

Carlo Toffalori
Division of Mathematics
University of Camerino School of Science and Technology
Camerino, Italy

Alex J. Wilkie
University of Manchester School of Mathematics
Manchester, United Kingdom

ISBN 978-3-642-54935-9 ISBN 978-3-642-54936-6 (eBook)
DOI 10.1007/978-3-642-54936-6
Springer Heidelberg New York Dordrecht London

Lecture Notes in Mathematics ISSN print edition: 0075-8434
ISSN electronic edition: 1617-9692

Library of Congress Control Number: 2014945978

Mathematics Subject Classification (2010): 03CXX; 03C45; 03C10; 03C60; 03C64; 12J20; 11U05; 11U09

Printed on acid-free paper

Springer is part of Springer Science+Business Media (www.springer.com)

Acknowledgements

CIME activity is carried out with the collaboration and financial support of the following Italian institutions: INdAM (Istituto Nazionale di Alta Matematica, National Institute of High Mathematics), MIUR (Ministero dell'Istruzione, dell'Università e della Ricerca, Ministry of Education, University and Research), and Ente Cassa di Risparmio di Firenze. This CIME Course is also sponsored and financially supported by the Italian research project FIR "New trends in the model theory of exponentiation".

Contents

Model Theory in Algebra, Analysis and Arithmetic: A Preface

Dugald Macpherson and Carlo Toffalori

1 Meeting Model Theory

Model theory is a branch of mathematical logic dealing with mathematical structures (models) from the point of view of first order logical definability. Although comparatively young, it is now well established, its major textbooks including [6, 17, 34, 43, 53]. A typical goal of model theory is to build, study and classify mathematical universes in which some given axioms (usually expressed in a first order way) are satisfied. Thus model theory has remote roots in the birth of non-Euclidean geometries and in the effort to realize them in suitable mathematical settings where their revolutionary (non-standard) assumptions are obeyed. Some major achievements of mathematical logic in the first half of the twentieth century, such as the Gödel Compactness Theorem around 1930, underpin modern model theory.

It seems that the name "model theory", more precisely "theory of models", was explicitly used for the first time by Alfred Tarski in 1954 in the papers [51, 52]. In the 1950s, thanks to the work of Tarski himself, Abraham Robinson, Vaught, and others, model theory started to emerge as a new independent topic in mathematical logic.

Today, model theory has a sophisticated and currently evolving internal abstract theory. It has also deepening and strengthening interactions with, for example, parts of algebra, combinatorics, geometry, number theory and analysis, often leading to striking applications—this relationship with other branches of mathematics dates

D. Macpherson (✉)
School of Mathematics, University of Leeds, Leeds LS2 9JT, UK
e-mail: H.D.Macpherson@leeds.ac.uk

C. Toffalori
School of Science and Technology, Division of Mathematics, University of Camerino, 62032 Camerino, Italy
e-mail: carlo.toffalori@unicam.it

L. van den Dries et al., *Model Theory in Algebra, Analysis and Arithmetic*, Lecture Notes in Mathematics 2111, DOI 10.1007/978-3-642-54936-6_1,

back to Tarski himself and the very beginning of model theory. These two features are often labelled ‘pure model theory’ and ‘applied model theory’. In the 1970s and early 1980s they were sometimes pursued rather independently, but the pure and applied aspects are now closely intertwined, feeding off and stimulating each other.

Central parts of the internal theory (e.g. types, saturation and homogeneity, ultraproducts, categoricity, along with parts of infinitary logic) arose in the 1950s and 1960s. The subject developed its modern flavour partly through the work of Morley [38] and then Shelah [48] and others, begun in the 1960s, which looks for abstract notions of classification, independence and dimension. Crucial concepts were singled out, such as stability, simplicity, o-minimality (see [42, 55, 56], for instance) and the NIP property, and others such as the NTP2 property and generic stability are just beginning to be developed. Stability theory is at the core.

Interactions of model theory with other parts of mathematics can be traced back to Tarski’s quantifier elimination for the complex and real field in the 1930s (see [54]). Connections are described in many sources, such as [7, 8, 15, 31, 32]. Among applications of model theory, we mention a few examples: o-minimality (and its close connection with real analysis and real analytic and algebraic geometry); Hrushovski’s work in Diophantine geometry [19], which yields new approaches to the Mordell-Lang and Manin-Mumford Conjectures with new results [5]; the recent work of Pila and Wilkie [41] on rational heights, together with further work of Pila [40] and others, which makes striking links between o-minimality and diophantine geometry; motivic integration as developed by Denef and Loeser and subsequently Cluckers, Hrushovski, and Kazhdan (see [29] for a general introduction and extended bibliographical references); and the rich ideas of Zilber around the logic of complex exponentiation [59] and Zariski geometries [60].

The present volume consists of four survey articles on contemporary themes in model theory. It is based on notes from lecture courses by the same authors at the meeting *Model Theory in Algebra, Analysis and Arithmetic* organized by CIME (Centro Internazionale Matematico Estivo), July 2–6, 2012, directed by the editors of the volume. The meeting continued the spirit of a much earlier CIME course *Model Theory and Applications* in 1975. The latter was coordinated by Piero Mangani,[1] with speakers H.J. Keisler, G. Sabbagh, G.E. Sacks, J.A. Makowski, M. Servi, C. Wood and A. Macintyre. Earlier, in 1968, a CIME course directed by Ettore Casari dealt with *Aspects of Mathematical Logic* and in particular with model theory, its speakers including A. Mostowski and A. Robinson himself.

The CIME 2012 course was intended particularly for young researchers (e.g. Ph.D. students and postdocs), both model theorists, and those from outside intrigued by recent applications and looking for an introduction to current model theory. It also had a very welcome participation of more senior researchers in other branches of mathematics, which encouraged cross-fertilization and was very beneficial for younger researchers. The course complemented other recent activities in Europe

---

[1]Piero Mangani died on April 4, 2013; the second author would like to devote this work to his memory.

in model theory and wider parts of logic. For example, between 2005 and 2008 a Marie Curie FP6 Research Training Network called MODNET greatly enhanced joint training and collaborations among the model theorists of Europe. The site

http://www.logique.jussieu.fr/modnet

witnesses what was done in this programme (conferences, research workshops, summer schools, publications) and hosts a current preprint server and worldwide list of meetings related to model theory. Many beginning model theorists were funded by MODNET, or by two other Marie Curie networks which trained PhD students in wider parts of logic: the FP6 Early Stage Training Network MATHLOGAPS (2004–2008), and the FP7 Initial Training Network MALOA (2009–2013).

2 The Present Volume

The CIME course focussed on both the purer and the more applied parts of model theory, with emphasis on the latter. The purpose was to present some deep and currently active topics in model theory and its applications to other branches of mathematics (such as algebraic geometry, diophantine geometry, real and complex analysis, transcendental number theory).

We briefly discuss below the wider context of the four main courses at the CIME meeting, and of the corresponding articles in this volume.

(1) *Some themes around first order theories without the independence property* (Anand Pillay, University of Leeds, now University of Notre Dame).

The course/article is an exposition of part of the internal structure theory in model theory, but is stimulated by recent connections to other parts of mathematics, such as ordered and valued fields, and the combinatorics and statistics of Vapnik-Chervonenkis classes. In Shelah's classification program (see e.g. [42,48]), stability was singled out as a first approximation of a dividing line between classifiable and unclassifiable first order theories. Shelah himself showed that stability is equivalent to the failure of two combinatorially-defined properties, the strict order property and the independence property. He developed a rich structure theory (for example, notions of independence) in the context of stability, and a major theme of recent model theory, perhaps *the* major theme, has been to extend parts of this to unstable contexts, where there is a great wealth of model-theoretically tractable and mathematically important examples. For example, the ordered field of reals is unstable (it has the strict order property) but does not have the independence property; likewise the field of p-adic numbers. On the other side, the random graph has the independence property but not the strict order property; likewise any non-principal ultraproduct of finite fields.

There has been considerable work extending the achievements of stability (in particular, the theory of forking) to more general classes of theories, such as o-minimal theories [55] (including the ordered field of reals), or simple theories [56] (including the random graph, and pseudofinite fields). In the last few years a rich fragment of stability theory has been developed for the so-called *NIP* theories (theories without the independence property, also called *dependent* theories). This emerged in work of Shelah [49], Adler, and then of Hrushovski, Peterzil, Pillay, Simon [23, 25, 26] who, over the last 4 years, worked out stability-theoretic ideas for NIP theories, revitalised a concept of measure due to Keisler, and applied the theory to prove some conjectures of Pillay about the relationship between groups definable in o-minimal structures and Lie groups [4]. Many others are now working in the field, and we mention three further developments.

First, Laskowski [28] observed a connection to 'Vapnik-Chervonenkis classes', which are important in statistical learning theory. Given a complete theory T, a formula $\phi(\bar{x}, \bar{y})$ does not have the independence property (or is 'NIP' or 'dependent') precisely if, in any model M of T, the family of uniformly definable subsets of $M^{l(\bar{y})}$ determined by ϕ is a Vapnik-Chervonenkis class. As noted in Pillay's paper in this volume, the Vapnik-Chervonenkis Theorem (a strengthening of the strong law of large numbers) has applications for measures in NIP theories. The Vapnik-Chervonenkis theory involves other invariants, such as VC *density*. This has connections to combinatorial geometry, and has been calculated in certain NIP theories [1]. This work is closely related to the UDTFS ('Uniform definability of types over finite sets') property of Laskowski. Pillay's article includes a sketch of the beautiful theorem of Chernikov and Simon, which uses a deep result of Matousek in combinatorial geometry to show that UDTFS is equivalent to NIP. This generalizes the fundamental fact that stability of a complete theory is equivalent to all types over all sets being definable.

Second, Shelah's philosophy to identify key tame/wild dividing lines in the class of first order theories (such as stable/unstable) has been rekindled. There are new results on other generalizations of stability, some identified by Shelah in the 1970s or early 1980s. One such is the NTP2 property, a common generalization of the NIP property and simplicity. For instance, Chernikov and Kaplan showed that working over a model in an NTP2 theory, forking is the same as dividing, and there are initial results on groups and fields in NTP2 theories. As an example, a non-principal ultraproduct (over primes) of p-adic fields has NTP2 theory [9]; it is not simple, due to the presence of the value group, or NIP, due to the presence of the residue field (which is pseudofinite, so simple but not NIP).

Third, stability is naturally a property of formulas (a theory is stable if and only if all formulas are stable). It is natural to investigate, in a possibly unstable theory, parts of a theory (formulas or types) which are stable, or at least have a stable trace. This led to the notion of *stably dominated type* in [16], of which the basic example is the generic type of the valuation ring in an algebraically closed valued field. Pillay's paper discusses the more general notion of *generically stable type*. This was introduced in [23], and relates also to the notion of compact domination from [25]. Generic stability makes sense also for measures [25, 26]. The theory develops particularly smoothly in the NIP context.

(2) *Lectures on the model theory of real and complex exponentiation* (Alex Wilkie, University of Manchester).

Topic 2 dates back to a problem posed by Tarski, concerning the decidability of the first order theory of the expansion $\mathbb{R}_{exp}$ of the ordered field of reals by the exponential function $x \to e^x$. A celebrated theorem of Wilkie in the 1990s [57] showed that $\mathbb{R}_{exp}$ is model complete and hence o-minimal, so inherits the rich model-theoretic properties of such structures (such as NIP). Wilkie's article, partly through a series of exercises, sketches the proof of this theorem.

Using Wilkie's model completeness of real exponentiation, Tarski's problem was positively answered by Macintyre and Wilkie [33], although only modulo Schanuel's Conjecture from transcendental number theory. Wilkie's work had rapid applications [47], but led also to a series of papers by van den Dries, Macintyre, and Marker on logarithmic-exponential power series, underpinning current work by Aschenbrenner, van den Dries and van der Hoeven on the field of transseries. It was generalized further by Wilkie's proof [57] that any Pfaffian expansion of the real field is o-minimal. Currently, stemming from the Pila-Wilkie work mentioned earlier, there is strong interest in quantifying, in terms of rational height, the number of rational solutions of systems of exponential-polynomial equations over the reals.

Complex exponentiation is even harder to approach, as ($\mathbb{C}$, *exp*) interprets the ring of integers and in this way inherits the Gödel incompleteness phenomena. On the other hand, the problem of understanding the complex exponential also involves ideas from pure model theory, which we now elaborate.

There is a central conjecture of Zilber on those first order structures [58] which are "strongly minimal"; these are structures in which, in all models of the theory, every definable subset of the domain is finite or cofinite. In a way prescribed by work of Baldwin and Lachlan [3] and developed by Zilber [58] they are the building blocks of uncountably categorical structures. Zilber's conjecture asserts that they all behave like three central examples: pure sets, vector spaces, and algebraically closed fields. However, Hrushovski [18] disproved this and built new, seemingly pathological, strongly minimal structures via partly classical amalgamation techniques; furthermore, he developed an axiomatic treatment of dimension functions, leading to a series of applications. Zilber's conjecture was shown valid [24] in the more restricted framework of "Zariski structures", i.e. strongly minimal structures equipped with a topology generalizing the Zariski topology on algebraic varieties. This last development was crucial in Hrushovski's application of model theory to the Mordell-Lang and Manin-Mumford Conjectures.

Zilber re-interpreted Hrushovski's counterexamples in a very natural way. In particular, Zilber discovered several dimension functions occurring "in nature". One such is easily described and just refers to complex exponentiation [59]: for X a finite subset of the complex field, let $l(X)$ denote the dimension of the vector space (over the rationals) which X spans, and let $et(X)$ (the exponential transcendence degree of X) denote the transcendence degree (again over the rationals) of the field generated by X and the exponentials of elements of X. Then the Zilber dimension of X, $Zd(X)$, is defined to be the integer $et(X) - l(X)$. The crucial point is whether this function is a dimension in the Hrushovski sense, in particular whether it is non-negative. Again this statement is closely related to Schanuel's conjecture in transcendental number theory. Zilber observed that the required properties of $Zd(X)$ are algebraically consistent (i.e. ignoring continuity and convergence properties of the actual exponential function) and was able to apply Hrushovski's method in order to build a surjective monomorphism $exp : (\mathbb{C}, +) \to (\mathbb{C} - \{0\}, \cdot)$ having a kernel isomorphic to $(\mathbb{Z}, +)$, and satisfying the Schanuel condition. Furthermore this "pseudoexponentiation" *exp* has other key model-theoretic properties, and is uniquely characterized by them. Zilber conjectures that the actual exponential function has these same properties. The topic is one of the most attractive in current model theory [2, 60].

This work is closely related to another conjecture of Zilber on complex exponentiation, generalizing the strong minimality of algebraically closed fields: he conjectures that the complex exponential field is *quasi-minimal*, that is, every definable subset of $\mathbb{C}$ is countable or has countable complement. The last section of Wilkie's article is motivated by this conjecture.

(3) *Model theory of valued fields* (Lou van den Dries, University of Illinois at Urbana-Champaign).

Valued fields are one of the main objects of classical applied model theory, emerging already in A. Robinson's 1956 proof of model completeness for algebraically closed valued fields [46]. Work by Ax, Kochen and independently Ershov led to the asymptotic solution of Artin's Conjecture on p-adic numbers, saying that every homogeneous polynomial with degree d and $n > d^2$ unknowns over the field $\mathbb{Q}_p$ has a non trivial zero in the field, provided p is large enough (see a discussion in [31]). Valued fields are not stable. The work of Ax-Kochen/Ershov gave methods, at least in residue characteristic 0, to reduce the model theory of a henselian valued field to that of its residue field and value group. This led to the p-adic quantifier elimination theorem of Macintyre [30], and to the beautiful work of Denef in the 1980s (e.g. [11]) which shows that certain power series associated with p-adic integrals are rational functions. This was extended in an analytic context by Denef and van den Dries [12], thereby verifying a conjecture of Serre and Oersterlé. This body of work was applied by Grunewald et al. [14], and later du Sautoy, to obtain rationality results for certain Poincaré series associated with groups. Denef's work also fed into a series of papers by Denef and Loeser, and more recently Cluckers and Loeser, on motivic integration.

Recent work by Haskell et al. [16] provided abstract model-theoretic tools allowing a satisfactory analysis of algebraically closed valued fields. These include stable domination, mentioned above. Aspects of this work have had further applications, for example in the work of Hrushovski and Kazhdan [20] in motivic integration, in results of Hrushovski and Loeser on Berkovich space in rigid analytic geometry (see [21]), and in work of Hrushovski and Martin on representation growth of finitely generated torsion-free nilpotent groups [22].

Many methods from the model theory of valued fields extend to an analytic context, in which a complete valued field (as a first order structure) is expanded by adding symbols for certain convergent power series functions. This emerged first in work of Denef and van den Dries [12] on analytic expansions of the p-adics, and was greatly extended by Lipschitz, Robinson and others, with a rather general setting described in [10]. The paper [12] also provides an approach to Gabrielov's model completeness of the o-minimal structure R_{an}, as described in the next article.

The van den Dries contribution, centred on the Ax-Kochen/Ershov theory, provides an extended introduction to the key basic ideas and insights of the subject.

(4) *Undecidability in number theory* (Jochen Koenigsmann, University of Oxford).

Topic 4 takes its origin in Hilbert's 10th Problem which in modern terminology can be stated as follows: find an algorithm (i.e. a Turing machine) to decide, given a polynomial equation $f(x_1, \ldots, x_n) = 0$ with coefficients in the ring of integers $\mathbb{Z}$, whether there exists a solution with $x_1, \ldots, x_n \in \mathbb{Z}$. In 1970 Matijasevic, following work of Davis, Putnam and Julia Robinson, proved that no such algorithm exists. Hence one says that Hilbert's Tenth Problem is undecidable. Since then, researchers have asked (see [13, 35, 39, 44, 45, 50]) what happens when $\mathbb{Z}$ is replaced by other commutative rings R of arithmetic interest. The most prominent open questions are when $R = \mathbb{Q}$ and when R is the ring of integers of an arbitrary number field. The question over $\mathbb{Q}$ is of fundamental interest to arithmetic geometers, because it is equivalent to the existence of an algorithm for deciding whether a variety over $\mathbb{Q}$ has a rational point. Attempts to reduce the question over $\mathbb{Q}$ to the question over $\mathbb{Z}$ come into conflict with conjectures of Mazur on the topology of rational points [36].

In a big advance, Koenigsmann [27] has proved that $\mathbb{Z}$ is universally definable in $\mathbb{Q}$: indeed there is a polynomial $g(t, x_1, \ldots, x_{418})$ with 419 unknowns and integer coefficients such that, for every rational number t, t is an integer if and only if $\forall x_1 \ldots \forall x_{418} \in \mathbb{Q}\ g(t, x_1, \ldots, x_{418}) \neq 0$. This implies that the $\forall\exists$-theory of the field of rationals is undecidable. Hilbert's 10th problem over $\mathbb{Q}$ is basically the question whether the $\exists$-theory of $\mathbb{Q}$ is decidable. As said, this is still open. However Koenigsmann also proved that $\mathbb{Z}$ is not existentially definable in $\mathbb{Q}$ under a rather mild arithmetic-geometric conjecture (much weaker than Mazur's).

For most number fields K, the question for the ring of integers of K is another open problem. By a clever use of elliptic curves Poonen has proved undecidability for certain of the rings between $\mathbb{Z}$ and $\mathbb{Q}$ where infinitely many primes are inverted (see Sect. 15 in [45]). Very recently Mazur and Rubin [37] have proved that Hilbert's Tenth Problem is unsolvable for the ring of integers of an arbitrary number field, assuming some standard conjectures about the Tate-Shafarevich group. This is a fast-moving and challenging area. Koenigsmann's paper illustrates and updates its development.

There are close links between the four chapters of this volume.

Valued fields provide key motivating examples of NIP theories—indeed, by a result of Belair in the Ax-Kochen/Ershov spirit, a henselian valued field of residue characteristic 0 is NIP if and only if its residue field is NIP. Likewise, o-minimal structures (such as the real exponential field) are NIP and, like (non-trivially) valued fields, are unstable. Several of the ideas described in Pillay's paper emerged through work on o-minimal structures and on algebraically closed valued fields. As sketched in Wilkie's paper, valuation-theoretic methods play a key role in the proof that the real exponential field is model complete and hence o-minimal. More generally, there are well-known analogies between the o-minimal ordered field of real numbers, and the valued field of p-adic numbers. Both are completions of the rationals, they share key model-theoretic properties such as NIP, but they differ crucially on connectedness properties of their topologies. Links are also emerging between quasi-minimality (which, as discussed in Wilkie's paper, is conjectured to hold for the complex exponential field) and some methods associated with NIP theories. Topic 4, Hilbert's 10th Problem, is slightly apart, due to its focus on the wilder non-Gödelian parts of model theory, and on the theme of decidability rather than definability. However, valued fields provide an important context for Hilbert's Tenth Problem, and decidability questions permeate the model theory of fields; examples are the Macintyre-Wilkie proof (relative to Schanuel's Conjecture) of the decidability of the real exponential field, and the still-open question whether the valued field $\mathbb{F}_p((t))$ of formal power series is decidable.

The 2012 CIME course also included other welcome contributions:

- A talk by Sergei Starchenko, University of Notre Dame, *A model theoretic proof of Bieri-Groves theorem*;
- An informal seminar kindly given by Umberto Zannier (SNS Pise) on certain diophantine problems related to model theory;
- A session of posters of young participants
  (1) Katharina Dupont (Konstanz), *Definable valuations and dependent fields*,
  (2) Antongiulio Fornasiero (Napoli 2), *Groups and rings definable in d-minimal structures*,
  (3) Robert Henderson (East Anglia), *Definable groups in exponential fields*,
  (4) Sonia L'Innocente (Camerino), *Diophantine sets of representations*,
  (5) Kota Takeuchi (Tsukuba), *Indiscernible objects in unstable theories*,
  (6) Giuseppina Terzo and Paola D'Aquino (Napoli 2), *From Schanuel's Conjecture to Shapiro's Conjecture*.

Acknowledgements We thank the four authors of the articles in this volume for their contributions. We also warmly thank

- the CIME director and vice-director, Pietro Zecca and Elvira Mascolo, and the whole CIME staff for the excellent organization of the course,
- all the speakers, and presenters of posters,
- all the participants (around 70, mostly young researchers),
- the Italian FIR *New trends in model theory of exponentiation* and its coordinator Sonia L'Innocente for their support.

We also thank the CIME Scientific Committee for accepting the proposal of this course. We hope that model theory may soon be the topic of other CIME meetings.

References

1. I.M. Aschenbrenner, A. Dolich, D. Haskell, D. Macpherson, S. Starchenko, Vapnik-Chervonenkis density in some theories without the independence property I. arXiv:1109.5438, 2011
2. J. Baldwin, *Categoricity*. University Lecture Series, vol. 50 (American Mathematical Society, Providence, 2009)
3. J.T. Baldwin, A.H. Lachlan, On strongly minimal sets. J. Symb. Log. **36**, 79–96 (1971)
4. A. Berarducci, M. Otero, Y. Peterzil, A. Pillay, A descending chain condition for groups definable in o-minimal structures. Ann. Pure Appl. Log. **134**, 303–313 (2005)
5. E. Bouscaren (ed.), *Model Theory and Algebraic Geometry*. Lecture Notes in Mathematics, vol. 1696 (Springer, New York, 1998)
6. C.C. Chang, H.J. Keisler, *Model Theory* (North Holland, Amsterdam, 1990)
7. Z. Chatzidakis, D. Macpherson, A. Pillay, A. Wilkie, *Model Theory with Applications to Algebra and Analysis I*. London Mathematical Society Lecture Note Series, vol. 349 (Cambridge University Press, Cambridge, 2008)
8. Z. Chatzidakis, D. Macpherson, A. Pillay, A. Wilkie, *Model Theory with Applications to Algebra and Analysis II*. London Mathematical Society Lecture Note Series, vol. 350 (Cambridge University Press, Cambridge, 2008)
9. A. Chernikov, Theories without the tree property of the second kind. Ann. Pure Appl. Log. **165**, 695–723 (2014)
10. R. Cluckers, L. Lipshitz, Fields with analytic structure. J. Eur. Math. Soc. **13**(4), 1147–1223 (2011)
11. J. Denef, The rationality of the Poincaré series associated to the p-adic points on a variety. Invent. Math. **77**, 1–23 (1984)
12. J. Denef, L. van den Dries, p-adic and real subanalytic sets. Ann. Math. (2) **128**, 79–138 (1988)
13. J. Denef, L. Lipshitz, T. Pheidas, J. Van Geel, *Hilbert's Tenth Problem: Relations with Arithmetic and Algebraic Geometry*. Contemporary Mathematics, vol. 270 (American Mathematical Society, Providence, 2000)
14. F.J. Grunewald, D. Segal, G.C. Smith, Subgroups of finite index in nilpotent groups. Invent. Math. **93**, 185–223 (1988)
15. D. Haskell, A. Pillay, C. Steinhorn (eds.), *Model Theory, Algebra and Geometry*. MSRI Publications (Cambridge University Press, Cambridge, 2000)
16. D. Haskell, E. Hrushovski, D. Macpherson, *Stable Domination and Independence in Algebraically Closed Valued Fields*. ASL Lecture Notes in Logic (Cambridge University Press, Cambridge, 2007)
17. W. Hodges, *Model Theory* (Cambridge University Press, Cambridge, 1993)
18. E. Hrushovski, A new strongly minimal set. Ann. Pure Appl. Log. **62**, 147–166 (1993)
19. E. Hrushovski, The Mordell-Lang conjecture for function fields. J. Am. Math. Soc. **9**, 667–690 (1996)
20. E. Hrushovski, D. Kazhdan, Integration in valued fields, in *Progress in Mathematics 850: Algebraic Geometry and Number Theory, in Honour of Vladimir Drinfeld's 50th Birthday* (Birkhäuser, Basel, 2006)
21. E. Hrushovski, F. Loeser, Non-archimedean tame topology and stably dominated types. Princeton Monograph Series (to appear)
22. E. Hrushovski, B. Martin, S. Rideau, with an appendix by R. Cluckers, Definable equivalence relations and zeta functions of groups, Math.LO/0701011v02, 2014
23. E. Hrushovski, A. Pillay, On *NIP* and invariant measures. J. Eur. Math. Soc. **13**, 1005–1061 (2011)
24. E. Hrushovski, B. Zilber, Zariski geometries. J. Am. Math. Soc. **9**, 1–56 (1996)
25. E. Hrushovski, Y. Peterzil, A. Pillay, Groups, measures and the *NIP*. J. Am. Math. Soc. **21**, 563–596 (2008)

26. E. Hrushovski, A. Pillay, P. Simon, Generically stable and smooth measures in NIP theories. Trans. Am. Math. Soc. **365**, 2341–2366 (2013)
27. J. Koenigsmann, Defining $\mathbb{Z}$ in $\mathbb{Q}$, in *Arithmetic of Fields*, Oberwolfach Reports (2009)
28. M.C. Laskowski, Vapnik-Chervonenkis classes of definable sets. J. London Math. Soc. (2) **45**, 377–384 (1992)
29. F. Loeser, Seattle lectures on motivic integration. http://www.math.ens.fr/~loeser/notes_seattle_09_04_2008.pdf, 2008
30. A. Macintyre, On definable subsets of p-adic fields. J. Symb. Log. **41**, 605–610 (1976)
31. A. Macintyre, Model completeness, in *Handbook of Mathematical Logic*, ed. by J. Barwise (North Holland, Amsterdam, 1977), pp. 139–180
32. A. Macintyre (ed.), *Connections between Model Theory and Algebraic and Analytic Geometry* (Quaderni di Matematica, Aracne, 2000)
33. A. Macintyre, A. Wilkie, On the decidability of the real exponential field, in *Kreiseliana* (A. K. Peters, Wellesley, 1996), pp. 441–467
34. D. Marker, *Model Theory: An Introduction*. Graduate Texts in Mathematics, vol. 217 (Springer, New York, 2002)
35. Y. Matiyasevich, *Hilbert's Tenth Problem*. Foundations of Computing Series (MIT Press, Cambridge, 1993)
36. B. Mazur, The topology of rational points. Exp. Math. **1**, 35–45 (1992)
37. B. Mazur, K. Rubin, Ranks of twists of elliptic curves and Hilbert's Tenth Problem. Invent. Math. **181**, 541–575 (2010)
38. M. Morley, Categoricity in power. Trans. Am. Math. Soc. **114**, 514–538 (1965)
39. T. Pheidas, K. Zahidi, Decision problems in algebra and analogues of Hilbert's tenth problem, in *Model Theory with Applications to Algebra and Analysis II*. London Mathematical Society Lecture Note Series, vol. 350 (Cambridge University Press, Cambridge, 2008)
40. J. Pila, O-minimality and the André-Oort conjecture for $\mathbb{C}^n$. Ann. Math. (2) **173**, 1779–1840 (2011)
41. J. Pila, A. Wilkie, The rational points of a definable set. Duke Math. J. **133**, 591–616 (2006)
42. A. Pillay, *Geometric Stability Theory* (Oxford University Press, Oxford, 1996)
43. B. Poizat, *A Course in Model Theory: An Introduction to Contemporary Mathematical Logic* (Springer, New York, 2000)
44. B. Poonen, Undecidability in number theory. Not. Am. Math. Soc. **55**, 344–350 (2008)
45. B. Poonen, Hilbert's Tenth problem over rings of number-theoretic interest (2010). http://math.mit.edu/~poonen/papers/aws2003.pdf
46. A. Robinson, *Complete Theories, Studies in Logic* (North-Holland, Amsterdam, 1956)
47. W. Schmid, K. Vilonen, Characteristic cycles of constructible sheaves. Invent. Math. **124**, 451–502 (1996)
48. S. Shelah, *Classification Theory* (North Holland, Amsterdam, 1990)
49. S. Shelah, Dependent first order theories, continued. Isr. J. Math. **173**, 1–60 (2009)
50. A. Shlapentokh, *Hilbert's Tenth Problem. Diophantine Classes and Other Extensions to Global Fields*. New Mathematical Monographs, vol. 7 (Cambridge University Press, Cambridge, 2007)
51. A. Tarski, Contributions to the theory of models I. Indag. Math. **16**, 572–581 (1954)
52. A. Tarski, Contributions to the theory of models II. Indag. Math. **16**, 582–588 (1954)
53. K. Tent, M. Ziegler, *A Course in Model Theory*. Lecture Notes in Logic, vol. 40 (Cambridge University Press, Cambridge, 2012)
54. L. van den Dries, Alfred Tarki's elimination theory for reals closed fields. J. Symb. Log. **53**, 7–19 (1988)
55. L. van den Dries, *Tame Topology and o-minimal Structures*. London Mathematical Society Lecture Notes Series, vol. 248 (Cambridge University Press, Cambridge, 1998)
56. F. Wagner, *Simple Theories* (Kluwer, Dordrecht, 2000)
57. A. Wilkie, Model completeness results for expansions of the ordered field of real numbers by restricted Pfaffian functions and the exponential function. J. Am. Math. Soc. **9**, 1051–1094 (1996)

58. B. Zilber, The structure of models of uncountably categorical theories, in *ICM-Varsavia 1983* (North Holland, Amsterdam, 1984), pp. 359–368
59. B. Zilber, Pseudoexponentiation on algebraically closed fields of characteristic 0. Ann. Pure Appl. Log. **132**, 67–95 (2004)
60. B. Zilber, *Zariski Geometries*. London Mathematical Society Lecture Note Series, vol. 360 (Cambridge University Press, Cambridge, 2010)

Some Themes Around First Order Theories Without the Independence Property

Anand Pillay

1 Model Theory

In this first section I discuss model theory and notation, in a form appropriate to the subsequent sections. In the subsequent sections I will draw a lot on the papers [7], [8] and [9] as well as on papers of Chernikov and Simon [3, 4].

I fix a "language" or "vocabulary" L consisting of relation symbols R_i (each of a given arity ≥ 1), function symbols f_j (each of a given arity ≥ 1) and constant symbols c_k. We always include a distinguished binary relation "$=$". If you want, take the language as being countable.

From these symbols, together with the logical connectives $\wedge, \vee, \rightarrow, \neg, \exists, \forall$, parentheses, and a collection of variables, we build up inductively the class of *first order L-formulas*. For a finite tuple $\bar{x} = (x_1, .., x_n)$ of variables, $\phi(\bar{x})$ denotes an L-formula whose free variables are among $x_1, .., x_n$. By an L-sentence σ I mean an L-formula with no free variables.

An L-structure M consists of an underlying set (often notationally identified with M), and interpretations $R_i(M)$, $f_j(M)$, $c_k(M)$ of the relation symbols, function symbols, constant symbols of L, where the equality symbol is interpreted as equality. So for example if R_i has arity n_i, then $R_i(M)$ is a subset of M^{n_i}. One could also work with a *many-sorted* language, where the symbols come together with sorts (so e.g. a n_i-ary relation symbol R_i comes equipped with an n_i-tuple of sorts). An L-structure will then be a collection of underlying sets indexed by the sorts and the symbols are interpreted accordingly. I will normally assume L to be 1-sorted, but there is a certain construction (the *eq* construction) taking us naturally into the many-sorted framework.

A. Pillay (✉)
University of Leeds, Leeds, UK

University of Notre Dame, Notre Dame, IN, USA
e-mail: apillay@nd.edu

L. van den Dries et al., *Model Theory in Algebra, Analysis and Arithmetic*, Lecture Notes in Mathematics 2111, DOI 10.1007/978-3-642-54936-6_2,

For an L-structure M, an L-formula $\phi(\bar{x})$ and an n-tuple $\bar{a}$ of elements of (the underlying set of) M, we obtain, by an inductive definition, the notion "$M \models \phi(\bar{a})$": $\phi(\bar{x})$ is true of $\bar{a}$ in M, or $\bar{a}$ satisfies $\phi(\bar{x})$ in M. When σ is an L-sentence, we say σ is true in M, or M is a model of σ, for $M \models \sigma$. If Σ is a collection of L-sentences, we write $M \models \Sigma$ for "$M \models \sigma$ for all $\sigma \in \Sigma$" and again say "M is a model of Σ".

I will not work with any notion of "formal proof", just "logical implication".

Definition 1.1. Let Σ be a set of L-sentences and τ an L-sentence. We write $\Sigma \models \tau$, and say Σ (logically) implies τ, if any model of Σ is a model of τ.

The most basic fact about first order logic is the *Compactness Theorem*.

Theorem 1.2. *Let Σ, τ be as above. Then $\Sigma \models \tau$ if and only if $\Sigma' \models \tau$ for some finite subset Σ' of Σ. Equivalently a collection Σ of sentences has a model iff every finite subset of Σ has a model.*

I sometimes say "Σ is consistent" for "Σ has a model".

Definition 1.3. (i) An L-theory is a consistent set of L-sentences, closed under logical implication. An L-theory is often denoted by T.

(ii) An L-theory T is said to be *complete* if for any sentence τ either $\tau \in T$ or $\neg\tau \in T$.

(iii) If K is a class of L-structures, then $Th(K)$ denotes the collection of sentences τ which are true in every $M \in K$. So $Th(K)$ is an L-theory.

(iv) When $K = \{M\}$ we write $Th(M)$ for $Th(K)$. Note that $Th(M)$ is a complete L-theory and every complete L-theory is of this form (for some M).

(v) Two L-structures M, N are said to be *elementarily equivalent* written $M \equiv N$, if they satisfy the same L-sentences.

Example 1.4. Take L to be the language of unitary rings $\{+, -, \times, 0, 1\}$.

(i) If K is the class of finite fields, $Th(K)$ (the theory of finite fields) is an incomplete L-theory. A compactness argument yields that $Th(K)$ has a infinite model, a so-called pseudofinite field.

(ii) The theory ACF of algebraically closed fields is an (incomplete) L-theory, but after fixing the characteristic p, is complete, and is called ACF_p. So two algebraically closed fields of the same characteristic are elementarily equivalent.

(iii) The theory RCF of real-closed fields is complete. This theory contains the axioms for ordered fields as well as axioms expressing the intermediate value property for polynomials.

(v) In each of the above cases the theories have a recursive "axiomatization" and are in fact decidable.

There are obvious notions of substructure and extension for L-structures (written $M \subseteq N$) generalizing subgroup, subring,... But the key model theoretic notion is *elementary substructure*.

Definition 1.5. Let M, N be L-structures.

(i) By a partial elementary map f from M to N we mean a map f from some subset of (the universe of) M to (the universe of) N such that for any L-formula $\phi(\bar{x})$ and tuple $\bar{a}$ from $dom(f)$, $M \models \phi(\bar{a})$ iff $N \models \phi(f(\bar{a}))$.
(ii) When $dom(f) = M$ we say that f is an elementary embedding of M into N.
(iii) When $dom(f) = M$ and f is the identity map we say that M is an elementary substructure of N, and N is an elementary extension of M, and write $M \prec N$.

By the cardinality $|M|$ of an L-structure M, I mean the cardinality of the underlying set of M. The compactness theorem, together with the "Downward Löwenheim Skolem Theorem" yield (assuming countability of L to avoid unnecessary complications):

Theorem 1.6. *Let M be an infinite L-structure. Then for $\kappa \geq |M|$, M has an elementary extension of cardinality κ. Also for any infinite $\kappa \leq |M|$, M has an elementary substructure of cardinality κ. In particular if the theory T has an infinite model then it has models in all infinite cardinalities.*

The proof of the first statement of the theorem involves adding constants c_m to the language for each element $m \in M$, as well as new constants $\{d_\alpha : \alpha < \kappa\}$ and showing that $Th(M)$ (with the new constants c_m) together with $\{d_\alpha \neq d_\beta : \alpha < \beta < \kappa\}$ is consistent, using the compactness theorem.

I now turn to definable sets and types.

Definition 1.7. (i) Let M be an L-structure. By a $\emptyset$-definable set in M I mean $\{\bar{a} \in M^n : M \models \phi(\bar{a})\}$ for some L-formula $\phi(\bar{x})$, written as $\phi(M)$.
(ii) Let M be an L-structure and B a subset of (the universe of) M. By a set definable in M *with parameters from* B I mean something of the form $\{\bar{a} \in M^n : M \models \phi(\bar{a}, \bar{b})\}$ where $\phi(\bar{x}, \bar{y})$ is an L-formula and $\bar{b}$ is a tuple from B. We consider $\phi(\bar{x}, \bar{b})$ as a formula *with parameters from* B, or *over* B.

For T a given L-theory, one can also think of a $\emptyset$-definable set with respect to T as a functor (from the category $Mod(T)$ of models of T to Set) taking M to $\phi(M)$ (for a given L-formula $\phi(\bar{x})$). The morphisms of $Mod(T)$ are the elementary embeddings. There are various formalisms for dealing with definable sets or formulas with parameters. One is to simply add new constants to the language for the relevant parameters, and then consider formulas in this expanded language. For example a formula over a structure M is a formula in the language L expanded by new constants c_m for $m \in M$ (and where M is tautologically expanded to a structure in this new language). Anyway I will leave it a bit loose. In any case an important aspect of the model-theoretic understanding or analysis of a given structure M (or theory T) is describing the "category" of definable sets (where the morphisms are the definable maps). One of the tools for accomplishing this in concrete instances is "quantifier elimination", abbreviated as QE: A theory T has quantifier elimination if for every L-formula $\phi(\bar{x})$ there is a quantifier-free L-formula $\psi(\bar{x})$ such that ϕ is equivalent to ψ modulo T in the sense that $\forall \bar{x}(\phi(\bar{x}) \leftrightarrow \psi(\bar{x})) \in T$. For example ACF has quantifier elimination, and RCF

has quantifier elimination after adding a new relation symbol for the nonnegative elements (already definable as the squares).

Now fix an L-theory T and let $\Sigma(\bar{x})$ be a collection of L-formulas $\phi(\bar{x})$. We say that $\Sigma(\bar{x})$ is *consistent with* T if there is a model M of T and a tuple $\bar{a}$ from M which *realizes* $\Sigma(\bar{x})$ in the sense that $M \models \phi(\bar{a})$ for all $\phi(\bar{x}) \in \Sigma$. Likewise if M is a structure, and $\Sigma(\bar{x})$ a collection of formulas over M, we say that $\Sigma(\bar{x})$ is consistent with M (or rather $Th(M)$ in the language expanded by constants for elements of M) if $\Sigma(\bar{x})$ is realized by a tuple in an elementary extension N of M. Equivalently, by compactness, every finite subset of Σ is realized by a tuple from M. A complete n-type of T is simply a maximal consistent (with T) collection of L-formulas $\phi(\bar{x})$ ($\bar{x} = (x_1, .., x_n)$). By virtue of the compactness theorem this is the same thing as an ultrafilter on the Boolean algebra of L-formulas $\phi(\bar{x})$ up to equivalence modulo T. We write the collection of complete n-types of T as $S_n(T)$. It is a compact Hausdorff totally disconnected space where a basic open is the set of types containing a given formula. Likewise we have $S_n(M)$, the space of complete n-types over a structure M, as well as $S_n(A)$ for A a subset of M. These are the fundamental compact spaces attached to a theory/structure.

For M a structure, $A \subseteq M$ and $\bar{a}$ an n-tuple from M, $tp_M(\bar{a}/A)$ is the set of formulas $\phi(\bar{x})$ over A such that $M \models \phi(\bar{a})$, an element of $S_n(A)$. When A is empty we just write $tp_M(\bar{a})$.

Definition 1.8. (i) An L-structure M is said to be κ-saturated if whenever $\Sigma(\bar{x})$ is a collection of formulas over a subset A of M of cardinality $< \kappa$ which is consistent with M then it is already realized by a tuple from M.
(ii) We say that M is saturated if it is $|M|$-saturated.

Under some set-theoretic assumptions, any theory T has saturated models in arbitrarily large cardinalities. On the other hand any uncountable algebraically closed field is saturated.

Proposition 1.9. *Suppose that M is saturated, of cardinality κ. Then*

(i) Any $N \equiv M$ with cardinality $\leq \kappa$ elementarily embeds in M.
(ii) Any partial elementary map f from M to itself such that $dom(f)$ has cardinality $< \kappa$ extends to an automorphism of M.
(iii) In particular if A is a subset of M of cardinality $< \kappa$ and $\bar{b}, \bar{c}$ are n-tuples from M, then $tp_M(\bar{b}/A) = tp_M(\bar{c}/A)$ if and only if there is an automorphism g of M which fixes A pointwise and takes $\bar{b}$ to $\bar{c}$.

Typically when studying a complete theory T we fix a large saturated model $\bar{M}$ of T, and work *inside* $\bar{M}$. So for example given a formula $\phi(\bar{x})$ over M and tuple $\bar{a}$ from $\bar{M}$ we just write $\models \phi(\bar{a})$ in place of $\bar{M} \models \phi(\bar{a})$. We will be studying study "type-definable sets" in $\bar{M}$. Such a set X is the set of common realizations in $\bar{M}$ of some collection $\Sigma(\bar{x})$ of formulas over a "small" (i.e. of cardinality $< |\bar{M}|$) subset A of $\bar{M}$. Also *tp* (without an index) denotes $tp_{\bar{M}}$.

From now on we will fix a complete first order theory T, and as mentioned above we will work in a big saturated model $\bar{M}$ which may be enlarged when appropriate.

By a *model* we will usually mean a "small" elementary substructure of $\bar{M}$. If A, B are subsets of $\bar{M}$ we sometimes write AB for $A \cup B$.

An important tool/notion in model theory is that of an *indiscernible sequence*.

Definition 1.10. (i) Let $(I, <)$ be a totally ordered set, and $(\bar{a}_i : i \in I)$ be a sequence of n-tuples (from $\bar{M}$) indexed by I. We say that the sequence is *indiscernible* if for each $i_1 < ... < i_k$, $j_1 < .. < j_k$ (k a positive integer) from I, $tp(\bar{a}_{i_1}, ..., \bar{a}_{i_k}) = tp(\bar{a}_{j_1}, ..., \bar{a}_{j_k})$. Likewise for indiscernible over a subset A of M. We sometimes write $\{\bar{a}_i : i \in I\}$ in place of $(\bar{a}_i : i \in I)$, hopefully without ambiguity. (ii) The *E-M type* (standing for Ehrenfeucht-Mostowski type) of the indiscernible sequence above is by definition the set of k-types $tp(\bar{a}_{i_1}, ..., \bar{a}_{i_k})$ above, as k varies.

By compactness:

Lemma 1.11. *Let $(\bar{a}_i : i \in I)$ be an indiscernible sequence with I infinite. Then for any small totally ordered set $(J, <)$ there is in $\bar{M}$ an indiscernible sequence $(\bar{b}_j : j \in J)$ with the same E-M type as $(\bar{a}_i : i \in I)$.*

There are various methods for producing indiscernibles. The first uses coheirs.

Lemma 1.12. *Let M be a model and M' an elementary extension of M. Let $p(\bar{x}) \in S_n(M)$. Let $p'(\bar{x}) \in S_n(M')$ be an extension of p which is* finitely satisfiable *in M. That is, every formula in p' is satisfied by a tuple from M. Let $\bar{b}$ realize p'. Let $\bar{b}_i \in M'$ for $i < \omega$ be such that $tp(\bar{b}/M \cup \{\bar{b}_0, ..., \bar{b}_{i-1}\}) = tp(\bar{b}_i/M \cup \{\bar{b}_0, ..., \bar{b}_{i-1}\})$. Then $(\bar{b}_i : i < \omega)$ is an indiscernible sequence over M.*

The second uses the Erdös-Rado Theorem.

Lemma 1.13. *There is a cardinal λ such that if M is an L-structure, and $(\bar{b}_i : i < \lambda)$ is a sequence of n-tuples from M, then in some elementary extension N of M there is an indiscernible sequence $(\bar{a}_i : i < \omega)$ such that for each nonnegative integer k, there are $j_0 < .. < j_k < \lambda$ such that $tp_N(\bar{a}_0, .., \bar{a}_k) = tp_M(\bar{b}_{j_0}, ..., \bar{b}_{j_k})$.*

The following special case uses Ramsey's Theorem.

Lemma 1.14. *Let $(b_i : i < \omega)$ be a sequence of n-tuples. For each $k < \omega$ let $\Sigma_k(\bar{x}_0, .., \bar{x}_k)$ be a collection of L-formulas (a so-called partial type) such that $\models \Sigma_k(\bar{b}_{i_0}, .., \bar{b}_{i_k})$ for all $i_0 < .. < i_k < \omega$. Then there is an indiscernible sequence $(\bar{c}_i : i < \omega)$ such that $\models \Sigma_k(\bar{c}_0, .., \bar{c}_k)$ for all k.*

Finally I mention the T^{eq} construction. First of all, at the semantic level, one is naturally interested not only in definable sets X, but also in quotients X/E by definable equivalence relations E. Such a set X/E is sometimes said to be *interpretable*, and elements of X/E are sometimes called "imaginaries". There is also a natural notion of a definable subset of X/E (a subset whose preimage in X is definable). One would like to speak of the type of an imaginary over a set of imaginaries. Again this can be done in an ad hoc manner. But the T^{eq} construction gives a nice formalism. For each L-formula $E(\bar{x}, \bar{y})$ which defines an equivalence relation on n-tuples, adjoin a new sort S_E to the language L, as well as a new function symbol f_E from n-tuples of the "home sort" to S_E. Adjoin a sentence σ_E saying $\forall \bar{x}\bar{y}(f(\bar{x}) = f(\bar{y}) \leftrightarrow E(\bar{x}, \bar{y}))$ and that f is onto S_E. Call the new language L^{eq}, and let T^{eq} be the L^{eq}-theory obtained from T by adjoining the sentences σ_E as E varies. Then any model M of T (in particular $\bar{M}$) has a unique "expansion" to a model M^{eq} of T^{eq}. No new structure is induced on the "home sort" in the sense that if $x_1, .., x_n$ are variables of L and $\phi(x_1, .., x_n)$ is an L^{eq} formula, then there is an L-formula $\psi(x_1, .., x_n)$ which is equivalent to $\phi(x_1, .., x_n)$ modulo T^{eq}. Also T^{eq} is complete (assuming completeness of T).

Remember that for a model M of T, and a subset A of M, the algebraic closure of A (definable closure of A) is the collection of $b \in M$ such that there is a formula $\phi(x)$ over A such that $M \models \phi(b)$ and there are only finitely many (exactly one) b' such that $M \models \phi(b')$. We write $acl(A)$, $dcl(A)$. When applied to M^{eq} we write $acl^{eq}(A)$, $dcl^{eq}(A)$.

From now on we feel free to work in $\bar{M}^{eq}$, and we no longer use bar notion for tuples, as finite tuples correspond to elements in $\bar{M}^{eq}$. Nevertheless 1-tuples from the home sort may play a distinguished role.

2 Stability and *NIP*

I continue with the conventions established at the end of Sect. 1. For those who are interested [12] is a reference for stability theory.
We fix T.

Definition 2.1. (i) The L-formula $\phi(x, y)$ has the *order property* or is unstable, if there are a_i, b_i for $i < \omega$ such that $\models \phi(a_i, b_j)$ iff $i \leq j$ for $i, j < \omega$. The formula $\phi(x, y)$ is stable if it is not unstable. T is stable if every formula $\phi(x, y)$ is stable.
(ii) The L-formula $\phi(x, y)$ has the *independence property* if there are a_i for $i \in \omega$ and b_s for $s \subseteq \omega$ such that $\models \phi(a_i, b_s)$ iff $i \in s$, for all i, s. The formula $\phi(x, y)$ has *NIP* (or is dependent) if it does not have the independence property. We say that T has the independence property if some formula $\phi(x, y)$ has the independence property, and otherwise we say that T is (or has) *NIP*.

It is rather important to know that T is stable (has *NIP*) just if any formula $\phi(x, y)$ where x is a 1-tuple variable of the home sort is stable (has *NIP*).

Example 2.2. (i) The theory of any infinite structure with a definable total ordering $<$ (such as $(\mathbb{Q}, <)$) is unstable, witnessed by the formula $x \leq y$.
(ii) Consider the 2-sorted structure $(\omega, P(\omega), \in)$ where $P(\omega)$ is the collection of subsets of ω. Then the theory of this structure has the independence property.
(iii) The theory of the random graph has the independence property.
(iv) Among stable (complete) theories are (apart from theories of finite structures), the theory of an infinite set in the empty language, ACF_p, and the theory of any commutative group, in the language $\{+, 0\}$.

(v) Among *NIP*, unstable theories, are *DLO* (the theory of dense linear orderings without endpoints), *RCF*, the theory of the p-adic field $\mathbb{Q}_p$ (in the language of rings), as well as (any completion of) the theory *ACVF* of algebraically closed valued fields (discussed later). For the latter we refer to the paper by Lou van den Dries in the current volume.

It is not hard to see that if $\phi(x, y)$ has the independence property then it has the order property. So T stable implies T has *NIP*. There has been an intensive study of stable theories, initiated by Shelah, involving counting the number of models, as well as describing definable sets, types, and their interaction, with many applications. "Generalized stability theory" attempts to adapt the stability machinery to more general first order theories. The case of *NIP* theories has been fairly intensively studied in the past 10 years (and from before that up to the present, the case of simple theories, which I will not define here). A basic "thesis" of Shelah is that the "theory" of *NIP* theories or structures should be what is common to stable theories and to *DLO*.

Lemma 2.3. *(i) The formula $\phi(x, y)$ is stable iff for any indiscernible sequence $(a_i : i \in I)$ and any c either $\{i \in I :\models \phi(a_i, c)\}$ is finite or $\{i \in I :\models \neg\phi(a_i, c)\}$ is finite. Moreover the relevant finite numbers depend only on $\phi(x, y)$. Also T is stable iff any indiscernible sequence is an* indiscernible set *(namely we have indiscernibility with respect to any ordering of the index set).*

(ii) $\phi(x, y)$ is NIP iff for any indiscernible sequence $\{a_i : i \in I\}$ and c there are at most finitely many (with the finite number depending on $\phi(x, y)$) of alternations of truth values of $\phi(a_i, c)$ as i goes through I.

(iii) In particular if $\phi(x, y)$ is NIP and $\{a_i : i < \omega\}$ is indiscernible, then $\{\phi(x, a_i)\Delta\phi(x, a_{i+1}) : i = 0, 2, 4, ..\}$ is inconsistent, i.e. has no common realization. (Here Δ denotes symmetric difference.)

An important notion for the remainder of the talks will be that of a *definable type*.

Definition 2.4. (i) Let A be a subset of $\bar{M}$. Let $p(x) = tp(b/A)$ for some $b \in \bar{M}$. Then p is *definable* if for every L-formula $\phi(x, y)$ there is a formula $\psi_\phi(y)$ over A such that for any (tuple) c from A, $\phi(x, c) \in p(x)$ (i.e. $\models \phi(b, c)$) iff $\models \psi_\phi(c)$.

(ii) A special case is when A is a model M. We say that $p \in S(M)$ is definable over $C \subseteq M$ if every $\psi_\phi(y)$ above has parameters from C. In any case such a defining schema $(\psi_\phi)_\phi$ gives rise to a complete type $p|B$ over any set B of parameters containing M, in particular to a complete type $p|\bar{M}$ over $\bar{M}$.

Example 2.5. Some easy examples:

(i) Take T to be the theory of an infinite set in the empty language (other than $=$). Let $b \neq a$ for all $a \in A$. Then $tp(b/A)$ is definable: via QE the only formula to be checked is $x = y$, and we have, for $a \in A$, $b = a$ iff $a \neq a$.

(ii) Let $M = (\mathbb{Q}, <)$ and $a \in M$. Let $p(x) = \{x > a\} \cup \{x < b : b \in M, a < b\}$. Then (by virtue of QE) p is a complete 1-type over M which is definable (over a). On the other hand if $\mathbb{Q} = A \cup B$, where $A < B$ and A has no greatest element, B no least element (i.e. a Dedekind cut corresponding to an irrational real) then the complete type over M: $\{x > a : a \in A\} \cup \{x < b : b \in B\}$ is not definable.

Let me now summarize the situation for stable theories. Assume T countable if you wish.

Proposition 2.6. *(i) T is stable iff for every $A \subseteq \bar{M}$ and $p(x) \in S_n(A)$, p is definable, iff for any model M of T, $|S_n(M)| \leq |M|^{\aleph_0}$.*

(ii) Assume T stable, Let $p(x) \in S(M)$ for some small model M. Then $p|\bar{M} = p'$ (from Definition 2.4 (ii)) is characterized by each of the following conditions:

(a) p' is finitely satisfiable in M,

(b) p' does not divide (fork) over M (see below),

(c) p' is invariant under $Aut(\bar{M}/M)$.

(iii) In the context of (ii) if $J = (a_i : i < \omega)$ is a Morley sequence *in p over M, namely a_0 realizes p in $\bar{M}$, a_1 realizes the restriction of p' to Ma_0, etc., then $(a_i : i < \omega)$ is indiscernible over M and p' is definable over J: given $\phi(x,y) \in L$ and $c \in \bar{M}$, $\phi(x,c) \in p'$ iff for some $i_0 < ... < i_N \leq 2N$, $\models \wedge_{j=i_0...i_N}\phi(a_j,c)$ (where N depends on ϕ as in Lemma 2.3 (i)).*

(iv) T is stable iff T^{eq} is stable.

Proof. I will not give the full proof, but just mention some parts of (ii). Note that p' is $Aut(\bar{M}/M)$-invariant. Let us now show (to exhibit the techniques) that p' is finitely satisfiable in M. Suppose not. So there is a formula $\phi(x,b) \in p'$ which is not satisfied by a tuple from M. As p' is definable over M there is formula $\psi(y)$ over M such that for all $b' \in \bar{M}$ (in particular in M), $\phi(x,b') \in p$ iff $\models \psi(b')$. We construct inductively $a_i, b_i \in M$ such that $\models \phi(a_i,b_j)$ iff $i > j$, and $\models \psi(b_i)$ for all i (contradicting that $\phi(x,y)$ does not have the order property). Suppose we have already found a_i, b_i for $i = 1,..,n$. Now $\bigwedge_{i=1,...,n}\phi(x,b_i) \in p(x)$. Hence there is $a_{n+1} \in M$ such that $\bigwedge_{i=1,...,n}\phi(a_{n+1},b_i)$. By our assumption that $\phi(x,b)$ is not satisfied in M, we have that $\models \bigwedge_{i=1,...,n}\neg\phi(a_i,b)$ and note that $\models \psi(b)$. As M is an elementary substructure of $\bar{M}$ there is $b_{n+1} \in M$ such that $\models \bigwedge_{i=1,...,n}\neg\phi(a_i,b_{n+1})$ and $\models \psi(b_{n+1}$. This completes the proof. □

An easy but important remark for later themes of this series is:

Lemma 2.7. *Assume that T is stable. Then complete types over sets are "uniformly definable". Namely for any L-formula $\phi(x,y)$ there is an L-formula $\psi(y,z)$ such that for any set of parameters A and any $p(x) \in S_n(A)$ there is c from A such that for all $b \in A$, $\phi(x,b) \in p(x)$ iff $\models \psi(b,c)$.*

Proof. Strictly speaking we should assume that the language contains some constant symbols so we can do definition by cases. Anyway add a new unary predicate symbol P to the language. Add also a new n-tuple d of constant symbols. Then the collection of sentences $T \cup \{\neg\exists w \in P(\forall y \in P(\phi(d,y) \leftrightarrow \psi(y,w))) : \psi(y,w) \in L\}$ is inconsistent (by Proposition 2.6 (i)). The compactness theorem yields that some finite subset has no model, giving finitely many possible ψ which as mentioned above can be taken to be one. □

The notions dividing and forking were mentioned in Proposition 2.6. These will again be important for the rest of the series of lectures.

Definition 2.8. Let $A \subset \bar{M}$ be small.

(i) A formula $\phi(x,b)$ (where we witness the parameters b) divides over A if there is some sequence $(b_i : i < \omega)$ indiscernible over A with $b_0 = b$ such that $\{\phi(x,b_i) : i < \omega\}$ is "inconsistent", namely has no common realization in $\bar{M}$.
(ii) $\phi(x,b)$ forks over A if it implies a finite disjunction of formulas each of which divides over A.
(iii) If $p(x) \in S_n(B)$ and $B \supseteq A$ then p divides (forks) over A if there is a formula $\phi(x,b) \in p$ which divides (forks) over A.

When $p(x) \in S_n(\bar{M})$ then dividing over A and forking over A coincide. For stable T it is a fact that forking and dividing coincide.

So Proposition 2.6 (ii) has a quite strong "uniqueness" content which in fact characterizes stable theories: Fix a small set A of parameters and tuple x of variables. Let I_A be the collection of formulas over $\bar{M}$ which divide (fork) over A. Then I_A is a proper ideal in the Boolean algebra of formulas $\psi(x)$ over $\bar{M}$ (up to equivalence). Fix $p(x) \in S(A)$. For $B \supseteq A$, call a realization d of p generic over B if it avoids the ideal I_A. Then there is a unique such type $tp(d/B)$, assuming $A = acl^{eq}(A)$.

Let us now continue with *NIP* theories. Shelah has a number of results trying to characterize *NIP* theories, or classify types in *NIP* theories, analogously to Proposition 2.6. See [15–17] for example. I will concentrate on part of this story, namely the interpretation and meaning of the theory of forking in the *NIP* context. It is NOT the case that in *NIP* theories, any type over a model does not fork over a "small" subset. Nevertheless types (over models) will have nonforking (invariant) extensions which play an important role, in applications too. The material in the rest of this section is taken from [8].

Note first (already alluded to in Sect. 1) that for any theory T, model M of T, $p(x) \in S(M)$, and $M \prec M'$, p has an extension $p'(x) \in S(M')$ which is *finitely satisfiable* in M. Then p' is called a coheir of p. Note that p' *does not split over* M in the sense that if $tp(b/M) = tp(c/M)$ $(b,c \in M')$ then $\phi(x,b) \in p'$ iff $\phi(x,c) \in p'$. Hence p' will be $Aut(M'/M)$-invariant. It was surprising for me to realize that the same is true for arbitrary "nonforking" extensions.

Proposition 2.9 (Assume that T has *NIP*). *Suppose that $p(x) \in S(\bar{M})$ and M is a small submodel. Then the following are equivalent:*

(i) p does not divide over M,
(ii) p does not fork over M,
(iii) p is $Aut(\bar{M}/M)$-invariant (i.e. p does not split over M).

Proof. The equivalence of (i) and (ii) is on the face of it because $\bar{M}$ is saturated, but in fact a result of Chernikov and Kaplan [2] states that for any M' such that $M \prec M' \prec \bar{M}$ and $p(x) \in S(M')$, p does not divide over M iff p does not fork over M.

Also clearly (iii) implies (i).

Let us now assume (i). Suppose $\phi(x,y) \in L$, $b,c \in \bar{M}$ and $tp(b/M) = tp(c/M)$. Assume $\phi(x,b) \in p(x)$ and we want to prove that $\phi(x,c) \in p(x)$. Using Lemma 1.12 we can find an infinite sequence $d_1, d_2, ...$ such that both $(b, d_1, d_2, ..)$ and $(c, d_1, d_2, ...)$ are indiscernible (over M). Now we claim that $\phi(x, d_1) \in p(x)$. If not, $\phi(x,b) \wedge \neg\phi(x,d_1) \in p(x)$. As $p(x)$ does not divide over M, $\{\phi(x,b) \wedge \neg\phi(x,d_1), \phi(x,d_2) \wedge \neg\phi(x,d_3), ...\}$ is consistent, and this contradicts Lemma 2.3 (iii). So $\phi(x,d_1) \in p(x)$. The same argument shows that $\phi(x,c) \in p(x)$, completing the proof. □

It follows from the proposition above that for an *NIP* theory T, if M is a model of T of cardinality κ and $p(x) \in S(M)$ then the number of nonforking extensions of p (to a given saturated model say) is bounded by 2^{2^κ}. It was asked by Adler whether this feature characterizes *NIP* theories. A counterexample appears in [5].

Let us consider the example of *DLO*. Take a small model M such as $(\mathbb{Q}, <)$ and a saturated model M' containing M. Suppose $p(x)$ is a complete 1-type over M' which does not fork over M (i.e. by the above is *Aut*(M'/M)-invariant). Then either (i) p is definable over M, or (ii) p is finitely satisfiable in M (i.e. is a coheir of $p|M$). A type of kind (i) is obtained by choosing a point $a \in M (= \mathbb{Q})$ (or maybe a is plus or minus infinity), and let p say that $x > a$ and $x < b$ for each $b \in \bar{M}$ such that $a < b$. A type of kind (ii) is obtained by again choosing $a \in M$, and let $p(x)$ say $x > a$ and $x < b$ for all $b \in M$ such that $a < b$, and also $x > c$ for all $c \in \bar{M}$ such that $a < c < b$ for all $b \in M$.

For a saturated model M', we will call $p(x) \in S(M')$ *invariant* if it is *Aut*(M'/M)-invariant for some "small" (compared to the saturation of M') submodel M. Given such p and $B \supseteq M'$ we can apply the "infinitary" definition of p to B to get an extension $p|B$ of p to a complete type over B. Given $p(x)$, $q(y)$, invariant types over M' we define $p(x) \otimes q(y)$ to be $tp(a,b/M')$ where b realizes q and a realizes $p|M'b$. For stable theories: $p(x) \otimes q(y)$ and $q(y) \otimes p(x)$ are equal (as types in variables x, y).

Definition 2.10. Let $p(x)$ be a global type (i.e. over $\bar{M}$) which is *Aut*$(\bar{M}/M)$-invariant. By a Morley sequence in p over M we mean (as in Proposition 2.6 (iii)) a sequence $J = (a_i : i < \omega)$ where a_i realizes the restriction of p to $Ma_0...a_{i-1}$.

It is again an easy exercise to see that such a Morley sequence is indiscernible over M and its type over M depends only on p. We could also do the same thing over $\bar{M}$ to obtain a type we call $p^{(\omega)}$ (obtained by iterating $\otimes$ on p itself ω-times). Anyway $tp(J/M)$ coincides with the restriction of $p^{(\omega)}$ to M. And this $tp(a_0, .., a_{n-1}/M)$ coincides with the restriction $p^{(n)}|M$ of $p^{(n)}$ to M.

Finally in this section:

Proposition 2.11 (Assume T has *NIP*). *Suppose $p(x) \in S(\bar{M})$ does not fork over small M. Then p is* strongly Borel definable *over M. Namely for any formula $\phi(x, y)$ there is a finite Boolean combination $\Gamma(y)$ of partial types over M such that for any $b \in \bar{M}$, $\phi(x, b) \in p(x)$ iff "$\models \Gamma(b)$".*

Proof. Let N be given for $\phi(x, y)$ by Lemma 2.3 (ii). Let $p^{(n)}|M$ denote $tp(a_0, .., a_{n-1}/M)$ where $(a_0, a_1, ...)$ is a Morley sequence for p over M.

Claim. For any $b \in \bar{M}$, $\phi(x, b) \in p$ if and only if for some $n \leq N$ the following holds: there is $(a_0, a_1, .., a_{n-1})$ realizing $p^{(n)}|M$ such that

$$(*)_n \models (\phi(a_i, b) \leftrightarrow \neg\phi(a_{i+1}, b))(i = 0, .., n-1)$$

and $\models \phi(a_{n-1}, b)$, but there is no $(a_0, .., a_n)$ realizing $p^{(n+1)}$ such that $(*)_{n+1}$ holds.

Proof of Claim. Suppose $\phi(x, b) \in p(x)$. Choose $(c_i : i < \omega)$ a realization of $p^{(\omega)}|M$ with maximal finite ($\leq N$) alternation of truth values of $\phi(c_i, b)$. Then eventually $\phi(c_i, b)$ holds. For otherwise letting c_ω realize $p|(M \cup \{b\} \cup (c_i : i < \omega))$ we contradict maximality.

The converse holds by applying the above proof to $\neg\phi(x, b)$. So the claim is proved and clearly yields strong Borel definability as required. □

3 Generically Stable Types

Here we discuss generically stable types, in the *NIP* context, because of their intrinsic interest but also because of their central role in the Hrushovski-Loeser approach to rigid analytic geometry (see [6]).

We fix T and again work in $\bar{M}$.

Definition 3.1. Let $\Sigma(x)$ be a partial type, over a small set A of parameters. Work over A (i.e. adjoin constants). We say that $\Sigma(x)$ is *stable* if it is not the case that there is a formula $\phi(x, y)$ and an indiscernible sequence $((a_i, b_i) : i < \omega)$ with a_i satisfying $\Sigma(x)$ (but no condition on the b_i) such that $\models \phi(a_i, b_j)$ iff $i \leq j$, for all i, j.

Remark 3.2. (i) The notion above is sometimes called "stable, stably embedded".
(ii) When x is a 1-tuple variable from the home sort, then stability of $x = x$ is equivalent to stability of T.
(iii) In *DLO* there are no stable partial types over any set of parameters other than algebraic formulas.
(iv) If T is some completion of the theory of algebraically closed valued fields the residue field is stable.

(v) Fixing a small set of parameters A, we define the stable part of T over A to be the union of all stable partial types over A.
(vi) If $\Sigma(x)$ (over $\emptyset$ say) is stable then all of Proposition 2.6 goes through for types of the form $tp(c/A)$ where c is a tuple of realizations of $\Sigma(x)$.

The notion of a generically stable type is different and weaker. Let us now assume that T has *NIP*.

Definition 3.3. Let $p(x) \in S(\bar{M})$ be a global type. We say that p is generically stable if for some small model M, p is both definable over M and finitely satisfiable in M.

Among the main results from [8] is:

Theorem 3.4. *Let $p(x) \in S(\bar{M})$. Then the following are equivalent:*

(i) p is generically stable (witnessed by the small model M).
(ii) p does not fork over M, and if $J = (a_i : i < \omega)$ is a Morley sequence of p over M, then J is an indiscernible set over M.
(iii) p does not fork over M and if $J = (a_i : i < \omega)$ is a Morley sequence of p over M then p is definable over J as in the stable case: for any $\phi(x,y)$ and $c \in \bar{M}$, $\phi(x,c) \in p$ iff $\models \wedge_{i \in w}\phi(a_i,c)$ for some $w \subset \{1,..,2N\}$ of size N (where N depends only on $\phi(x,y)$ not even on p).
(iv) p is the unique nonforking extension of $p|B$ (the restriction of p to B) for any $M \subseteq B$. (In fact p is the unique nonforking extension of the restriction of p to $acl^{eq}(A)$ whenever p does not fork over A.)
(v) For all n, $p^{(n)}$ is the unique (global) nonforking extension of $p^{(n)}|M$.

Remark 3.5. The condition in (ii) on total indiscernibility of the Morley sequence J is easily seen to be equivalent to requiring that $p(x_1) \otimes p(x_2) = p(x_2) \otimes p(x_1)$, as complete global types in variables x_1, x_2. Here we say p is *symmetric*.

Now some comments on the proof of Theorem 3.4.

(i) implies (ii) is an easy inductive argument.

For (ii) implies (iii), note that if J is an indiscernible set then by Lemma 2.3 (ii) and *NIP*, for any $\phi(x,y)$ there is an N such that for any c either $\{i :\models \phi(a_i,c)\}$ or $\{i :\models \neg\phi(a_i,c)\}$ has cardinality $< N$. Then, given J a Morley sequence as in (ii), we claim that for any c, $\phi(x,c) \in p(x)$ iff $\{i :\models \phi(a_i,c)\}$ has cardinality $\geq N$ (equivalently has co-cardinality $< N$). If $\phi(x,c) \in p(x)$ we can continue the Morley sequence "over c", to obtain $a_\omega, a_{\omega+1}, ...$ such that $\models \phi(a_\alpha,c)$ for the new α. So there had better be already infinitely many $i < \omega$ such that $\models \phi(a_i,c)$.

For (iv) implies (iii): Fix a Morley sequence $J = (a_i : i < \omega)$ of p over M. By virtue of Lemma 2.3 (ii) we can form the "average type of J over $\bar{M}$", namely $\{\phi(x,c) : \phi(x,y) \in L, c \in \bar{M}$ and eventually $\models \phi(a_i,c)\}$. This is a global complete type which is *Aut*($\bar{M}/MJ$)-invariant, so equals p. This is true for every such Morley sequence. As p is *Aut*($\bar{M}/M$)-invariant, for any formula $\phi(x,y)$ and type $q(y)$ over M such that $\phi(x,c) \in p$ for c realizing q we have, by compactness some N such that for all J, $\models \phi(a_i,c)$ for all but N $a_i \in J$. Another compactness argument yields an N such that for all Morley sequences J and all c either $\{i :\models \phi(a_i, c)\}$ or $\{i :\models \neg\phi(a_i, c)\}$ has cardinality $< N$, which suffices.

If $p(x)$ is a complete type over a small model N we say p is *generically stable* if p has a global $Aut(\bar{M}/M)$-invariant extension $p'(x) \in S(\bar{M})$ which is generically stable.

Example 3.6. (i) Any complete type over a model which extends a stable partial type, is generically stable.

(ii) There are no (nonalgebraic) generically stable types in *DLO*.

(iii) Let M be a model of *ACVF*. Let $p(x)$ be the complete 1-type over M saying that $v(x) \geq 0$ (i.e. x is in the valuation ring), and $v(x - a) < \alpha$ for all $a \in M$ and $\alpha \in \Gamma(M)$ with $\alpha > 0$. Then p is a complete generically stable type although it does satisfy (i) above.

Let me point out that condition (iv) in Theorem 3.4 cannot be weakened to saying that p is the unique nonforking extension of $p|M$. We can again find it in *ACVF*, and even in the more basic "C-minimal" theory T consisting of a family of equivalence relations E_α indexed (definably) by a dense linear ordering I such that $\alpha < \beta$ implies that each E_α-class breaks into infinitely many E_β-classes (plus some other axioms). In the case of *ACVF*, $E_\alpha(x, y)$ is $v(x - y) \geq \alpha$. By a definable ball we mean some E_α-class. By a type-definable ball we mean an intersection of definable balls. We can find a model M, and a type-definable over M ball, say B, containing no proper definable (over M) ball. Let $p(x) \in S(M)$ say that $x \in B$. Then p has a unique global $Aut(\bar{M}/M)$-invariant extension p', saying that $x \in B$ and x is in no proper definable (over M) subball. But p' is not definable over M (because B is only type-definable).

In Example 3.6 (iii), the generic stability of p is explained by "stable domination".

Definition 3.7. Let $p(x) = tp(a/M)$. We say that p is stably dominated if there is some partial M-definable function f, defined at a such that with $b = f(a)$ we have:

(i) $q(y) = tp(b/M)$ extends a stable partial type, and

(ii) for every formula $\phi(x, c)$, if b' realizes the nonforking extension of q over M, c then EITHER for all realizations a' of p such that $f(a') = b'$, $\models \phi(a', c)$, OR for all realizations a' of p such that $f(a') = b'$, $\models \neg\phi(a', c)$.

It can be shown fairly quickly that a stably dominated type is generically stable. Moreover as is proved in [6]:

Lemma 3.8. *In ACVF the generically stable types coincide with the stably dominated types.*

Going back to Example 3.5 (iii) where $p(x)$ is the "generic type over M of the valuation ring", stable domination of p is witnessed by the residue map. Namely if a realizes p take $f(a) = \bar{a} = a/\mathcal{M}$, where $\mathcal{M}$ is the maximal ideal.

Then $tp(f(a)/M)$ is the "generic type" of the residue field, and clause (ii) of Definition 3.7 can be proved.

Finally in this section/lecture, we discuss the "definable space" of generically stable types in *NIP* theories. We fix a sort or variable x and want to show that the collection of generically stable types $p(x) \in S(\bar{M})$ can be viewed as a "generalized definable set". First, given a definable type $p(x) \in S(\bar{M})$ we have for each $\phi(x, y) \in L$, the ϕ-definition of p, a formula with parameters and free variable y, which I will call $d_p(\phi)(y)$. Such a formula (or the set it defines) can be viewed as an element of $\bar{M}^{eq}$. (In fact given a formula with parameters $\psi(y, c)$ let $E(z_1, z_2)$ be the equivalence relation $\forall x(\psi(x, z_1) \leftrightarrow \psi(x, z_2))$. Then the set defined by $\psi(x, c)$ can be identified with the imaginary c/E, of sort S_E). So we can identify the type p with the countable sequence $(d_p(\phi)(y) : \phi(x, y) \in L)$ of imaginaries. However for another generically stable type $q(x) \in S(\bar{M})$, $d_p(\phi)(y)$ and $d_q(\phi)(y)$ will be imaginaries in different sorts, unless $d_p(\phi)(y)$ and $d_q(\phi)(y)$ have the form $\psi(y, c)$ and $\psi(y, d)$ respectively, for some L-formula $\psi(y, z)$. Namely we want to have "uniform definability" of generically stable types. This turns out to be the case.

Lemma 3.9. *(i) Let $\phi(x, y)$ be an L-formula. Then there is an L-formula $\psi(y, z)$ such that for all generically stable types $p(x) \in S(\bar{M})$ there is c such that $\psi(y, c)$ is the ϕ-definition of p (i.e. $\psi(x, c)$ is $d_p(\phi)(y)$).*

(ii) Moreover, in the context of (i), there is a partial type $\Sigma(z)$ over $\emptyset$, such that for any d, $\models \Sigma(d)$ if and only if $\psi(y, d)$ is the ϕ-definition of some generically stable global type.

Proof. (i) In fact clause (iii) of Theorem 3.4 provides such a formula: Take $\psi(y, z)$ to be $\vee_{w \subseteq 2N, |w|=N} \wedge_{i \in w} \phi(x_i, y)$ where $z = (x_i : i < 2N)$ and N is determined by ϕ.

(ii) Let $\Sigma(z)$ say that there is a totally indiscernible set $\{a_i : i < \omega\}$ such that $\psi(y, z)$ is equivalent to $\vee_{w \subseteq 2N, |w|=N} \wedge_{i \in w} \phi(a_i, y)$. The point is that this totally indiscernible sequence $(a_i)_i$ has to be the Morley sequence of some global generically stable type. □

By a $*$-definable set (over A) we mean a set X of infinite tuples $(a_1, a_2, ...)$ where a_i is of a given sort S_i say, such that for some collection $\Sigma(x_1, x_2, ...)$ of formulas over A, $X = \{(a_i)_i :\models \phi((a_i)_i) \text{ for each } \phi((x_i)_i) \in \Sigma\}$.

Corollary 3.10. *Identifying a global generically stable type $p(x)$ with its sequence of definitions, the collection of global generically stable types is a $*$-definable set over $\emptyset$.*

Remark 3.11. In the case of (some completion of) *ACVF*, and saturated model $M = (K, +, ...)$ and irreducible subvariety V of K^n, then the definable space of global generically stable types, concentrating on V, is the Hrushovski-Loeser version of the Berkovich space attached to V [6]. It is also equipped with a topology, which is simply the restriction to the set of generically stable types of the "strict spectral topology" on the type space $S_x(M)$: a basic closed set of types is the set of types containing a finite collection of polynomial equations and weak valuation inequalities.

4 Keisler Measures

The material in this section is essentially from [8] and [9].

The first definition is outside the model-theoretic context.

Definition 4.1. Let X be a set and $\mathcal{C}$ a family of subsets of X. $\mathcal{C}$ is said to have finite VC-dimension if there is a finite bound on the cardinality of finite subsets A of X which are "shattered by $\mathcal{C}$"; here A is said to be shattered by $\mathcal{C}$, if $\{A \cap C : C \in \mathcal{C}\}$ is the full power set of A.

VC denotes Vapnik-Chervonenkis. See [18].

Remark 4.2. Let M an L-structure, and $\phi(x, y)$ an L-formula ($x = (x_1, .., x_n)$ say). Then $\phi(x, y)$ has *NIP* for $T = Th(M)$, if and only if $\{\phi(x, b)(M) : b \in M\}$ as a collection of subsets of M^n, has finite VC-dimension.

Now we again fix a complete L-theory T.

Definition 4.3. Let M be a model. A (Keisler) measure $\mu(x)$ over M is a finitely additive probability measure on the collection of formulas $\phi(x, m)$, $m \in M$ (or on the corresponding definable sets).

So $\mu(x)$ has values in the real unit interval [0, 1]. When $\mu(x)$ has values only in $\{0, 1\}$ μ is a complete type $p(x)$ over M. As with types we say that $\mu(x)$ is a global Keisler measure if it is over $\bar{M}$. In a stable theory, a Keisler measure is essentially a weighted average of types. The study of Keisler measures in stable theories is more or less the content of Keisler's paper [10].

A Keisler measure $\mu(x)$ over M is the "same thing" as a regular probability measure on the Stone space $S_x(M)$ (complete types over M in variables x). Regular means that for any Borel subset B of the space, and $\epsilon > 0$, there are closed C and open U such that $C \subseteq B \subseteq U$ and $\mu(U \setminus C) < \epsilon$. In the context of a totally disconnected space such as $S_x(M)$, regularity in turn implies that for a closed C, $\mu(C)$ is the infimum of the $\mu(C')$ for C' a clopen containing C. So this says that a regular probability measure on $S_x(M)$ is determined by its restriction to the clopens, which is precisely a Keisler measure.

Lemma 4.4. *Suppose that $\mu(x)$ is a global Keisler measure, $\phi(x, y) \in L$, $\epsilon > 0$, and $(b_i : i < \omega)$ is an indiscernible sequence that $\mu(\phi(x, b_i)) \geq \epsilon$ for all b_i. Then $\{\phi(x, b_i) : i < \omega\}$ is consistent.*

The closeness of Keisler measures to types in the *NIP* context is witnessed by:

Proposition 4.5 (Assume T has *NIP*). *Any Keisler measure has "bounded support". Namely there is κ depending on the cardinality of T such that for any model*

M (including $\bar{M}$) and Keisler measure $\mu(x)$ over M, the set of formulas $\phi(x,m)$ over M, up to $\sim_\mu$ has cardinality at most κ. (Where $\sim_\mu$ is the equivalence relation on formulas of the symmetric difference being μ-measure 0.)

Proof. If not, one can find a formula $\phi(x,y) \in L$ and a large set $\{b_i : i \in I\}$ such that $\mu(\phi(x,b_i)\Delta\phi(x,b_j)) > 0$ for all $i \neq j$. Some Ramsey-style arguments yield $\epsilon > 0$ and an indiscernible sequence $(b_i : i < \omega)$ such that (without loss of generality) $\mu(\phi(x,b_{2i}) \wedge \neg\phi(x,b_{2i+1})) \geq \epsilon$ for all $i < \omega$. By Lemma 4.4, there is c realizing $\phi(x,b_{2i}) \wedge \neg\phi(x,b_{2i+1})$ for all i. This contradicts *NIP*. (See Lemma 2.3 (iii).) □

Let us just note for the future that $S_{(x,y)}(M)$ is not the same thing as $S_x(M) \times S_y(M)$. Nevertheless for $\mu(x)$ a Keisler measure over M, let μ^k denote the product measure (on $S_x(M) \times ... \times S_x(M)$).

Proposition 4.6 (*NIP*).

(i) *Suppose $\mu(x)$ is a Keisler measure over M, $\phi(x,y) \in L$, and $\epsilon > 0$. Then there is k, and $B \subset S_x(M)^k$ of positive μ^k measure, such that for all $c \in M$, and $(p_1,..,p_k) \in B$, $\mu(\phi(x,c))$ is within ϵ of $Fr(\phi(x,c),p_1,..,p_k)$. (Here $Fr(\phi(x,c),p_1,..,p_k)$ denotes the proportion of the p_i's which contain $\phi(x,c)$.)*

(ii) *Assuming $M = \bar{M}$ and μ is $Aut(\bar{M}/M_0)$-invariant (equivalently every formula of positive measure does not divide over M_0) for some small model M_0, then in (i) we can also insist that the types in B are $Aut(\bar{M}/M_0)$-invariant (do not divide over M_0).*

Explanation. (i) is a direct consequence of the VC (Vapnik-Chervonenkis) theorem [18], applied to $(S_x(M),\mu)$. This VC theorem is a "uniform law of large numbers" which I now describe. Let (X,Ω,μ) be a probability space. Let $\mathcal{C} \subset \Omega$ be a family of "events" which has finite VC-dimension. Given $A \in \Omega$, k and $p_1,..,p_k \in X$, let $g_{A,k}(p_1,..,p_k) = |\mu(A) - Fr(A,p_1,..,p_k)|$. Fix $\epsilon > 0$. THEN $\mu^k(\{\bar{p} \in X^k : sup_{A\in\mathcal{C}} g_{A,k}(\bar{p}) > \epsilon\}) \to 0$ as $k \to \infty$.

(One assumes μ^k-measurability of the set in the parentheses, so some additional fine tuning is required to apply the VC-theorem to the situation at hand.)

Definition 4.7. We call a global Keisler measure *invariant* if it is $Aut(\bar{M}/M_0)$-invariant for some small model M_0.

From Proposition 4.6 (ii) together with Proposition 2.11 (Borel definability of invariant types) we obtain:

Proposition 4.8 (*NIP*). *Any invariant global Keisler measure $\mu(x)$ over $\bar{M}$ is Borel definable (over M_0 if $\mu(x)$ is $Aut(\bar{M}/M_0)$-invariant). Namely, for any L-formula $\phi(x,y)$, and closed $C \subseteq [0,1]$, $\{q(y) \in S_y(M_0) : \mu(\phi(x,b)) \in C$ for some (any) b realizing $q\}$ is a Borel subset of $S_y(M_0)$.*

Via Proposition 4.8 we can form "nonforking amalgams" of global Keisler measures (generalizing the case $p \otimes q$ for types mentioned earlier). So let $\mu(x)$

be a global invariant Keisler measure, and $\lambda(y)$ any global Keisler measure. We want to form the "nonforking amalgam", $\mu(x) \otimes \lambda(y)$, a global Keisler measure on formulas of the form $\phi(x, y)$ over M. Fix $\phi(x, y)$ over $\bar{M}$. Let M_0 be a small model containing the parameters from ϕ and such that μ is $Aut(\bar{M}/M_0)$-invariant. We define

$$(\mu(x) \otimes \lambda(y))(\phi(x, y)) = \int_{S_y(M_0)} \mu(\phi(x, q)) d\lambda | M_0$$

Some explanations of the formula above are in order. We write $\mu(\phi(x, q))$ because by M_0-invariance, for $b \in \bar{M}$, $\mu(\phi(x, b))$ depends only on $tp(b/M_0)$. And $\lambda|M_0$ means the restriction of λ to formulas $\psi(y)$ over M_0, equivalently as mentioned above to $S_y(M_0)$. The integral makes sense, because by Borel definability of μ (Proposition 4.8) the function taking $q \in S_y(M_0)$ to $\mu(\phi(x, q))$ is Borel.

Note that when μ, λ are global types $p(x)$, $q(y)$, respectively, and p is invariant then the definition above yields $p(x) \otimes q(y)$ as described earlier, i.e. $tp(a, b/\bar{M})$ where b realizes q and a realizes $p|(\bar{M}, b)$.

If $\lambda(y)$ is also invariant, we can form $\lambda(y) \otimes \mu(x)$ (again as a measure on formulas in variables x, y), and it makes sense to ask whether $\mu(x) \otimes \lambda(y) = \lambda(y) \otimes \mu(x)$. Of course this need not be the case in general.

Definition 4.9. Suppose $\mu(x)$ is an invariant global Keisler measure. We call $\mu(x)$ *symmetric* if $\mu(x_1) \otimes \mu(x_2) = \mu(x_2) \otimes \mu(x_1)$.

For an invariant measure $\mu(x)$ we can form the n-fold nonforking amalgam, $\mu^{(n)}(x_1, .., x_n)$ inductively: $\mu^{(n+1)}(x_1, x_2, .., x_{n+1}) = \mu(x_{n+1}) \otimes \mu^{(n)}(x_1, .., x_n)$. $\mu^{(\omega)}(x_1, x_2, ...)$ is the union. Symmetry of μ is equivalent to "total indiscernibility" of $\mu^{(\omega)}$ (in the obvious sense).

An elaboration of the VC theorem (see Theorem 3.2 (iii) from [9]) yields:

Proposition 4.10 (Assume ***NIP*****).** *Let $\mu(x)$ be a global $Aut(\bar{M}/M)$-invariant measure, which is symmetric. Let $\phi(x, y) \in L$ and $\epsilon > 0$. Then for sufficiently large m there are formulas $\theta_m(x_1, .., x_m)$ over M, such that $\mu^{(m)}(\theta_m(x_1, .., x_m) \to 1$ as $m \to \infty$, and for all $(a_1, .., a_m)$ realizing θ_m, and all $b \in \bar{M}$, $\mu(\phi(x, b))$ is within ϵ of $Fr(\phi(x, b), a_1, .., a_m)$.*

Global invariant symmetric Keisler measures are what we have called *generically stable measures*, and coincide with (global) generically stable types when μ is a type. Their ubiquity in *NIP* theories is still to be made precise. In [14] it is pointed out how these measures are involved in characterizing strongly dependent theories. In the case of suitable definable groups G (so called *fsg*-groups), there is a unique G-invariant global Keisler measure, which happens to be generically stable. There is also a basic construction from indiscernible segments, giving rise to generically stable measures: given an indiscernible segment $J = (a_i : i \in [0, 1])$, define the global average measure μ of J by: $\mu(\phi(x, b))$ is the Lebesgue measure of $\{i \in [0, 1] : \models \phi(a_i, b)\}$ (which by *NIP* is a finite union of intervals and points). Then μ

is generically stable (being both definable over and finitely satisfiable in J). After my lecture in the summer school Berarducci, Mamino, and Viale pointed out that any Borel probability measure on [0, 1] will work in place of Lebesgue measure.

5 Uniform Definability of Types Over Finite Sets (UDTFS)

In this final section I will describe a very recent result by Chernikov and Simon [4] which gives a characterization of *NIP* in terms of definability of types, analogous in a sense with the stable case.

The context is really that of a one-sorted theory T. Note first that any complete type $p(x)$ over a finite set A is definable (in any theory T). Chris Laskowski asked whether in *NIP* theories, we have *uniform definability* as $p(x)$ and A vary.

Definition 5.1. The theory T has UDTFS (uniform definability of types over finite sets), if given variable x, for any L-formula $\phi(x, y)$ there is an L-formula $\psi(y, z)$ such that for any finite set A and complete type $p(x)$ over A, there is some tuple d from A such that for all tuples a from A (of length that of y) $\phi(x, a) \in p(x)$ iff $\models \psi(a, d)$.

Remark 5.2. UDTFS implies *NIP*.

Proof. Assume T has UDTFS. Let $\phi(x, y)$ be an L-formula, and $\psi(y, z)$ the L-formula given by UDTFS. Suppose z is an ℓ-tuple. Then for any finite set A, the collection of subsets of A picked out by formulas $\phi(b, y)$ as b varies, has cardinality at most $|A|^{\ell}$. This clearly shows that $\phi(x, y)$ has *NIP*. See also Remark 4.2. □

We aim towards the converse of Remark 5.2 (see Theorem 5.7 below).

An important result of Shelah [15], (see also [13]) is:

Fact 5.3. *Suppose T has NIP, and let $M \models T$. Let M_{ext} be the expansion of M by adding predicates for all "externally definable" subsets of M^n for all n (where $X \subseteq M^n$ is externally definable if for some formula $\phi(x)$ with parameters from $\bar{M}$, $X = \{a \in M^n : \bar{M} \models \phi(a)\}$). Then $Th(M_{ext})$ has quantifier elimination and also has NIP.*

I was interested in a kind of characterization of *NIP* theories in terms related to this quantifier elimination result. An elaboration of ideas in the proof of the above (together with work of Guingona) led Chernikov and Simon to prove the related, and in fact stronger result [4], which I state now. As a matter of notation, for $A \subset \bar{M}$ and $\theta(x, c)$ a formula with parameters c from $\bar{M}$, $\theta(x, c)(A)$ denotes the set of tuples from A which satisfy the formula $\theta(x, c)$ in $\bar{M}$.

Lemma 5.4 (Assume *NIP***).** *Let $A \subset \bar{M}$. Given a formula $\phi(x, c)$ with parameters c from $\bar{M}$ there is a formula $\theta(x, w)$ (w a tuple of variables) such that for any finite subset $A_0 \subseteq A$ contained in $\phi(x, c)(A)$ there is some tuple b_0 from A such that $A_0 \subseteq \theta(x, b_0)(A) \subseteq \phi(x, c)(A)$.*

Let us note how *QE* for $Th(M_{ext})$ follows from Lemma 5.4. We apply Lemma 5.4 to the case where $A = M$. Consider for example a formula $\phi(x, y, c)$ where x, y are single variables, and c is from $M_1 > M$. We want to show that the projection of $\phi(x, y, c)(M)$ on the x-coordinate is also externally definable. Let $\theta(x, y, w)$ be given by Lemma 5.4. The Lemma and compactness yields an elementary extension (M_1', M') of the pair (M_1, M) and b_0 from M' such that

(i) $\theta(x, y, b_0)(M) = \phi(x, y, c)(M)$ and
(ii) $(M_1', M') \models \forall x, y \in M'(\theta(x, y, b_0) \to \phi(x, y, c))$.

Claim. $\{x \in M : \exists y \in M(\phi(x, y, c)\}$ coincides with $\{x \in M : M' \models \exists y(\theta(x, y, b_0))\}$ and hence is externally definable.

Proof of Claim. Clearly the left hand side is contained in the right hand side. Suppose now that $a \in M$ and $M' \models \exists y(\theta(a, y, b_0))$. By (ii), $(M_1', M') \models \exists y \in M'(\phi(a, y, c))$. So as (M_1, M) is an elementary substructure of (M_1', M'), we have that $(M_1, M) \models \exists y \in M(\phi(a, y, c))$ as required.

Compactness (i.e. adding a predicate for A) and Lemma 5.4 yields:

Corollary 5.5. *Let us pick, for each formula $\theta(x, w)$ an integer n_θ. Then for any formula $\phi(x, y)$ there are $\theta_1(x, w_1)$,.., $\theta_k(x, w_k)$ such that for any subset A of $\bar{M}$ and tuple c from $\bar{M}$, there is $i = 1, .., k$ such that whenever A_0 is a subset of $\phi(x, c)(A)$ of cardinality $\leq n_{\theta_i}$ there is a tuple b_0 from A such that $A_0 \subseteq \theta_i(x, b_0)(A) \subseteq \phi(x, c)(A)$.*

The other ingredient needed comes from discrete geometry. Before stating it, let me recall that the *VC*-codimension of a family $\mathcal{C}$ of subsets of a set X, is the greatest d such that some subset of $\mathcal{C}$ of size d is shattered by X (in the obvious sense). $\mathcal{C}$ has finite *VC*-dimension iff it has finite *VC*-codimension.

Theorem 5.6 ([1, 11]). *Let X be a set, and $\mathcal{F}$ a family of subsets of X of finite VC-dimension. Then for sufficiently large k ($\geq$ the VC-codimension of $\mathcal{F}$), for any $p \geq k$ there is N such that if $\mathcal{G} \subseteq \mathcal{F}$ is a finite subfamily and among any p elements of $\mathcal{G}$ some k of them have nonempty intersection, then there is an N-element subset Y of X such that every set in $\mathcal{G}$ meets Y. Moreover N depends only on p, k (not $\mathcal{F}$).*

Now we state and prove the main result of this section:

Theorem 5.7 (Chernikov-Simon). *T has NIP iff T has UDTFS.*

Proof. We already saw right implies left (5.2).

Left implies right: The reader should be warned that the roles of x, y in Definition 5.1 are interchanged in this proof. We take x a single variable for simplicity. For each formula $\theta(x, w)$ let n_θ be the *VC*-dimension of the family of sets defined by $\theta(x, b)$ as b varies (which is finite by *NIP*).

Now fix an L-formula $\phi(x, y)$. Let $\theta_1(x, w_1)$,.., $\theta_k(x, w_k)$ be the formulas given by Corollary 5.5. For any finite set A and tuple a (for y) there is $i = 1, .., k$ such that the conclusion of Corollary 5.5 holds for i. We will produce a uniform definition for $\{x \in A :\models \phi(x, a)\}$ depending on which i works. This will of course give a uniform definition, by usual coding tricks (as long as $|A| \geq 2$ which can clearly be assumed). Hence there is no loss in assuming $i = 1$ and writing $\theta(x, w)$ for $\theta_1(x, w_1)$. Let $n = n_\theta$. Suppose w is an ℓ-tuple of variables. Consider a finite set A and tuple a from $\bar{M}$. Let B be the set of $b \in A^\ell$ such that $\theta(x, b)(A) \subseteq \phi(x, a)(A)$. Let $\mathcal{F}$ be the family of nonempty subsets $\theta(d, w)(B)$ of B as d varies in A. Then $\mathcal{F}$ has VC-codimension $\leq n$. Let N be given by Theorem 5.6 for $p = k = n$.

Claim 1. The intersection of any k members of $\mathcal{F}$ is nonempty.

Proof of Claim 1. Let $\theta(d_1, w)(B)$,.., $\theta(d_k, w)(B)$ be in $\mathcal{F}$. Let $A_0 = \{d_1, .., d_k\}$ (a subset of A). By definition of B, $A_0 \subseteq \phi(x, a)(A)$ and has size $k = n$ (VC-dimension of $\theta(x, w)$). For example $\models \theta(d_1, b)$ for some $b \in B$, so $\theta(x, b)(A) \subseteq \phi(x, a)(A)$ whereby $\models \phi(d_1, a)$. So we can apply the conclusion of Corollary 5.5 to obtain $b_0 \in A^\ell$ such that $\models \theta(d_i, b_0)$ for $i = 1, .., k$, and such that $\theta(x, b_0)(A) \subseteq \phi(x, a)(A)$. But then, by definition of B again, $b_0 \in B$, and yields Claim 1.

By Claim 1, Theorem 5.6 (with $\mathcal{G} = \mathcal{F}$), and the choice of N,
(*) there are $b_1, .., b_N \in B$ such that any set $\theta(d, w)(B)$ in $\mathcal{F}$ contains some b_i.

Claim 2. For all $d \in A$, $\models \phi(d, a)$ iff $\models \bigvee_{i=1,...,N} \theta(d, b_i)$.

Proof of Claim 2. If $\models \phi(d, a)$ then by definition $\theta(d, w)(B) \in \mathcal{F}$, hence by (*) $\models \theta(d, b_i)$ for some $i = 1, .., N$. Conversely, if $\models \theta(d, b_i)$ for some i, then by definition of $b_i \in B$, we have $\models \phi(d, a)$.

Claim 2 gives us our uniform definition, as θ and N depend only on $\phi(x, y)$. □

Acknowledgements Supported by EPSRC grant EP/I002294/1.

References

1. N. Alon, D.J. Kleitman, Piercing convex sets and the Hadwiger-Debrunner (p, q)- problem. Adv. Math. **96**, 103–112 (1992)
2. A. Chernikov, I. Kaplan, Forking and dividing in *NTP*$_2$ theories. J. Symb. Log. **77**, 1–20 (2012)
3. A. Chernikov, P. Simon, Externally definable sets and dependent pairs. Isr. J. Math. **194**, 409–425 (2013)
4. A. Chernikov, P. Simon, Externally definable sets and dependent pairs II. Trans. AMS (to appear)
5. A. Chernikov, I. Kaplan, S. Shelah, On nonforking spectra (preprint, 2012)
6. E. Hrushovski, F. Loeser, *Non-archimedean Tame Topology, and Stably Dominated Types*. Princeton Monograph Series (to appear)
7. E. Hrushovski, Y. Peterzil, A. Pillay, Groups, measures and the *NIP*. J. AMS **21**, 563–596 (2008).
8. E. Hrushovski, A. Pillay, On NIP and invariant measures. J. Eur. Math. Soc. **13**, 1005–1061 (2011)
9. E. Hrushovski, A. Pillay, P. Simon, On generically stable and smooth measures in *NIP* theories. Trans. AMS **365**, 2341–2366 (2013)
10. H.K. Keisler, Measures and forking. Ann. Pure Appl. Log. **45**, 119–169 (1987)

11. J. Matousek, Bounded VC-dimension implies a fractional Helly theorem. Discrete Comput. Geom. **31**, 251–255 (2004)
12. A. Pillay, *Geometric Stability Theory* (Oxford University Press, Oxford, 1996)
13. A. Pillay, Externally definable sets and a theorem of Shelah, in *Felgner Festchrift* (College Publications, London, 2007)
14. A. Pillay, On weight and measure in *NIP* theories. Notre Dame J. Formal Log. (to appear)
15. S. Shelah, Dependent first order theories, continued. Isr. J. Math. **173**, 1–60 (2009)
16. S. Shelah, Dependent dreams and counting types (preprint 2012)
17. S. Shelah, Strongly dependent theories. Isr. J. Math. (to appear)
18. V. N. Vapnik, A.Y. Chervonenkis, On the uniform convergence of relative frequencies of events to their probabilities. Theory Prob. Appl. **16**, 264–280 (1971)

Lectures on the Model Theory of Real and Complex Exponentiation

A.J. Wilkie

1 Introduction

We shall be interested in the structures $\mathbb{R}_{exp}$ and $\mathbb{C}_{exp}$, the expansions of the rings of real and complex numbers respectively, by the corresponding exponential functions and, in particular, the problem of describing mathematically their definable sets.

In the real case I shall give the main ideas of a proof of the theorem stating that the theory T_{exp} of the structure $\mathbb{R}_{exp}$ is *model complete*.

Definition 1. A theory T in a language L is called *model complete* if it satisfies one of the following equivalent conditions:

(i) for every formula $\phi(x)$ of L, there exists an existential formulas $\psi(x)$ of L such that $T \models \forall x(\phi(x) \leftrightarrow \psi(x))$;
(ii) for all models $\mathbb{M}_0$, $\mathbb{M}$ of T with $\mathbb{M}_0 \subseteq \mathbb{M}$ we have that $\mathbb{M}_0 \preccurlyeq_1 \mathbb{M}$ (i.e. existential formulas are absolute between $\mathbb{M}_0$ and $\mathbb{M}$);
(iii) for all models $\mathbb{M}_0$, $\mathbb{M}$ of T with $\mathbb{M}_0 \subseteq \mathbb{M}$ we have that $\mathbb{M}_0 \preccurlyeq \mathbb{M}$ (i.e. all formulas are absolute between $\mathbb{M}_0$ and $\mathbb{M}$);
(iv) for all models $\mathbb{M}$ of T, the $L_{\mathbb{M}}$-theory $T \cup \text{Diagram}(\mathbb{M})$ is complete.

(In (iv), $L_{\mathbb{M}}$ is the language extending L by a constant symbol naming each element of the domain of $\mathbb{M}$, and Diagram($\mathbb{M}$) denotes the set of all quantifier-free sentences of $L_{\mathbb{M}}$ that are true in $\mathbb{M}$.)

This is, of course, a theorem. The actual definition that gives rise to the name is (iv).

I shall prove that (ii) holds for $T = T_{exp}$ and in this case the task can be further reduced:

A.J. Wilkie (✉)
School of Mathematics, The Alan Turing Building, University of Manchester, Manchester M13 9PL, UK
e-mail: alex.wilkie@manchester.ac.uk

L. van den Dries et al., *Model Theory in Algebra, Analysis and Arithmetic*, Lecture Notes in Mathematics 2111, DOI 10.1007/978-3-642-54936-6_3,

Exercise 1. Suppose that all pairs of models $\mathbb{M}_0$, $\mathbb{M}$ of T_{exp} with $\mathbb{M}_0 \subseteq \mathbb{M}$ have the property that any quasipolynomial (see below) with coefficients in $\mathbb{M}_0$ and having a solution (i.e. a zero) in $\mathbb{M}$, also has a solution in $\mathbb{M}_0$. Prove that T_{exp} is model complete.

Definition 2. Let $\mathbb{M}_0 = \langle M_0, ...\rangle$ be a substructure of a model $\mathbb{M} = \langle M, ...\rangle$ of T_{exp}. Then a function (from M^n to M) that can be written in the form

$$\langle x_1, \ldots, x_n\rangle \mapsto P(x_1, \ldots, x_n, \exp(x_1), \ldots, \exp(x_n)),$$

where P is a polynomial (in $2n$ variables) with coefficients in M_0, is called a *quasipolynomial with coefficients in* $\mathbb{M}_0$.

In fact I shall only prove, in the notation of Exercise 1, that a quasipolynomial with solutions in M^n has one, $\langle b_1, \ldots, b_n\rangle$ say, that is $\mathbb{M}_0$-bounded, i.e. for some $a \in M_0$ with $a > 0$, we have that $-a \leq b_i \leq a$ for $i = 1, \ldots, n$.

To go on to find a solution in M_0^n requires a separate argument, which I shall not go into in these notes. The method is fairly routine and appeared several years before the eventual proof of the model completeness of T_{exp} (see [15]). The case $n = 1$ of both arguments is reasonably straightforward and serves as a good introduction to the general case:

Exercise 2. Let $\mathbb{M}_0 = \langle M_0, ...\rangle$ and $\mathbb{M} = \langle M, ...\rangle$ be models of T_{exp} with $\mathbb{M}_0 \subseteq \mathbb{M}$. Use the two step method discussed above to show that every zero in M of a one variable (nonzero) quasipolynomial with coefficients in $\mathbb{M}_0$, actually lies in M_0. [That is, first show that such zeros are $\mathbb{M}_0$-bounded, and then show that they lie in M_0.]

Of course it is (i) in Definition 1 that is the desired conclusion, but by establishing it via the model theoretic statement (ii) we get no information on how to *effectively* find $\psi(x)$ from $\phi(x)$. We do know that this can be done in principle if Schanuel's Conjecture is true, but even with this assumption a transparent (say, primitive recursive) algorithm is still lacking.

In the complex case, where we know that model completeness fails (see [7]), we discuss the conjecture of Zilber stating that every definable subset of $\mathbb{C}$ (in the structure $\mathbb{C}_{exp}$) is either countable or cocountable. This property (of an expansion of the complex field say) is called *quasi-minimality*. This is a weakening of a well known model-theoretic notion known as *m*inimality: a structure is called just *minimal* if every definable subset of its domain is either finite or cofinite.

When discussing Zilber's conjecture I shall be assuming that the reader has studied some complex analysis as well as model theory. Many of the arguments will combine the two. Here is an example.

Exercise 3. Let $\mathbb{M}$ be an expansion of the ring of complex numbers in which an entire function $f : \mathbb{C} \to \mathbb{C}$ is definable. Prove that if $\mathbb{M}$ is minimal then f is a polynomial.

I shall study these rather specific topics in the more general context of expansions of the underlying fields (real or complex) by analytic functions.

2 Analytic Functions

In this section $\mathbb{K}$ denotes either $\mathbb{R}$ or $\mathbb{C}$.

Let U be an open neighbourhood of a point $\omega \in \mathbb{K}^n$ (for the usual topology) and let $f : U \to \mathbb{K}$ be an infinitely differentiable function. Then we may form the Taylor series of f at ω:

$$Tf := \sum_{\alpha \in \mathbb{N}^n} \frac{f^{(\alpha)}(\omega)}{\alpha!} \cdot (x - \omega)^\alpha.$$

It is important to realise that in the real case this is just a formal series in the variables $x = \langle x_1, \ldots, x_n \rangle$. It may not converge for any values of the variables (other than ω itself) and even if it does, the sum may bear no relation to $f(x)$. If there exists an open neighbourhood V of ω with $V \subseteq U$ such that the series converges to $f(x)$ for each $x \in V$, then we say that the function f is *analytic* at ω. If it is analytic at all points ω in its domain U then we just say that f is *analytic*. In the complex case, the continuous differentiability of f on U is in fact sufficient to guarantee that f is analytic.

I now state some theorems that will be important when we come to discuss quantifier elimination and model completeness for expansions of $\mathbb{K}$ by analytic functions.

Throughout these notes, $\Delta_{\mathbb{K}}^{(n)}(\omega; r)$ denotes the *box neighbourhood* (also called the *polydisk*, in the case that $\mathbb{K} = \mathbb{C}$) centred at $\omega \in \mathbb{K}^n$ and polyradius $r \in \mathbb{R}^n_{>0}$ defined by

$$\Delta_{\mathbb{K}}^{(n)}(\omega; r) := \{x \in \mathbb{K}^n : |x_i - \omega_i| < r_i \text{ for } i = 1, \ldots, n\}.$$

The subscript $\mathbb{K}$ will be omitted if it is clear from the context, and the superscript also in the case that $n = 1$.

For polyradii r and s we write $s < r$, or $s \le r$, if corresponding coordinates are so ordered. In the following statements we assume that $n > 1$ and make the convention that primed variables are $(n-1)$-tuples. Further, if x is already designated as an n-tuple, then x' is its initial $(n-1)$-tuple.

Theorem 1 (The Implicit Function Theorem for One Dependent Variable). *Suppose that $F : \Delta_{\mathbb{K}}^{(n)}(0; r) \to \mathbb{K}$ is analytic. Suppose further that*

$$F(0) = 0 \ne \frac{\partial F}{\partial x_n}(0).$$

Then there exists $s \in \mathbb{R}^n_{>0}$ with $s \le r$ and an analytic function $\phi : \Delta_{\mathbb{K}}^{(n-1)}(0'; s') \to \mathbb{K}$ such that $\phi(0') = 0$ and for all $x' \in \Delta_{\mathbb{K}}^{(n-1)}(0'; s')$ we have $F(x', \phi(x')) = 0$. Further, for each such x', $\phi(x')$ is the only $y \in \mathbb{K}$ satisfying $|y| < s_n$ and $F(x', y) = 0$.

The uniqueness statement here is important for definability issues. It guarantees that (at least, in the real case) if the data is definable then so is the implicit function ϕ. In the following more general result the non-singularity condition is relaxed. However, it suffers from the disadvantage that the functions asserted to exist are not necessarily definable from the data.

Theorem 2 (The Weierstrass Preparation Theorem). *Suppose that* $F : \Delta_{\mathbb{K}}^{(n)}(0;r) \to \mathbb{K}$ *is analytic. Suppose further that* $p \geq 1$ *and*

$$F(0) = \frac{\partial F}{\partial x_n}(0) = \cdots = \left(\frac{\partial}{\partial x_n}\right)^{p-1} F(0) = 0 \neq \left(\frac{\partial}{\partial x_n}\right)^{p} F(0).$$

Then there exists $s \in \mathbb{R}^n_{>0}$ *with* $s \leq r$*, analytic functions* $\phi_1, \ldots, \phi_p : \Delta_{\mathbb{K}}^{(n-1)}(0'; s') \to \mathbb{K}$ *and* $u : \Delta_{\mathbb{K}}^{(n)}(0;s) \to \mathbb{K}$ *such that* $\phi_1(0') = \ldots = \phi_p(0') = 0$*,* u *does not vanish, and for all* $x \in \Delta_{\mathbb{K}}^{(n)}(0;s)$

$$F(x) \;=\; u(x)\cdot(x_n^p + \phi_1(x')\cdot x_n^{p-1} + \cdots + \phi_p(x')).$$

Theorem 2 forms the foundation for the local theory of analytic functions and analytic sets (= zero sets of analytic functions). For the geometric theory the non-singularity hypothesis constitutes no loss of generality:

Exercise 4. Let $F : \Delta_{\mathbb{K}}^{(n)}(0;r) \to \mathbb{K}$ be an analytic function. Suppose that $F(0) = 0$ but that F is not identically zero. Prove that after a linear change of coordinates (which may be taken to have an integer matrix) the hypothesis of Theorem 2 holds (for some $p \geq 1$ and possibly a smaller r).

Unfortunately, as we shall see in the next section, this is not good enough for model theoretic considerations: one cannot just, e.g., permute variables when one is considering a fixed projection of, say, the zero set of an analytic function into lower dimensions. One has to have some form of preparation theorem that deals with the case that $\left(\frac{\partial}{\partial x_n}\right)^p F(0) = 0$ for all p, or, equivalently, when the function $F(0', \cdot)$ is identically zero.

Theorem 3 (The Denef-van den Dries Preparation Theorem). *Let* $F : \Delta_{\mathbb{K}}^{(n)}(0;r) \to \mathbb{K}$ *be an analytic function. Then there exist* $d \geq 1$*,* $s \leq r$ *and analytic functions* $\phi_j : \Delta_{\mathbb{K}}^{(n-1)}(0';s') \to \mathbb{K}$*, and* $u_j : \Delta_{\mathbb{K}}^{(n)}(0;s) \to \mathbb{K}$ *(for* $j \leq d$*) with the* u_j*'s nonvanishing, such that for all* $x \in \Delta_{\mathbb{K}}^{(n)}(0;s)$

$$F(x) = \sum_{j=0}^{d} \phi_j(x')\cdot x_n^j \cdot u_j(x).$$

Finally in this section I state the many variable version of the Implicit Function Theorem. The role of the derivative is played by the *Jacobian*:

Definition 3. Let $F : U \times V \to \mathbb{K}^m$ be an analytic map (i.e. its coordinate functions are analytic functions) where $U \subseteq \mathbb{K}^n$ and $V \subseteq \mathbb{K}^m$ are open sets. Write x for the U-variables and y for the V-variables. Then the *Jacobian* J_F of F with respect to y is given by the determinant of the matrix

$$\left(\frac{\partial F_i}{\partial y_j}\right)_{1 \le i,j \le m}.$$

It is an analytic function from $U \times V$ to $\mathbb{K}$.

Theorem 4 (The Implicit Function Theorem for Several Dependent Variables). *Let $F : U \times V \to \mathbb{K}^m$ be as in the definition. Let $\langle a, b \rangle \in U \times V$ be such that $F(a,b) = 0$ and $J_F(a,b) \neq 0$. Then there exists an open neighbourhood $U_a \times V_b \subseteq U \times V$ of $\langle a, b \rangle$ and an analytic map $\Phi : U_a \to V_b$ such that for all $x \in U_a$ we have $F(x, \Phi(x)) = 0$. Further, for each such x, $\Phi(x)$ is the unique $y \in V_b$ satisfying $F(x,y) = 0$.*

It hardly seems worth giving a reference for the Implicit Function Theorem—just consult your favourite analysis text. The same is almost true for the Weierstrass Preparation Theorem though for this (and for the formal case) I would recommend [10]. As for Theorem 3, I give a sketch of the proof in [17] but for more details you will have to consult the original paper [1].

3 The Structure $\mathbb{R}_{an}$ and Its Reducts

In this and the next three sections we are only concerned with the case that $\mathbb{K} = \mathbb{R}$.

The local theory of real analytic functions and real analytic sets is captured by the definability theory of the structure $\mathbb{R}_{an}$. This is the expansion of the ordered ring of real numbers $\bar{\mathbb{R}} := \langle \mathbb{R}; +, \cdot, -, 0, 1, < \rangle$ by all $r \in \mathbb{R}$ as distinguished elements and all *restricted* analytic functions:

Definition 4. Let $n \ge 1$, $\omega \in \mathbb{R}^n$, $r \in \mathbb{R}^n_{>0}$ and $f : \Delta^{(n)}(\omega; r) \to \mathbb{R}$ be an analytic function. Let s be a polyradius with $s < r$. Then the function $\tilde{f} : \mathbb{R}^n \to \mathbb{R}$ defined by

$$\tilde{f}(x) = \begin{cases} f(x) & \text{if } x \in \Delta^{(n)}(\omega; s), \\ 0 & \text{otherwise.} \end{cases}$$

is called a *restricted analytic function*. Of course it is not analytic on $\mathbb{R}^n$, only on $\Delta^{(n)}(\omega; s)$.

Definition 5. (i) $\mathbb{R}_{an} := \langle \bar{\mathbb{R}}, \{r\}_{r \in \mathbb{R}}, \{\text{all restricted analytic functions}\} \rangle$.
(ii) $T_{an} := Th(\mathbb{R}_{an})$.
(iii) $\mathbb{R}^D_{an} := \langle \mathbb{R}_{an}, D \rangle$, where $D : \mathbb{R}^2 \to \mathbb{R}$ is defined by

$$D(x, y) = \begin{cases} x/y & \text{if } y \neq 0 \text{ and } |x| \leq |y|, \\ 0 & \text{otherwise.} \end{cases}$$

(iv) $T_{an}^D := Th(\mathbb{R}_{an}^D)$.

We have the following results.

Theorem 5 (Gabrielov). *T_{an} is model complete.*

Theorem 6 (Denef-van den Dries). *T_{an}^D admits elimination of quantifiers.*

Since both the graph of the function D and its complement in $\mathbb{R}^3$ are existentially definable in the language $\mathcal{L}(\mathbb{R}_{an})$, we see immediately that the two structures $\mathbb{R}_{an}^D$ and $\mathbb{R}_{an}$ have the same definable (= existentially definable) sets, and that Theorem 6 implies Theorem 5. The theory T_{an} does not have quantifier elimination (in the language $\mathcal{L}(\mathbb{R}_{an})$) even locally (see Exercise 6 below). However, our approach to the proof of Theorem 6 will still involve investigating projections of those subsets of $\mathbb{R}^n$ that are quantifier-free definable in the language $\mathcal{L}(\mathbb{R}_{an})$ and the following exercise reduces the problem to studying the bounded situation.

Exercise 5. Let $A \subseteq \mathbb{R}^{n+m}$ be quantifier-free definable in the language $\mathcal{L}(\mathbb{R}_{an})$ and let $\pi : \mathbb{R}^{n+m} \to \mathbb{R}^n$ be the projection map onto the first n coordinates.

Prove that there exists $R \in \mathbb{R}_{>0}^{n+m}$ and a set $C \subseteq \mathbb{R}^n$ that is quantifier-free definable in the language $\mathcal{L}(\mathbb{R}_{an})$ such that

$$\pi[A] = \pi[A \cap \Delta^{(n+m)}(0; R)] \cup C.$$

It is rather a deep fact that any subset of $\mathbb{R}^2$ which is quantifier free definable in the language $\mathcal{L}(\mathbb{R}_{an}^D)$ is in fact quantifier free definable in the language $\mathcal{L}(\mathbb{R}_{an})$. However:

Exercise 6 (Hard, the result is due to Osgood). Find a set $A \subseteq \mathbb{R}^3$ which is quantifier-free definable in the language $\mathcal{L}(\mathbb{R}_{an}^D)$ and has the property that for no $\epsilon > 0$ is $A \cap \Delta^{(3)}(0; \epsilon)$ quantifier-free definable in the language $\mathcal{L}(\mathbb{R}_{an})$.

The following exercises are set with a view towards the proof of Theorem 6.

Exercise 7. Let $A \subseteq \mathbb{R}^n$ be a quantifier-free definable set in the language $\mathcal{L}(\mathbb{R}_{an})$. Prove that for each $\omega \in \mathbb{R}^n$ there exists $s_\omega \in \mathbb{R}_{>0}^n$ such that the set $A \cap \Delta^{(n)}(\omega; s_\omega)$ is a finite union of sets of the form

$$\{x \in \Delta^{(n)}(\omega; s_\omega) : f(x) = 0, g_1(x) > 0, \ldots, g_N(x) > 0\} \qquad (*)$$

for some analytic functions $f, g_1, \ldots, g_N : \Delta^{(n)}(\omega; s_\omega) \to \mathbb{R}$.

Exercise 8. Let A be as in Exercise 7 and assume that $n \geq 2$. Suppose that for all $\omega \in \mathbb{R}^n$, each f and g_j arising in the conclusion of Exercise 7 has the (pleasant, but not usually realised) property that either it does not vanish at ω, or else it is *regular* *in* x_n at ω. That is there is some $p \geq 1$ (depending on the function) such that it satisfies the hypothesis of Theorem 2. Prove that the projection, $\{x' \in \mathbb{R}^{n-1} : \exists x_n \in \mathbb{R}, \langle x', x_n \rangle \in A\}$, of A onto the first $n-1$ coordinates, is quantifier-free definable in the language $\mathcal{L}(\mathbb{R}_{an})$. [Hint: First consider the case when A has the form (*). As well as Exercise 7 and Theorem 2 (translated to arbitrary points in $\mathbb{R}^n$), you will need Tarski's Theorem on the quantifier elimination for $\bar{\mathbb{R}}$. Then use Exercise 5 and the compactness of bounded closed subsets of $\mathbb{R}^n$.]

If you managed to do these, then you should not have too much trouble with Exercise 9. This time you will need Theorem 3 as well as the following fact: if $f : \Delta(0; r) \to \mathbb{R}$ is an analytic function and $f(0) = 0$ then there is an analytic function $g : \Delta(0; r) \to \mathbb{R}$ such that for all $x \in \mathbb{R}$, $f(x) = x \cdot g(x)$.

Exercise 9. Let $A \subseteq \mathbb{R}^2$ be a quantifier-free definable set in the language $\mathcal{L}(\mathbb{R}_{an})$. Prove that the projection of A onto the first coordinate is also quantifier-free definable in the language $\mathcal{L}(\mathbb{R}_{an})$.

Having studied these special cases let us now consider the general situation, where Theorem 3 is needed and where there is no possibility of dividing the coefficients ϕ_j by a common factor and thereby reducing to the situation of Exercise 8 (as you should have done in solving Exercise 9).

As in Exercise 8, it is sufficient to consider sets of the form $(*)$, and for simplicity I consider the projection of the set

$$A := \{x \in \Delta^{(n)}(0; r) : F(x) = 0\}$$

onto the first $n-1$ coordinates. We may obviously assume that the analytic function $F : \Delta^{(n)}(0; r) \to \mathbb{R}$ is not identically zero and the difficult case is when the function $F(0', \cdot)$ is identically zero.

We choose a representation of F as displayed in Theorem 3. The idea is to divide by the largest (in modulus) of the ϕ_j's and then appeal to Weierstrass. To this end, let

$$S := \{c \in [-1, 1]^{d+1} : \text{at least one coordinate of } c \text{ is } 1\}$$

Let $v = \langle v_0, \ldots, v_d \rangle$ be new variables, let $c \in S$ and consider the analytic function $F_c : (-1, 1)^{d+1} \times \Delta^{(n)}(0; s) \to \mathbb{R}$ given by

$$F_c(v, x) := \sum_{j=0}^{d} (c_j + v_j) \cdot x_n^j \cdot u_j(x).$$

Now if p is minimal such that $c_p \neq 0$, then, by direct calculation,

$$(\frac{\partial}{\partial x_n})^p(0, 0) = p! \cdot c_p \cdot u_p(0) \neq 0.$$

So we are in the Weierstrass situation (of Exercise 8) with respect to the variables v, x and hence we obtain a quantifier-free formula of $\mathcal{L}(\mathbb{R}_{an})$, $\Theta_c(v, x')$ say, which defines the projection of the zero set of F_c onto the $\langle v, x'\rangle$- coordinates, at least in some sufficiently small box neighbourhood $\Delta^{((d+1)+(n-1))}(\langle 0, 0'\rangle; \langle t_c, s'_c\rangle)$ of $\langle 0, 0'\rangle$.

Now S is compact. So there is a finite set $\sigma \subseteq S$ such that S is covered by the collection $\{\Delta^{(d+1)}(c; t_c) : c \in \sigma\}$. Choose $\tau' \in \mathbb{R}^{n-1}_{>0}$ such that $\tau' < s'_c$ for all $c \in \sigma$.

It follows that a point $x' \in \Delta^{(n-1)}(0; \tau')$ lies in our projection of A (at least, for sufficiently small x_n) if and only if:

either $\phi_j(x') = 0$ for $j = 0, \ldots, d$,
or else for some j_0 with $0 \leq j_0 \leq d$, $\phi_{j_0}(x') \neq 0$,
and $|\phi_{j_0}(x')| \geq |\phi_j(x')|$ for $j = 0, \ldots .d$,
and for some $c \in \sigma$ we have $|\phi_j(x') - c_j \cdot \phi_{j_0}(x')| < (t_c)_j \cdot |\phi_{j_0}(x')|$ for $j = 0, \ldots, d$,
and $\Theta_c(h_0(x'), \ldots, h_d(x'), x')$ holds, where $h_j(x')$ denotes the term $D(\phi_j(x'), \phi_{j_0}(x')) - c_j$ of $\mathcal{L}(\mathbb{R}^D_{an})$ for each $j = 0, \ldots, d$.

One now has to do another compactness argument to cover all $x_n \in [-r_n, r_n]$.

Now, you will have noticed that we started with a quantifier-free formula of $\mathcal{L}(\mathbb{R}_{an})$ and we ended up with a quantifier-free formula of $\mathcal{L}(\mathbb{R}^D_{an})$, so it would appear that we cannot proceed further to eliminate more existentially quantified variables. However, this is easily dealt with. One first proves a generalization of Theorem 3 in which x_n is replaced by a tuple $x_n, \ldots, x_{n+m-1}$ of variables. This is done by induction on m, the representation of $F(x', x_n, \ldots, x_{n+m-1})$ looking the same as in Theorem 3 except that the subscript j becomes a multi-index $\alpha \in \mathbb{N}^m$ with $|\alpha| \leq d$. And now the point is that in successively eliminating the quantifiers $\exists x_{n+m-1}, \exists x_{n+m-2}, \ldots, \exists x_n$ (as above) we apply the D function to terms involving the variables x' only. There is one extra complication which I should point out, but about which I will not go into detail, namely that, even after dividing by the largest $\phi_\alpha(x')$, we might not have guaranteed the regularity of F with respect to any of the variables $x_n, \ldots, x_{n+m-1}$. This can be achieved, however, by a suitable (polynomial) invertible transformation of these variables (leaving x' fixed).

For our purposes in these notes, one of the most important consequences of Theorem 6 is contained in the following.

Exercise 10. Prove that every subset of $\mathbb{R}$ that is definable in the language $\mathcal{L}(\mathbb{R}_{an})$ is a finite union of intervals and points.

It might now hardly seem worth commenting on the (very difficult) proof of Theorem 5. However, apart from pointing out the fact that [2] appeared 20 years before [1], it turned out that Gabrielov's arguments were much more suited to dealing with reducts of $\mathbb{R}_{an}$ than those of Denef-van den Dries and in [3] he proves the following result.

Theorem 7 (Gabrielov). *Let $\tilde{\mathbb{R}}$ be a reduct of $\mathbb{R}_{an}$ having the property that for each restricted analytic function $\tilde{f}$ appearing in its signature, and for each j, $\widetilde{\frac{\partial f}{\partial x_j}}$*

also appears. (Or, more generally, that $\widetilde{\frac{\partial f}{\partial x_j}}$ *is the interpretation of some term of the language* $\mathcal{L}(\tilde{\mathbb{R}})$.*) Then* $Th(\tilde{\mathbb{R}})$ *is model complete in its language* $\mathcal{L}(\tilde{\mathbb{R}})$.

The reason that the Denef-van den Dries method fails here is essentially due to the remark I made just before the statement of Theorem 2 (but see [12]).

The importance of Theorem 7 for us is the

Corollary 1. *Let* $\mathbb{R}_{rexp} := \langle \bar{\mathbb{R}}; \exp \restriction (0,1)\rangle$ *where* $\exp : \mathbb{R} \to \mathbb{R}, x \mapsto e^x$ *is the usual exponential function. Then* $Th(\mathbb{R}_{rexp})$ *is model complete. So is the theory of the structure* $\mathbb{R}_{rexp,rtrig} := \langle \bar{\mathbb{R}}; \exp \restriction (0,1), \sin \restriction (0, 2\pi), \cos \restriction (0, 2\pi)\rangle$.

4 Some Topics in o-minimality

Let $\mathbb{M} = \langle M, <, \ldots\rangle$ be a structure, where $<$ is a dense, total ordering without endpoints of its domain M. Then $\mathbb{M}$ is called *o-minimal* if every parametrically definable subset of M is a finite union of open intervals and points.

Clearly any reduct of an o-minimal structure is o-minimal, so we immediately deduce the following fact from Exercise 10.

Corollary 2. *Any reduct of* $\mathbb{R}_{an}$*, in particular* $\mathbb{R}_{rexp}$ *and* $\mathbb{R}_{rexp,rtrig}$*, is o-minimal.*

In this section I shall be listing those properties of o-minimal structures that I shall be using in the sequel. Unless otherwise stated, all proofs may be found in the excellent book [14]. From now on, notions of definability are always with reference to the structure under consideration and are *without* parameters. We thereby achieve *uniformity in parameters* in our results.

Theorem 8 (The Monotonicity Theorem). *Let* $\mathbb{M} = \langle M, <, \ldots\rangle$ *be an o-minimal structure and suppose that* $f : M \to M$ *is a definable function. Then there are points* $a_1 < a_2 < \ldots < a_p$ *in* M *such that (setting* $a_0 = -\infty$ *and* $a_{p+1} = +\infty$*) for each* $j = 0, \ldots, p$, f *is either constant or strictly monotone and continuous (for the order topology) on the interval* (a_j, a_{j+1}).

Exercise 11. Prove that the a_j's in Theorem 8 may be taken to be definable (even without knowing the proof of Theorem 8).

Theorem 9 (Existence of Definable Skolem Functions). *Let* $\mathbb{M} = \langle M, <, +, 0, \lambda, \ldots\rangle$ *be an o-minimal structure where* $\langle M, <, +, 0\rangle$ *is an ordered abelian group and* λ *is some nonzero element of* M. *Then* $Th(\mathbb{M})$ *admits definable Skolem functions. This means that for any definable set* $A \subseteq M^{n+1}$*, there exists a definable function* $f : M^n \to M$ *such that for all* $a \in M^n$*, if there is some* $b \in M$ *such that* $\langle a, b\rangle \in A$*, then* $\langle a, f(a)\rangle \in A$.

Henceforth, all o-minimal structures will be assumed to be equipped with an ordered abelian group structure as in Theorem 9.

Let us fix one such, $\mathbb{M} = \langle M, <, +, 0, \ldots\rangle$ say.

It follows from Theorem 9 that for any subset S of M, $Dcl(S)$ is the domain of an elementary substructure of $\mathbb{M}$ which we denote by $\widetilde{Dcl(S)}$. Here $Dcl(S)$ denotes the *definable closure* of S:

$$Dcl(S) := \{f(s_1, \ldots, s_n) : n \geq 0, s_1, \ldots, s_n \in S,\ f \text{ a definable function}\}.$$

By convention, a 0-place definable function is a definable element of M, so $\widetilde{Dcl(\emptyset)}$ is an isomorphic copy of the unique prime model of $Th(\mathbb{M})$.

I should mention at this point the foundational theory of o-minimality developed by Pillay and Steinhorn in [8] and [6]. They prove the Cell Decomposition Theorem which has the consequence that any structure elementarily equivalent to an o-minimal structure is itself o-minimal. (This certainly does not follow from the definition which involves sets defined *with* parameters.) I do not need these results explicitly in these lectures (though a slightly more detailed account would certainly need to discuss them) so I return to properties of the $Dcl(\cdot)$ operator.

Exercise 12. Prove that $Dcl(\cdot)$ is a pre-geometry on M. [Hint: all is clear apart from Exchange. For this use Theorem 8 and Exercise 11.]

So by Exercise 12 we may assign a cardinal number (usually finite in our applications) to any subset $S \subseteq M$:

$$rank(S) := \max\{|I| : I \subseteq S, I \text{ is } Dcl(\cdot)\text{-independent}\}.$$

Thus, for example, $rank(M)$, also written $rank(\mathbb{M})$, is the smallest number of elements of M required to generate M under the definable functions. Exercise 12 guarantees that this is well-defined in the sense that any two minimal sets of generators have the same cardinality.

Exercise 13. Let $\mathbb{M} = \langle M, <, +, \cdot, 0, 1\rangle$ be a real closed, ordered field (i.e. $\mathbb{M} \equiv \bar{\mathbb{R}}$). Prove that for any $S \subseteq M$, $rank(S)$ is the transcendence degree (over $\mathbb{Q}$) of the subfield of $\mathbb{M}$ generated by S.

5 Some Valuation Theory for o-minimal Structures

Let $\tilde{\mathbb{R}}$ be an o-minimal expansion of $\bar{\mathbb{R}}$ and let $\mathbb{M} \equiv \tilde{\mathbb{R}}$. The set of *finite* and *infinitesimal* elements of M (= the domain of $\mathbb{M}$) are defined, respectively, by:

$$Fin(\mathbb{M}) := \{a \in M : |a| < N \text{ for some } N \in \mathbb{Q}\},$$

$$\mu(\mathbb{M}) := \{a \in M : |a| < q \text{ for all } q \in \mathbb{Q}_{>0}\},$$

where we have identified $\mathbb{Q}$ with the prime subfield of $\mathbb{M}$.

Exercise 14. Prove that $Fin(\mathbb{M})$ is (the domain of) a subring of $\mathbb{M}$ and that $\mu(\mathbb{M})$ is the unique maximal ideal of $Fin(\mathbb{M})$.

Thus $Fin(\mathbb{M})/\mu(\mathbb{M})$ is a field, called the *residue field* of $\mathbb{M}$, and is denoted $Res(\mathbb{M})$.

Exercise 15. Let $\mathcal{S}$ be the collection of all elementary substructures $\mathbb{A} = \langle A, \ldots \rangle$ of $\mathbb{M}$ such that $A \subseteq Fin(\mathbb{M})$. Prove that $\mathcal{S}$ satisfies the hypotheses of Zorn's Lemma (in particular, that $\mathcal{S} \neq \emptyset$). Deduce that there exists $\mathbb{A}_0 = \langle A_0, \ldots \rangle \in \mathcal{S}$ such that for each $a \in Fin(\mathbb{M})$, there exists a unique $b \in A_0$ (the "standard part" of a) such that $|a - b| \in \mu(\mathbb{M})$.

Thus, the field $Res(\mathbb{M})$ can be expanded to an $\mathcal{L}(\mathbb{M})$-structure in such a way that it has an isomorphic copy, $\mathcal{R}(\mathbb{M})$ say, in $\mathbb{M}$ (contained in $Fin(\mathbb{M})$) and, further, we have that $\mathcal{R}(\mathbb{M}) \preccurlyeq \mathbb{M}$. We call $rank(\mathcal{R}(\mathbb{M}))$ the *residue rank* of $\mathbb{M}$ and denote it by $resrank(\mathbb{M})$.

Obviously $resrank(\mathbb{M}) \leq rank(\mathbb{M})$ and we wish to investigate the deficiency $rank(\mathbb{M}) - resrank(\mathbb{M})$, which, in some sense, measures the number of "degrees of infinity" in $\mathbb{M} \setminus Fin(\mathbb{M})$. The following is completely standard algebra, and not particularly special to the situation here.

Exercise 16. Prove that there exists a unique (up to isomorphism) ordered $\mathbb{Q}$-vector space $\langle \Gamma, <, +, 0 \rangle$ (called the *value group* of $\mathbb{M}$) and a unique surjective function $\nu : M \setminus \{0\} \to \Gamma$ (called the *valuation map* of $\mathbb{M}$) having the following properties.

For all $a, b \in M \setminus \{0\}$,

(i) $\nu(a \cdot b) = \nu(a) + \nu(b)$;
(ii) $\nu(a + b) \geq \min\{\nu(a), \nu(b)\}$;
(iii) $\nu(a) = 0$ if and only if $a \in Fin(\mathbb{M}) \setminus \mu(\mathbb{M})$.

[Hint: $Fin(\mathbb{M}) \setminus \mu(\mathbb{M})$ is a multiplicative subgroup of $\mathbb{M} \setminus \{0\}$. Then Γ is the quotient group written additively.]

We continue to use the notation of Exercise 16 throughout this section.

Exercise 17. Prove that if $a, b \in M \setminus \{0\}$ and $\nu(a) > \nu(b)$, then $\nu(a + b) = \nu(b)$.

You should now be able to establish the following classical inequality.

Exercise 18. Let $a_1, \ldots, a_n \in \mathcal{R}(\mathbb{M})$ be (field-theoretically) algebraically independent (over $\mathbb{Q}$) and let $b_1, \ldots, b_m \in \mathbb{M} \setminus \{0\}$ be such that $\nu(b_1), \ldots, \nu(b_m)$ are linearly independent (over $\mathbb{Q}$) elements of Γ. Prove that $a_1, \ldots, a_n, b_1, \ldots, b_m$ are algebraically independent (over $\mathbb{Q}$). Deduce that in the case $\tilde{\mathbb{R}} = \bar{\mathbb{R}}$ (see Exercise 13), we have the inequality

$$rank(\mathbb{M}) \geq resrank(\mathbb{M}) + dim_{\mathbb{Q}}(\Gamma).$$

The main step in the proof of the model completeness of T_{exp} is to generalize Exercise 18 to the case that $\tilde{\mathbb{R}}$ is $\mathbb{R}_{rexp}$. In fact we have the following

Theorem 10 (The Valuation Inequality). *Let $\tilde{\mathbb{R}}$ be any reduct of $\mathbb{R}_{an}$ that expands $\bar{\mathbb{R}}$. Then for any $\mathbb{M} \equiv \tilde{\mathbb{R}}$ with rank($\mathbb{M}$) finite, we have the inequality*

$$rank(\mathbb{M}) \geq resrank(\mathbb{M}) + dim_{\mathbb{Q}}(\Gamma).$$

I can say very little here about the proof of Theorem 10 except that one uses the method of Exercise 18 after approximating analytic functions by polynomials via their Taylor expansions. Various proofs of Theorem 10 now exist in the literature for much wider classes of o-minimal structures, and I refer you to [11] for a very readable account.

Actually, I will need a relativized version of Theorem 10 which follows fairly easily from it. I now assume that $Th(\mathbb{M})$ is model complete.

Firstly, and in general, for $\mathbb{A} = \langle A, <, \ldots \rangle$ and $\mathbb{B} = \langle B, <, \ldots \rangle$ o-minimal structures in the same language with $\mathbb{A} \preccurlyeq \mathbb{B}$, let us write $rank_{\mathbb{A}}(\mathbb{B})$, the *rank of* $\mathbb{B}$ *over* $\mathbb{A}$, for the rank (as defined above) of the expansion $\langle \mathbb{B}, a \rangle_{\{a \in A\}}$ of $\mathbb{B}$, where each element of A is distinguished.

To return to situation at hand, let $\mathbb{M}_0 = \langle M_0, <, \ldots \rangle$ be an elementary substructure of $\mathbb{M}$. Then one may easily modify the Zorn's Lemma argument of Exercise 15 to show that the copies of the residue fields may be chosen in such a way that $\mathcal{R}(\mathbb{M}_0) \preccurlyeq \mathcal{R}(\mathbb{M})$. (Recall that both are models of $Th(\mathbb{M})$, and it is at this point that one needs model completeness.) So we may define $resrank_{\mathbb{M}_0}(\mathbb{M}) := rank_{\mathcal{R}(\mathbb{M}_0)}(\mathcal{R}(\mathbb{M}))$.

One can also easily show that Γ_0 (the value group of $\mathbb{M}_0$) is a sub-$\mathbb{Q}$-vector space of Γ.

Then in this situation we have

Theorem 11 (The Relativized Valuation Inequality). *If $rank_{\mathbb{M}_0}(\mathbb{M})$ is finite then*

$$rank_{\mathbb{M}_0}(\mathbb{M}) \geq resrank_{\mathbb{M}_0}(\mathbb{M}) + dim_{\mathbb{Q}}(\Gamma/\Gamma_0).$$

6 The Model Completeness of T_{exp}

In this section I give a proof of the model completeness of T_{exp}, at least insofar as I indicated after the statement of Definition 2 in Sect. 1. But let us proceed for the moment as if we are trying to establish the hypotheses of Exercise 1 as stated.

So consider models $\mathbb{M}_0$, $\mathbb{M}$ of T_{exp} with $\mathbb{M}_0 \subseteq \mathbb{M}$. Let their domains be M_0 and M respectively and suppose we are given a quasipolynomial $f : M^n \to M$ say, with coefficients in M_0 and having a zero, b say, in M^n. The first step in finding such a zero in M_0^n is a *reduction to the non-singular case*:

Lemma 1. *It is sufficient to show that whenever $F : M^n \to M^n$ is a quasipolynomial map with coefficients in M_0, and $b \in M^n$ is a* non-singular *solution to the equation $F(x) = 0$, i.e.*

$$F(b) = 0 \text{ and } J_F(b) \neq 0,$$

then $b \in M_0^n$.

(Here, J_F *is the Jacobian with respect to all the variables* $x = x_1, \ldots, x_n$.*)*

There is now a simpler argument than that given in [16] of a much more general result than this, and I refer you to [4] for the proof of this lemma.

It is perhaps overdue for me also to mention Khovanski's paper [5]. There it is shown that quasipolynomial maps as in the lemma have only finitely many non-singular zeros. Khovanski only works over the reals, but his (completely effective) upper bound for the number of such zeros is independent of any parameters occurring as coefficients, and hence one obtains the finiteness in all models of T_{exp}.

Now let $n \geq 1$ be given and assume, for a contradiction, that we have a counterexample, i.e. for some quasipolynomial map $F : M^n \to M^n$ (with coefficients in M_0) and some $b \in M^n$, we have $F(b) = 0$ and $J_F(b) \neq 0$ but $b \notin M_0^n$. Suppose further that these have been chosen to maximise the number r of coordinates of b that lie in $Fin(\mathbb{M})$. We may assume that these are the first r coordinates, $b_1, \ldots, b_r$, of b. Choose $N \in \mathbb{N}$ such that $-N < b_j < N$ for $j = 1, \ldots, r$.

Case 1. $r = n$.

Now we may consider $\mathbb{M}_0$ and $\mathbb{M}$ as models of T_{rexp} (:= $Th(\mathbb{R}_{rexp})$) by simply restricting their exponential functions to (0, 1). Let us call these structures $\mathbb{M}_0^*$ and $\mathbb{M}^*$ respectively. Since the exponentiation of $\mathbb{M}$ restricted to the interval $(-N, N)$ is definable in $\mathbb{M}^*$, the map F restricted to $(-N, N)^n$ is definable in $\mathbb{M}^*$ using parameters from $\mathbb{M}_0^*$, and we have, by Khovanski's theorem, that

$$\mathbb{M}^* \models \exists^{=k} x \in (-N, N)^n \, (F(x) = 0 \wedge J_F(x) \neq 0)$$

for some $k \geq 1$.

But $\mathbb{M}_0^* \preccurlyeq \mathbb{M}^*$ (by Corollary 1) and hence

$$\mathbb{M}_0^* \models \exists^{=k} x \in (-N, N)^n \, (F(x) = 0 \wedge J_F(x) \neq 0),$$

whence $b \in M_0^n$, a contradiction.

Case 2. $r < n$.

Then $b_1, \ldots, b_r \in Fin(\mathbb{M})$ and $b_{r+1}, \ldots, b_n \in \mathbb{M} \setminus Fin(\mathbb{M})$.

In fact, it is easy to see that if C is an $(n-r) \times (n-r)$ invertible matrix with rational entries, and $\gamma \in M_0^{n-r}$, then the n-tuple $\langle b_1, \ldots, b_r, \gamma + \langle b_{r+1}, \ldots, b_n \rangle C \rangle$ also constitutes a counterexample (for a suitably transformed map F).

Hence, in particular, for all $\gamma \in M_0$ and $q_{r+1}, \ldots, q_n \in \mathbb{Q}$ (not all zero) we have

$$\gamma + q_{r+1} \cdot b_{r+1} + \cdots + q_n \cdot b_n \notin Fin(\mathbb{M}) \qquad (*)$$

(by the maximality of r).

We now work in the theory T_{rexp} and consider its models $\mathbb{M}_0^*$ and $\mathbb{M}^*$ as discussed in Case 1. Let M_1 be the definable closure in $\mathbb{M}^*$ of the set $M_0 \cup \{b_1, \ldots, b_n, \exp(b_{r+1}), \ldots, \exp(b_n)\}$ and let $\mathbb{M}_1^*$ be the elementary substructure of $\mathbb{M}$ with domain M_1.

Now each coordinate function F_j of the map F has the form

$$F_j : M^n \to M, \langle x_1, \ldots, x_n \rangle \mapsto P_j(x_1, \ldots, x_n, \exp(x_1), \ldots, \exp(x_n))$$

where P_j is a polynomial (in $2n$ variables) with coefficients in M_0. Let us consider the function G_j given by

$$G_j : M^{n+(n-r)} \to M, \langle x_1, \ldots, x_n, y \rangle \mapsto P_j(x_1, \ldots, x_n, \exp(x_1), \ldots, \exp(x_r), y),$$

where $y = \langle y_{r+1}, \ldots, y_n \rangle$

Notice that these functions are definable in $\mathbb{M}^*$ (with parameters from M_0) if the variables $x_1, \ldots x_r$ are constrained to the interval $(-N, N)$.

It is immediate that $\langle b_1, \ldots, b_n, \exp(b_{r+1}), \ldots, \exp(b_n) \rangle$ is a zero of the map $G := \langle G_1, \ldots, G_n \rangle : M^{n+(n-r)} \to M^n$ and one may easily check, by direct calculation, that it is a non-singular zero with respect to some sub-n-tuple of the variables $x_1, \ldots, x_n, y_{r+1}, \ldots, y_n$. It now follows from Khovanski's theorem that the coordinates of the corresponding sub-n-tuple of $\langle b_1, \ldots, b_n, \exp(b_{r+1}), \ldots, \exp(b_n) \rangle$ are all contained in the definable closure (over the parameters M_0) of the remaining $n-r$ coordinates. (We are using here the fact that in totally ordered structures, "algebraic closure = definable closure".). We conclude that $rank_{\mathbb{M}_0^*}(\mathbb{M}_1^*) \leq n - r$.

However,

$$\nu(\exp(b_{r+1})), \ldots, \nu(\exp(b_n))$$

are $\mathbb{Q}$-linearly independent elements of the vector space Γ_1^* / Γ_0^* (where Γ_1^* and Γ_0^* denote the value groups of $\mathbb{M}_1^*$ and $\mathbb{M}_0^*$ respectively). Since if not, then

$$\beta \cdot \exp(q_{r+1}b_{r+1} + \ldots + q_n b_n) \in Fin(\mathbb{M}_1^*) \setminus \mu(\mathbb{M}_1^*)$$

for some $\beta \in M_0$, and some $q_{r+1}, \ldots, q_n \in \mathbb{Q}$ (not all zero).

We may assume that $\beta > 0$ and since $\mathbb{M}_0 \models T_{exp}$, there is some $\gamma \in M_0$ such that $\exp(\gamma) = \beta$. But then

$$\exp(\gamma + q_{r+1}b_{r+1} + \ldots + q_n b_n) \in Fin(\mathbb{M}_1^*) \setminus \mu(\mathbb{M}_1^*)$$

whence

$$\gamma + q_{r+1}b_{r+1} + \ldots + q_n b_n \in Fin(\mathbb{M})$$

which contradicts $(*)$.

So $dim_{\mathbb{Q}}(\Gamma_1^*/\Gamma_0^*) \geq n - r$.

Since we have already established that $rank_{\mathbb{M}_0^*}(\mathbb{M}_1^*) \leq n - r$, it follows from Theorem 11 that $Resrank_{\mathbb{M}_0^*}(\mathbb{M}_1^*) = 0$, so

$$\mathcal{R}(\mathbb{M}_0^*) = \mathcal{R}(\mathbb{M}_1^*),$$

and that

$$\nu(\exp(b_{r+1})), \ldots, \nu(\exp(b_n))$$

generate Γ_1^* (as a $\mathbb{Q}$-vector space) over Γ_0^*.

It now takes a very easy valuation theoretic argument (together with the fact that M_0 is closed under taking logarithms of positive elements) to establish the following claim, and I leave the details of the proof to you.

Claim. For all nonzero $a \in \mathbb{M}_1^*$ there exists $\gamma \in M_0$ and a homogeneous linear form with rational coefficients $\lambda : M^{n-r} \to M$ and an infinitesimal $\epsilon \in \mu(\mathbb{M}_1^*)$ such that

$$a = (1 + \epsilon) \cdot \exp(\gamma + \lambda(b_{r+1}, \ldots, b_n)).$$

In order to obtain our contradiction we use the claim in the obvious way to define inductively a sequence $a_1, a_2, \ldots$ of elements of M_1, where a_1 is an arbitrary positive element of $M_1 \setminus Fin(\mathbb{M}_1^*)$, and, for each $j \geq 1$,

(i) $a_j = (1 + \epsilon_j) \cdot \exp(a_{j+1})$ for some $\epsilon_j \in \mu(\mathbb{M}_1)$, and
(ii) $a_{j+1} = \gamma_{j+1} + \lambda_{j+1}(b_{r+1}, \ldots, b_n)$ for some $\gamma_{j+1} \in M_0$ and some homogeneous linear form λ_{j+1} with rational coefficients.

It is clear that $\langle a_j : j \geq 1 \rangle$ is a decreasing sequence of positive elements of M_1 and for all $l, m, j \geq 1$,

$$l < a_{j+1}^m < a_j \qquad (**).$$

Now since the $\mathbb{Q}$-vector space of $(n - r)$-variable rational linear forms is finite dimensional it follows from (ii) that for some $p \geq 1$ and rational numbers $q_1, \ldots, q_p$ (not all zero), we have

$$q_1 a_1 + \cdots + q_p a_p \in M_0.$$

Thus by $(**)$ there certainly exists some $\gamma \in M_0$ such that $0 < a_{j_0} < \gamma$, where j_0 is the least j such that $q_j \neq 0$. If $j_0 > 1$ we may set $j = j_0 - 1$ in (i) and (using the fact that M_0 is closed under exponentiation) deduce that there also exists some $\gamma' \in M_0$ such that $0 < a_{j_0 - 1} < \gamma'$. Continuing in this way we eventually arrive at some $\gamma'' \in M_0$ such that $0 < a_1 < \gamma''$. Since a_1 was an arbitrary positive infinite element of M_1 this shows that every element of M_1 is $\mathbb{M}_0$-bounded. This is true, in particular, for $b_1, \ldots, b_n$ and this is all that I claimed we would show here.

7 The Complex Exponential Field and Analytic Continuation for Definable Functions

We now consider the case that $\mathbb{K} = \mathbb{C}$.

As mentioned in the introduction, this section is motivated by the conjecture of Zilber stating that every $\mathbb{C}_{exp}$-definable subset of $\mathbb{C}$ is either countable or co-countable. As far as I know, even sets of the form

$$\{z \in \mathbb{C} : \exists w \in \mathbb{C}\ F(z, w) = 0\} \qquad (*)$$

where $F(z, w)$ is a (two variable) term of the language $\mathcal{L}(\mathbb{C}_{exp})$ have not been shown to satisfy Zilber's conjecture.

Our approach to this particular case is as follows. Let us suppose that

$$F(0,0) = 0 \neq \frac{\partial F}{\partial w}(0,0).$$

Then by the Implicit Function Theorem there exists $\epsilon > 0$ and a complex analytic function $\phi : \Delta(0;\epsilon) \to \mathbb{C}$ such that for all $z \in \Delta(0;\epsilon)$, we have $F(z, \phi(z)) = 0$. We must show that the set $(*)$ is co-countable and it seems reasonable to conjecture that the function ϕ has an analytic continuation (which necessarily preserves the equation $F(z, \phi(z)) = 0$) to all but countably many points in the complex plane. Indeed, one can fairly easily show that if one proves a suitably generalized version of this *analytic continuation conjecture* (in which w is allowed to be an n-tuple of variables and F an n-tuple of terms in the $1 + n$ variables z, w, and where the countably many exceptional points have a certain specific form) then Zilber's conjecture (even for subsets of $\mathbb{C}$ defined by formulas of the language $\mathcal{L}_{\omega_1,\omega}(\mathbb{C}_{exp})$) would follow.

Let us now consider issues of definability. The approach to Zilber's conjecture suggested above transcends $\mathcal{L}(\mathbb{C}_{exp})$-definability (at least, if Zilber's conjecture is true!): one cannot define *restricted* functions $\phi : \Delta(0;\epsilon) \to \mathbb{C}$ without the resource of the real line and the usual metric. So we follow the Peterzil-Starchenko idea of doing complex analysis definably in a suitable o-minimal structure via the usual identifications $\mathbb{C} \sim \mathbb{R} \oplus i\mathbb{R} \sim \mathbb{R} \times \mathbb{R}$. (Actually, we will only be considering a fixed o-minimal expansion $\tilde{\mathbb{R}}$ of the ordered field of real numbers $\overline{\mathbb{R}}$, so many of the subtleties of [9] will not be required here.)

In order to study the complex exponential function in this way we require an o-minimal structure in which it is definable (considered as a function from $\mathbb{R} \times \mathbb{R}$ to $\mathbb{R} \times \mathbb{R}$).

Exercise 19. Prove that there is no such structure.

However, Miller and van den Dries generalized the proof of the model completeness of T_{exp} to show that the expansion $\mathbb{R}_{an,exp}$ of the structure $\mathbb{R}_{an}$ by the (unrestricted) real exponential function is model complete and o-minimal (see [13]).

Exercise 20. Prove that for any $N \in \mathbb{N}$ the restriction of the complex exponential function to the strip $\{x + iy \in \mathbb{C} : -N < y < N\}$ is definable in the structure $\mathbb{R}_{an,exp}$.

Exercise 21. Prove that any (analytic) branch of the complex logarithm function restricted to the slit plane $\{x + iy \in \mathbb{C} : y \neq 0 \text{ or } x > 0\}$ is definable in the structure $\mathbb{R}_{an,exp}$.

So some useful complex analytic functions are definable in the structure $\mathbb{R}_{an,exp}$. For many others see [18].

To return to the approach to Zilber's conjecture suggested above, I propose the program of investigating analytic continuation for functions definable in the structure $\mathbb{R}_{an,exp}$ and I conclude these notes with a first step in this direction.

We first require the analytic cell decomposition theorem for $\mathbb{R}_{an,exp}$ (see [13]). Since we only need it for subsets of the complex plane I state it only in the two (real) dimensional case. Definability here, and for the remainder of these notes, is with reference to the structure $\mathbb{R}_{an,exp}$.

Theorem 12 (van den Dries-Miller). *Let $\mathcal{A}$ be a finite collection of definable subsets of $\mathbb{R}^2$. Then there are points $a_1 < a_2 < \cdots < a_p$ in $\mathbb{R}$ and, for each $j = 0, \ldots, p$, a finite collection $\mathcal{F}_j$ of definable (real) analytic functions with domain (a_j, a_{j+1}) (where we have set $a_0 = -\infty$ and $a_{p+1} = \infty$ and we also include the two functions with constant value $\infty, -\infty$ in $\mathcal{F}_j$) such that*

(i) for f, g distinct functions in $\mathcal{F}_j$, either for all $x \in (a_j, a_{j+1})$, $f(x) < g(x)$ (written $f \prec g$) or else for all $x \in (a_j, a_{j+1})$, $f(x) > g(x)$;

(ii) for each $A \in \mathcal{A}$ and each $f \in \mathcal{F}_j$, either $graph(f) \subseteq A$ or else $graph(f) \cap A = \emptyset$;

(iii) for each $f, g \in \mathcal{F}_j$ with $f \prec g$ and for which there is no $h \in \mathcal{F}_j$ with $f \prec h \prec g$, and for each $A \in \mathcal{A}$, we have that either $(f, g) \subseteq A$ or else $(f, g) \cap A = \emptyset$, where $(f, g) := \{\langle x, y\rangle \in \mathbb{R}^2 : a_j < x < a_{j+1}, f(x) < y < g(x)\}$.

The sets graph(f) and (f, g) mentioned above are called *1-cells* and *2-cells* respectively and the collection of all of them is called an *analytic cell decomposition compatible with* $\mathcal{A}$. (Strictly speaking we should also include, for each $j = 0, \ldots, p$, a compatible partition of the ordinates $\{a_j\} \times \mathbb{R}$ into open intervals and points.)

The results contained in the following two exercises will be required for the proof of our analytic continuation theorem. They both require a little o-minimal theory, and the second some elementary complex analysis (Cauchy's Theorem) as well.

We use the notation $\bar{X}$ to denote the closure of a set $X \subseteq \mathbb{C}$ in the usual topology on $\mathbb{C}$.

Exercise 22. In the notation of Theorem 12, suppose that $A \in \mathcal{A}$ and that A is a (nonempty) *regular open* set (i.e. A is the interior of its closure in $\mathbb{R}^2$). Prove that there exists a finite set $S \subseteq \bar{A} \setminus A$ with the property that for all $a \in \bar{A} \setminus (A \cup S)$, there is a *unique* 2-cell, V_a say, in the cell decomposition such that $V_a \subseteq A$ and $a \in \bar{V_a} \setminus V_a$.

Exercise 23. Let $F : \bar{C} \to \mathbb{C}$ be a definable continuous function, where $C \subseteq \mathbb{C}$ is an analytic 2-cell, and assume that F has infinitely many zeros. Suppose further that $F \restriction C$ is (complex) analytic. Prove that F is identically zero.

Theorem 13 (See [18]). *Let $U \subseteq \mathbb{C}$ be a definable, regular open set and let $\phi : U \to \mathbb{C}$ be a definable (complex) analytic function. Then there exists a finite set $T \subseteq \bar{U} \setminus U$ such that for all $w \in \bar{U} \setminus (T \cup U)$, ϕ has an analytic continuation to some open set containing w.*

Proof. Consider an analytic cell decomposition compatible with the collection $\mathcal{A} = \{U, \{z \in U : |\phi(z)| \leq 1\}\}$. If C is a 2-cell of this decomposition and $C \subseteq U$ then either $|\phi(z)| \leq 1$ throughout C or else $|\phi(z)| > 1$ throughout C. Now to prove the theorem it follows from Exercise 22 that we may assume that U is in fact such a 2-cell (exercise). Say

$$U = C = \{x + iy : a < x < b, f(x) < y < g(x)\},$$

where $f, g : (a, b) \to \mathbb{R}$ are definable real analytic functions.

I show that ϕ has an analytic continuation across all but finitely many points of $graph(g)$ (assumed $\neq \infty$), other cases being dealt with similarly.

Consider first the case that ϕ is bounded on C. Then using o-minimality one can show that ϕ has a continuous extension, $\bar{\phi}$ say, to all but finitely many points of $graph(g)$, and by a further use of analytic cell decomposition we may assume that $\bar{\phi} \circ g$ is real analytic at all but finitely many points of (a, b). So, avoiding such points, let $x_0 \in (a, b)$ and choose $\epsilon > 0$ so that $a < x_0 - \epsilon < x_0 + \epsilon < b$ and, further, so that the (real) Taylor series of both g and $\bar{\phi} \circ g$ at x_0 define complex analytic functions $G : \Delta(x_0; \epsilon) \to \mathbb{C}$ and $\Phi : \Delta(x_0; \epsilon) \to \mathbb{C}$ respectively. We have to show that ϕ has an analytic continuation to some open set containing $g(x_0)$.

Define the complex analytic function $H : \Delta(x_0; \epsilon) \to \mathbb{C}$ by $H(z) := z + iG(z)$.

Since the Taylor coefficients of G are real it follows that $H'(x_0) \neq 0$ and hence (by reducing ϵ if necessary) that H is a holomorphic homeomorphism from $\Delta(x_0; \epsilon)$ onto an open set, V say. Further, H maps the interval $(x_0 - \epsilon, x_0 + \epsilon)$ onto $graph(g \restriction (x_0 - \epsilon, x_0 + \epsilon))$. We may also suppose, by reducing ϵ further, that H is definable (its real and imaginary parts being restricted analytic functions (when extended by 0) of two real variables).

Now consider the function

$$\Psi := \phi - \bar{\Phi} \circ H^{-1} : (C \cup graph(g \restriction (x_0 - \epsilon, x_0 + \epsilon))) \cap V \to \mathbb{C}.$$

By our construction Ψ is definable, continuous, holomorphic on $C \cap V$, and identically zero on the curve $graph(g \restriction (x_0-\epsilon, x_0+\epsilon))$ which obviously contains infinitely many points of the boundary of $C \cap V$. It now easily follows from Exercise 23 that Ψ is identically zero throughout $(C \cup graph(g \restriction (x_0 - \epsilon, x_0 + \epsilon))) \cap V$ and hence that $\Phi \circ H^{-1}$ provides an analytic continuation of ϕ to V, as required.

The case that $|\phi(z)| > 1$ for all $z \in C$ is dealt with by applying the above argument to the function $1/\phi$ and then inverting the analytic continuation. (The proof actually shows that ϕ is necessarily locally bounded at all but finitely many points of $graph(g)$.)

This completes the proof of Theorem 13. □

References

1. J. Denef, L. van den Dries, p-Adic and real subanalytic sets. Ann. Math. **128**, 79–138 (1988)
2. A. Gabrielov, Projections of semi-analytic sets. Funct. Anal. Appl. **2**, 282–291 (1968)
3. A. Gabrielov, Complements of subanalytic sets and existential formulas for analytic functions. Inv. Math. **125**, 1–12 (1996)
4. G.O. Jones, A.J. Wilkie, Locally polynomially bounded structures. Bull. Lond. Math. Soc. **40**(2), 239–248 (2008)
5. A. Khovanski, On a class of systems of transcendental equations. Sov. Math. Dokl. **22**, 762–765 (1980)
6. J.F. Knight, A. Pillay, C. Steinhorn, Definable sets in ordered structures, II. Trans. Am. Math. Soc. **295**, 593–605 (1986)
7. D. Marker, A remark on Zilber's pseudoexponentiation. J Symb. Log. **71**(3), 791–798 (2006)
8. A. Pillay, C. Steinhorn, Definable sets in ordered structures I. Trans. Amer. Math. Soc. **295**, 565–592 (1986)
9. Y. Peterzil, S. Starchenko, Expansions of algebraically closed fields in o-minimal structures. Sel. Math. New Ser. **7**, 409–445 (2001)
10. J.M. Ruiz, *The Basic Theory of Power Series* (Edizioni ETS, Pisa, 2009)
11. P. Speissegger, in *Lectures on o-minimality*, ed. by B. Hart, M. Valeriote. Lectures on Algebraic Model Theory, Fields Institute Monographs (American Mathematical Society, Providence, 2000)
12. L. van den Dries, On the elementary theory of restricted elementary functions. J. Symb. Log. **53**, 796–808 (1988)
13. L. van den Dries, C. Miller, On the real exponential field with restricted analytic functions. Isr. J. Math **85**, 19–56 (1994)
14. L. van den Dries, *Tame Topology and o-minimal Structures*. London Mathematical Society Lecture Note Series, vol. 248 (Cambridge University Press, Cambridge, 1998)
15. A.J. Wilkie, On the theory of the real exponential field. Ill. J. Math. **33**(3), 384–408 (1989)
16. A.J. Wilkie, Model completeness results for expansions of the ordered field of real numbers by restricted Pfaffian functions and the exponential function. J. Am. Math. Soc. **9**(4), 1051–1094 (1996)
17. A.J. Wilkie, Model theory of analytic and smooth functions, in *Models and Computability*, ed. by S.B. Cooper, J.K. Truss. LMS Lecture Note Series, vol. 259, pp. 407–419 (1999)
18. A.J. Wilkie, Some results and problems on complex germs with definable Mittag-Leffler stars. Notre Dame J. Formal Logic **54**(3-4), 603–610 (2013)

Lectures on the Model Theory of Valued Fields

Lou van den Dries

1 Introduction

The subject originates in the 1950s with Abraham Robinson when he established the model completeness of the theory of algebraically closed valued fields. In the 1960s Ax & Kochen and, independently, Ershov, proved a remarkable theorem on henselian valued fields, with applications to p-adic number theory. These results and their refinements and extensions remain important in more recent developments like motivic integration.

The AKE-theorems—with AKE abbreviating Ax, Kochen, Ershov—were the starting point for further work by many others. These lectures (given in the fall semester of 2004 at the University of Illinois at Urbana-Champaign, and augmented on the occasion of my talks at the CIME meeting in July 2012) have the modest goal of giving a transparent treatment of the original AKE-theorems and of Robinson's result, including the relevant background on local rings and valuation theory. We assume only very basic knowledge: from algebra some familiarity with rings and fields (including some Galois theory) and from model theory: compactness, saturation, back-and-forth, model-theoretic tests for quantifier elimination. We do go beyond the original AKE-results by also paying attention to definable sets, which are at the center of present applications.

In some of the key arguments in Chaps. 4 and 5 we use the technique of pseudocauchy sequences. This old technique originates with Ostrowski and was also used by AKE. There are ways to avoid this and give more constructive proofs, but in recent extensions to suitable valued difference fields and valued differential fields, it was this approach, suitably elaborated, that was successful. It is also very much in the spirit of model theory.

L. van den Dries (✉)
Department of Mathematics, University of Illinois, Urbana, IL, USA
e-mail: vddries@math.uiuc.edu

L. van den Dries et al., *Model Theory in Algebra, Analysis and Arithmetic*, Lecture Notes in Mathematics 2111, DOI 10.1007/978-3-642-54936-6_4,

With just the techniques from the present notes we can do much more. With sharper model theoretic tools one can even go far beyond the scope of these notes; see for example [30] and [28]. As a partial compensation for our limited material, we give at the end more historical background and a list of references with comments. This includes pointers to developments closely related to the more classical material presented here. It should also be clear from the other lectures at this meeting that valuation theory currently plays a significant role in a variety of model-theoretic topics.

I thank the referee and the editors for their careful reading of the manuscript. Implementing their suggestions improved the final product considerably.

Conventions. Throughout, m, n range over $\mathbb{N} = \{0, 1, 2, \ldots\}$, the set of natural numbers. Unless specified otherwise, "ring" means "commutative ring with 1" and given a ring R we let $U(R) := \{x \in R : xy = 1 \text{ for some } y \in R\}$ be its multiplicative group of units, and for $a_1, \ldots, a_n \in R$ we denote the ideal $a_1R + \cdots + a_nR$ of R also by $(a_1, \ldots, a_n)R$ or just by $(a_1, \ldots, a_n)$. A ring is considered as an L-structure for the language $L = \{0, 1, +, -, \cdot\}$ of rings. Given a ring R and distinct indeterminates $t_1, \ldots, t_n$, we let $R[t_1, \ldots, t_n]$ and $R[[t_1, \ldots, t_n]]$ denote the corresponding polynomial ring and formal power series ring, with

$$R \subseteq R[t_1, \ldots, t_n] \subseteq R[[t_1, \ldots, t_n]].$$

Note that for every $f \in R[[t_1, \ldots, t_n]]$ we have

$$f = a + t_1 f_1 + \cdots + t_n f_n$$

where $a \in R$ and each $f_i \in R[[t_1, \ldots, t_i]]$. The element a is uniquely determined by f and is called the *constant term of* f, and denoted by $f(0)$. Throughout, $\boldsymbol{k}, \boldsymbol{k}_1, \boldsymbol{k}_2$ are fields, and we put $\boldsymbol{k}^\times := \boldsymbol{k} \setminus \{0\} = U(\boldsymbol{k})$. Given a prime number p the field $\mathbb{Z}/p\mathbb{Z}$ of p elements is usually denoted by $\mathbb{F}_p$. We let $\boldsymbol{k}^{\text{ac}}$ be the algebraic closure of a field $\boldsymbol{k}$.

As to model-theory, we assume only familiarity with the basics, and use the following standard notations for L-structures $\mathcal{M}$ and $\mathcal{N}$, with L any language:

(i) $\mathcal{M} \equiv \mathcal{N}$ means that $\mathcal{M}$ and $\mathcal{N}$ are elementarily equivalent (that is, satisfy the same L-sentences),
(ii) $\mathcal{M} \subseteq \mathcal{N}$ means that $\mathcal{M}$ is a substructure of $\mathcal{N}$,
(iii) $\mathcal{M} \preccurlyeq_1 \mathcal{N}$ means that $\mathcal{M}$ is an existentially closed substructure of $\mathcal{N}$,
(iv) $\mathcal{M} \preccurlyeq \mathcal{N}$ means that $\mathcal{M}$ is an elementary substructure of $\mathcal{N}$.

For example, given rings R and S we mean by $R \subseteq S$ that R is a subring of S; in writing $R \subseteq R[t_1, \ldots, t_n] \subseteq R[[t_1, \ldots, t_n]]$ earlier, we already used this meaning tacitly. In Chaps. 4, 5, and 7 we often construe valued fields as 2-sorted or 3-sorted structures, and then we use the above notations accordingly.

The goal of this course is the famous Ax-Kochen-Ershov Theorem from the 1960s. We shall prove this result in Chap. 5 in various stronger forms, and discuss applications. We begin by stating some very special cases of the theorem. Let t be a single indeterminate.

1.1. *Suppose* $\boldsymbol{k}_1$ *and* $\boldsymbol{k}_2$ *have characteristic* 0. *Then, as rings,*

$$\boldsymbol{k}_1[[t]] \equiv \boldsymbol{k}_2[[t]] \iff \boldsymbol{k}_1 \equiv \boldsymbol{k}_2.$$

The direction $\Rightarrow$ is easy and holds also when $\boldsymbol{k}_1$ and $\boldsymbol{k}_2$ have characteristic $p > 0$, but the direction $\Leftarrow$ lies much deeper, and is not yet known to be true when $\boldsymbol{k}_1$ and $\boldsymbol{k}_2$ have characteristic $p > 0$. (This is one of the main open problems in the subject.) Recall that $\mathbb{Q}^{\mathrm{ac}} \equiv \mathbb{C}$, as fields, where $\mathbb{Q}^{\mathrm{ac}}$ is the algebraic closure of $\mathbb{Q}$, so as a special case of 1.1 we have $\mathbb{Q}^{\mathrm{ac}}[[t]] \equiv \mathbb{C}[[t]]$. On the other hand, $\mathbb{Q}^{\mathrm{ac}}[[t_1, t_2]] \not\equiv \mathbb{C}[[t_1, t_2]]$ and $\mathbb{Q}^{\mathrm{ac}}[t] \not\equiv \mathbb{C}[t]$. This highlights the fact that 1.1 is a *one-variable* phenomenon about *power series rings*: it fails for two or more variables, and also fails for polynomial rings instead of power series rings. A stronger version of 1.1 is in terms of sentences:

1.2. *For every sentence* σ *in the language of rings, there is a sentence* $\overline{\sigma}$ *in that language, such that for all* $\boldsymbol{k}$ *of characteristic* 0,

$$\boldsymbol{k}[[t]] \models \sigma \iff \boldsymbol{k} \models \overline{\sigma}.$$

□

For the next result, let $\boldsymbol{k}[t_1, \dots, t_n]^{\mathrm{alg}}$ be the subring of $\boldsymbol{k}[[t_1, \dots, t_n]]$ consisting of all $f \in \boldsymbol{k}[[t_1, \dots, t_n]]$ that are algebraic over $\boldsymbol{k}[t_1, \dots, t_n]$.

1.3. *If* $\boldsymbol{k}$ *has characteristic* 0, *then* $\boldsymbol{k}[t]^{\mathrm{alg}} \preccurlyeq \boldsymbol{k}[[t]]$.

For any field $\boldsymbol{k}$ we have a weak version of 1.3 for n variables:

$$\boldsymbol{k}[t_1, \dots, t_n]^{\mathrm{alg}} \preccurlyeq_1 \boldsymbol{k}[[t_1, \dots, t_n]].$$

This is part of the Artin Approximation Theorems, [3]. Again, if $n \geq 2$, then $\boldsymbol{k}[t_1, \dots, t_n]^{\mathrm{alg}}$ is never an elementary substructure of $\boldsymbol{k}[[t_1, \dots, t_n]]$.

Next, let $\mathbb{C}[[t_1, \dots, t_n]]_{\mathrm{conv}}$ be the subring of $\mathbb{C}[[t_1, \dots, t_n]]$ consisting of the power series $f \in \mathbb{C}[[t_1, \dots, t_n]]$ that converge absolutely in some neighborhood of the origin in $\mathbb{C}^n$ (with the neighborhood depending on the series). For the case of a single variable t the AKE-theorems give:

1.4. $\mathbb{C}[[t]]_{\mathrm{conv}} \preccurlyeq \mathbb{C}[[t]]$.

Artin approximation [2] gives $\mathbb{C}[[t_1, \dots, t_n]]_{\mathrm{conv}} \preccurlyeq_1 \mathbb{C}[[t_1, \dots, t_n]]$ for all n, but for $n > 1$ we cannot replace here $\preccurlyeq_1$ by $\preccurlyeq$.

A key algebraic property of the power series rings $\boldsymbol{k}[[t_1, \dots, t_n]]$ is that they are *henselian local rings*. This property is closely linked to the theorems above, and distinguishes these power series rings from the polynomial rings $\boldsymbol{k}[t_1, \dots, t_n]$. From the model-theoretic point of view it is significant that "henselian local" is an elementary (first-order) property of a ring, in contrast to higher order properties such as "noetherian" and "complete".

Definition 1.5. A ring R is *local* if it has exactly one maximal ideal. Given a local ring R, we denote its maximal ideal by $\mathfrak{m}_R$, or just $\mathfrak{m}$ if R is clear from the context, and we let $\boldsymbol{k} = R/\mathfrak{m}$ be the residue field, with residue map

$$x \mapsto \bar{x} = x + \mathfrak{m} : R \to \boldsymbol{k}.$$

Local rings are exactly the models of a (finite) set of axioms in the language of rings, because a ring R is local iff its set of non-units is an ideal. (Even simpler, a ring R is local iff $1 \neq 0$ and its set of non-units is closed under addition.) Note also that an element x in a local ring is a unit iff $\bar{x} \neq 0$.

Examples.

(1) Fields are exactly the local rings with $\mathfrak{m} = 0$.
(2) $\boldsymbol{k}[[t_1, \ldots, t_n]]$ is a local ring with $\mathfrak{m} = (t_1, \ldots, t_n)$. This follows by the above decomposition of any $f \in R$ as a sum $f(0)+t_1 f_1+\cdots+t_n f_n$, and by noting that $f \in U(R)$ iff $f(0) \neq 0$. The kernel of the ring morphism $f \mapsto f(0) : R \to \boldsymbol{k}$ is $\mathfrak{m}$, so this morphism induces a field isomorphism $R/\mathfrak{m} \to \boldsymbol{k}$. We shall usually identify the residue field $R/\mathfrak{m}$ with $\boldsymbol{k}$ via this isomorphism.
(3) Let p be a prime number and k a positive integer. Then $\mathbb{Z}/p^k\mathbb{Z}$ is a local ring. The integers $a_0 + a_1 p + \cdots + a_{k-1} p^{k-1}$ with $a_i \in \{0, 1, \ldots, p-1\}$ are in one-to-one correspondence with their images in $\mathbb{Z}/p^k\mathbb{Z}$. The maximal ideal of this ring is generated by the image of p. The elements of this ring behave somewhat like power series in the "variable" p truncated at p^k.
(4) Letting $k \to \infty$ in (3) we get the ring $\mathbb{Z}_p$ of p-adic integers. Its elements can be represented as infinite series $\sum_{i=0}^{\infty} a_i p^i$, where all $a_i \in \{0, 1, \ldots, p-1\}$. A precise definition is given in the next section.

All local rings in these examples are henselian, as we shall see in the next section. The localizations

$$\mathbb{Z}_{p\mathbb{Z}} = \{\frac{a}{b} : a, b \in \mathbb{Z}, b \notin p\mathbb{Z}\} \subseteq \mathbb{Q}, \quad (p \text{ a prime number}),$$

are examples of local rings that are not henselian.

Definition 1.6. Let R be a local ring. We say that R is *henselian* if for any polynomial $f(X) \in R[X]$ and any $\alpha \in R$ such that

$$f(\alpha) \in \mathfrak{m}, \qquad f'(\alpha) \notin \mathfrak{m},$$

there is $a \in R$ with $f(a) = 0$ and $a \equiv \alpha \mod \mathfrak{m}$.

Remark 1.7. By Lemma 2.1 below there can be at most one such a.
The henselian property means that non-singular zeros in the residue field can be lifted to the ring itself: let R be a local ring, $f(X) \in R[X]$, $\alpha \in R$; then

$$f(\alpha) \in \mathfrak{m} \text{ and } f'(\alpha) \notin \mathfrak{m} \iff \bar{f}(\bar{\alpha}) = 0 \text{ and } \bar{f}'(\bar{\alpha}) \neq 0$$
$$\iff \bar{\alpha} \text{ is a non-singular zero of } \bar{f}(X).$$

(Here we let $\bar{f}(X)$ be the image of $f(X)$ in $\boldsymbol{k}[X]$ obtained by replacing each coefficient of f by its residue class.) Note that the henselian local rings are exactly the models of a certain set of sentences in the language of rings: this set consists of the axioms for local rings and has in addition for each $n \geq 1$ an axiom expressing the henselian property for polynomials of degree $\leq n$.

The AKE-theorem concerns a particular class of henselian local rings, namely henselian valuation rings. The rings $\boldsymbol{k}[[t]]$, $\boldsymbol{k}[[t]]^{\mathrm{alg}}$, $\mathbb{C}[[t]]_{\mathrm{conv}}$, and $\mathbb{Z}_p$ are indeed valuation rings, unlike $\boldsymbol{k}[[t_1, \dots, t_n]]$ for $n > 1$, and $\mathbb{Z}/p^k\mathbb{Z}$ for $k > 1$.

Definition 1.8. A *valuation ring* is a domain R whose set of ideals is linearly ordered by inclusion.

Exercises. Prove the following:

1. Each valuation ring is a local ring.
2. Let R be a domain with fraction field K. Then R is a valuation ring iff for every $a \in K^\times$ either $a \in R$ or $a^{-1} \in R$.
3. $\boldsymbol{k}[[t]]$ is a valuation ring.
4. $\boldsymbol{k}[[t_1, \dots, t_n]]$ is not a valuation ring if $n > 1$.

2 Henselian Local Rings

In this chapter we establish basic facts about henselian local rings, prove that complete local rings are henselian (Hensel's Lemma), show that under certain conditions the residue field of a henselian local ring can be lifted, and give a baby version of the Ax-Kochen Principle.

2.1 Some Elementary Results

Lemma 2.1. *Let R be a local ring, $f(X) \in R[X]$, and $\alpha \in R$, such that*

$$f(\alpha) \in \mathfrak{m}, \qquad f'(\alpha) \notin \mathfrak{m}.$$

Then there is at most one $a \in R$ such that $f(a) = 0$ and $a \equiv \alpha \mod \mathfrak{m}$.

Proof. Suppose $a \in R$ satisfies $f(a) = 0$ and $a \equiv \alpha \mod \mathfrak{m}$. Then $f'(a) \equiv f'(\alpha) \mod \mathfrak{m}$, hence $f'(a)$ is a unit. Taylor expansion in $x \in \mathfrak{m}$ around a gives

$$f(a+x) = f(a) + f'(a)x + bx^2 = f'(a)x + bx^2 \quad (b \in R)$$
$$= f'(a)x\left[1 + f'(a)^{-1}bx\right].$$

Since $f'(a)\left[1 + f'(a)^{-1}bx\right] \in U(R)$, this gives: $f(a+x) = 0$ iff $x = 0$. □

Lemma 2.2. *Let R be a local ring. The following are equivalent.*

(1) *R is henselian.*
(2) *Each polynomial $1 + X + cd_2X^2 + \cdots + cd_nX^n$, with $n \geq 2$, $c \in \mathfrak{m}$ and $d_2, \ldots, d_n \in R$, has a zero in $U(R)$.*
(3) *Each polynomial $Y^n + Y^{n-1} + cd_2Y^{n-2} + \cdots + cd_n$, with $n \geq 2$, $c \in \mathfrak{m}$ and $d_2, \ldots, d_n \in R$, has a zero in $U(R)$.*
(4) *Given a polynomial $f(X) \in R[X]$, $\alpha \in R$, and $c \in \mathfrak{m}$ such that $f(\alpha) = cf'(\alpha)^2$, there is $a \in R$ such that $f(a) = 0$ and $a \equiv \alpha \mod cf'(\alpha)$.*

The "Newton version" (4) gives extra precision in the henselian property when $f'(\alpha) \notin \mathfrak{m}$, but (4) is devised to deal also with the case $f'(\alpha) \in \mathfrak{m}$.

Proof. (1)⇒(2). Assume (1) and let $f(X) = 1 + X + cd_2X^2 + \cdots + cd_nX^n$ with $c \in \mathfrak{m}$ and $d_2, \ldots, d_n \in R$. Then for $\alpha = -1$ we have: $f(\alpha) \equiv 0 \mod \mathfrak{m}$ and $f'(\alpha) \equiv 1 \mod \mathfrak{m}$. Thus $f(X)$ has a zero in $-1 + \mathfrak{m} \subseteq U(R)$, by (1).

(2)⇔(3). Use the substitution $X = 1/Y$.

(2)⇒(4). For f, α, c as in the hypothesis of (4), let $x \in R$ and consider the expansion:

$$f(\alpha + x) = f(\alpha) + f'(\alpha)x + \sum_{i \geq 2} b_i x^i$$

where the $b_i \in R$ do not depend on x

$$= cf'(\alpha)^2 + f'(\alpha)x + \sum_{i \geq 2} b_i x^i$$

Set $x = cf'(\alpha)y$ where $y \in R$. Then

$$f(\alpha + cf'(\alpha)y) = cf'(\alpha)^2\left[1 + y + \sum_{i \geq 2} cd_i y^i\right],$$

where the $d_i \in R$ do not depend on y. Assuming (2), choose $y \in R$ such that

$$1 + y + \sum cd_i y^i = 0.$$

This yields an $a = \alpha + cf'(\alpha)y$ as required.

(4)⇒(1). Clear. □

Exercise. Let R be a henselian local ring, $m \geq 1$, and suppose char $\boldsymbol{k}$ does not divide m (this includes the case char $\boldsymbol{k} = 0$). Show that for $a \in U(R)$:

$$a \text{ is an } m\text{-th power in } R \iff \bar{a} \text{ is an } m\text{-th power in } \boldsymbol{k}.$$

2.2 Hensel's Lemma

Hensel's Lemma says that *complete* local rings are henselian. Here is the idea of the proof. Let R be a local ring, and think of the elements in $\mathfrak{m}$ as *infinitesimals* (very small). Let $f(X) \in R[X]$ and $\alpha \in R$ be such that $f(\alpha) \in \mathfrak{m}$ and $f'(\alpha) \notin \mathfrak{m}$, so $f(\alpha)$ is tiny compared to $f'(\alpha)$. If $f(\alpha) \neq 0$ we can apply Newton's method to perturb α by a tiny amount x to make $f(\alpha + x)$ much smaller than $f(\alpha)$: Taylor expansion in $x \in R$ around α yields

$$\begin{aligned} f(\alpha + x) &= f(\alpha) + f'(\alpha)x + \text{terms of higher degree in } x \\ &= f'(\alpha)\left(f'(\alpha)^{-1} f(\alpha) + x + \text{terms of higher degree in } x\right) \end{aligned}$$

Setting $x = -f'(\alpha)^{-1} f(\alpha)$ yields $x \in \mathfrak{m}$ and

$$f(\alpha + x) = f'(\alpha)[\text{multiple of } f(\alpha)^2]$$

By the occurrence of $f(\alpha)^2$ as a factor in the above expression, this choice of x will make $f(\alpha + x)$ much smaller than $f(\alpha)$, provided we have a suitable norm on R by which to measure size. The process above can now be repeated with $\alpha + x$ in the role of α. Iterating this process indefinitely and assuming R is *complete* with respect to our norm, we can hope to obtain a sequence converging to some $a \in R$ such that $f(a) = 0$ and $a \equiv \alpha \mod \mathfrak{m}$. A precise version of this limit process occurs in the proof of Hensel's Lemma 2.7.

Definition 2.3. Let R be a ring. A *norm* on R is a function $|\ | : R \to \mathbb{R}^{\geq 0}$ such that for all $x, y \in R$:

(i) $|x| = 0 \iff x = 0, \quad |1| = |-1| = 1$,
(ii) $|x + y| \leq |x| + |y|$,
(iii) $|xy| \leq |x||y|$.

It follows easily that $|-x| = |x|$ for all $x \in R$, and thus we obtain a metric $d(x, y) = |x - y|$ on R; we consider R as a metric space with this metric if the norm $|\ |$ is clear from context. Note that then the operations

$$+, \cdot, - : R^2 \to R \quad \text{and} \quad |\ | : R \to \mathbb{R}$$

are continuous. We say that R is *complete* if it is complete as a metric space. If condition (ii) holds in the strong form

$$|x + y| \leq \max\{|x|, |y|\},$$

then we call the norm an *ultranorm* (or ultrametric norm, or nonarchimedean norm). Note that then $|x+y| = \max\{|x|, |y|\}$ if $|x| \neq |y|$. When our norm is an ultranorm, then convergence of infinite series $\sum a_n$ is an easy matter:

Lemma 2.4. *Suppose the ring R is complete with respect to the ultranorm $| \ |$. Let (a_n) be a sequence in R. Then*

$$\lim_{n\to\infty} \sum_{i=0}^{n} a_i \text{ exists in } R \iff \lim_{n\to\infty} |a_n| = 0.$$

We leave the proof as an exercise. If the limit on the left exists, it is denoted by $\sum_{n=0}^{\infty} a_n$.

Example. Let A be a ring with $1 \neq 0$, and $R := A[[t_1, \ldots, t_n]]$. Each $f \in R$ has a unique representation

$$f = \sum a_{i_1 \ldots i_n} t_1^{i_1} \cdots t_n^{i_n}$$

where the formal sum is over all $(i_1, \ldots, i_n) \in \mathbb{N}^n$ and the coefficients $a_{i_1 \ldots i_n}$ lie in A. For such f we define $\operatorname{ord} f \in \mathbb{N} \cup \{\infty\}$ by

$$\operatorname{ord} f = \min\{i_1 + \cdots + i_n : a_{i_1 \ldots i_n} \neq 0\} \quad \text{if } f \neq 0,$$

and $\operatorname{ord} 0 = \infty$. This order function satisfies:

- $\operatorname{ord} 1 = 0$,
- $\operatorname{ord}(f + g) \geq \min\{\operatorname{ord}(f), \operatorname{ord}(g)\}$,
- $\operatorname{ord}(fg) \geq \operatorname{ord}(f) + \operatorname{ord}(g)$.

It follows that $|f| := 2^{-\operatorname{ord} f}$ defines an ultranorm on R.

Exercise. Show that R is complete with respect to this norm, and has the polynomial ring $A[t_1, \ldots, t_n]$ as a dense subring.

Remark 2.5. Let the ring R be complete with respect to the norm $| \ |$, and let $x \in R$ satisfy $|x| < 1$. Then $\sum_{i=0}^{\infty} x^i := \lim_{N\to\infty} \sum_{i=0}^{N} x^i$ exists in R, and

$$(1 - x) \sum_{i=0}^{\infty} x^i = 1.$$

Lemma 2.6. *Assume the ring R is complete with respect to the ultranorm $|\ |$ and $|x| \leq 1$ for all $x \in R$. Then the set $\mathfrak{M} = \{x \in R : |x| < 1\}$ is an ideal. With $\overline{R} := R/\mathfrak{M}$, we have for all $a \in R$,*

$$a \in U(R) \iff \bar{a} \in U(\overline{R}), \quad \text{where } \bar{a} := a + \mathfrak{M}.$$

Proof. The direction $\Rightarrow$ is obvious. Conversely, let $a \in R$ be such that $\bar{a} \in U(\overline{R})$. Take $b \in R$ such that $ab = 1 - x$ with $x \in \mathfrak{M}$. Then $1 - x$ is a unit in R by the above remark. Hence $a \in U(R)$ and $a^{-1} = b \sum_{i=0}^{\infty} x^i$. □

Hensel's Lemma 2.7. *Assume that the ring R is complete with respect to the ultranorm $|\ |$ and that $|x| \leq 1$ for all $x \in R$. Let $f(X) \in R[X]$, and suppose $\alpha \in R$ is such that*

$$|f(\alpha)| < 1, \qquad f'(\alpha) \in U(R).$$

Then there is a unique $a \in R$ such that $f(a) = 0$ and $|a - \alpha| < 1$.

Proof. We shall obtain a as the limit of a sequence $\{a_n\}$ in R. Put $a_0 := \alpha$ and $\varepsilon := |f(\alpha)| < 1$. Suppose $a_n \in R$ is such that

$$|f(a_n)| \leq \varepsilon^{2^n}, \qquad |a_n - \alpha| \leq \varepsilon.$$

(Clearly this is true for $n = 0$.) Note that $a_n \equiv \alpha \mod \mathfrak{M}$, so $f'(a_n)$ is a unit because $f'(\alpha)$ is. Now, put

$$a_{n+1} := a_n + h, \quad \text{where} \quad h = -f'(a_n)^{-1} f(a_n).$$

Then:

$$\begin{aligned} f(a_{n+1}) &= f(a_n) + f'(a_n)h + \text{multiple of } h^2 \\ &= 0 + \text{multiple of } f(a_n)^2, \end{aligned}$$

hence

$$|f(a_{n+1})| \leq |f(a_n)|^2 \leq (\varepsilon^{2^n})^2 = \varepsilon^{2^{n+1}}.$$

Also,

$$|a_{n+1} - \alpha| \leq \max\{|a_{n+1} - a_n|, |a_n - \alpha|\} \leq \varepsilon,$$

so the induction step is complete. We have

$$|a_{n+1} - a_n| = |h| \leq |f(a_n)| \leq \varepsilon^{2^n},$$

so $\{a_n\}$ is a Cauchy sequence. Let $a \in R$ be its limit. Then

$$f(a) = f(\lim a_n) = \lim f(a_n) = 0.$$

The condition $|a_n - \alpha| \leq \varepsilon$ for all n insures that $|a - \alpha| \leq \varepsilon < 1$. The uniqueness of a follows exactly as in the proof of Lemma 2.1. □

Exercise. Assume the ring R is complete with respect to the ultranorm $|\ |$ and $|x| \leq 1$ for all $x \in R$. Let $f_1, \ldots, f_N \in R[X]$ where $X = (X_1, \ldots, X_N)$ is a tuple of distinct indeterminates. Let $\alpha \in R^N$ be a "near zero" of $f_1, \ldots, f_N$, that is,

$$|f_i(\alpha)| < 1 \text{ for } i = 1, \ldots, N, \text{ and } \det\left(\frac{\partial f_i}{\partial X_j}(\alpha)\right)_{i,j} \text{ is a unit in } R.$$

Then there is a unique $a \in R^N$ such that $f_1(a) = \cdots = f_N(a) = 0$, and $a \equiv \alpha \bmod \mathfrak{M}$ componentwise.

Hint. Put $f = (f_1, \ldots, f_N) \in R[X]^N$. Check that in $R[X]^N$ we have an identity

$$f(\alpha + X) = f(\alpha) + f'(\alpha)X + \text{terms of degree} \geq 2 \text{ in } X,$$

where $f'(\alpha)$ is the matrix $\left(\frac{\partial f_i}{\partial X_j}(\alpha)\right)_{i,j}$, and in the matrix product $f'(\alpha)X$ we view X as an $N \times 1$ matrix. Verify that the proof of Hensel's Lemma goes through.

2.3 *Completion*

Let $(R, |\ |)$ be a normed ring, that is, R is a ring, and $|\ |$ is a norm on R. As a metric space, R has a completion $\hat{R}$, and the ring operations $+, -, \cdot : R^2 \to R$ extend uniquely to continuous operations $+, -, \cdot : \hat{R}^2 \to \hat{R}$. With these extended operations $\hat{R}$ is again a ring. The norm $|\ | : R \to \mathbb{R}$ also extends uniquely to a continuous function $\widehat{|\ |} : \hat{R} \to \mathbb{R}$, and $\widehat{|\ |}$ is a norm on the ring $\hat{R}$ whose corresponding metric is the metric of the complete metric space $\hat{R}$. We have now a complete normed ring $(\hat{R}, \widehat{|\ |})$ in which R is a dense subring. If $(R', |\ |')$ is a second complete normed ring extending the normed ring $(R, |\ |)$ such that R is dense in R', then there is a unique normed ring isomorphism $(\hat{R}, \widehat{|\ |}) \cong (R', |\ |')$ that is the identity on R. Thus we have the right to call $(\hat{R}, \widehat{|\ |})$ *the* completion of the normed ring $(R, |\ |)$; to keep notations simple we usually write $|\ |$ for the norm $\widehat{|\ |}$ on $\hat{R}$.

If $|xy| = |x||y|$ for all $x, y \in R$, then we call $|\ |$ an *absolute value*, and in that case R is a domain, and $|\ |$ extends uniquely to an absolute value on the fraction field. If K is a field with absolute value $|\ |$, then the map $x \mapsto x^{-1}$ on $K^\times$ is continuous, and the completion $\hat{K}$ of K with respect to $|\ |$ is a field, and the extended norm is again an absolute value. In this case, for any subring $R \subseteq K$, the closure of R in $\hat{K}$ is a subring, and is thus the completion of R with respect to the restricted norm.

Suppose that $R = A[[t_1, \dots, t_n]]$ with A a domain. Then its norm given by $|f| = 2^{-\mathrm{ord}(f)}$ is an absolute value. Note that R is complete with respect to this norm. If $A = \boldsymbol{k}$ is a field, then the ring $R = \boldsymbol{k}[[t]]$ is a valuation ring, and its fraction field K is also complete with respect to the extended absolute value. In this fraction field we have $R = \{f \in K : |f| \leq 1\}$. One can identify K in this case with the field of Laurent series $\boldsymbol{k}((t))$.

The Field of p-adic Numbers. Let p be a prime number. For $a \in \mathbb{Z}$ we define $v_p(a) \in \mathbb{N} \cup \{\infty\}$ as follows: $v_p(0) = \infty$, and if $a \neq 0$, then $v_p(a)$ is the natural number such that $a = p^{v_p(a)} b$ where $b \in \mathbb{Z}$ and $p \nmid b$. It is clear that then for all $a, b \in \mathbb{Z}$:

1. $v_p(a + b) \geq \min\{v_p(a), v_p(b)\}$,
2. $v_p(ab) = v_p(a) + v_p(b)$,
3. $v_p(1) = 0$.

This yields an absolute value $| \ |_p$ on $\mathbb{Z}$ by setting

$$|a|_p = p^{-v_p(a)}.$$

(The reason we defined $|a|_p$ in this way with base p, instead of setting $|a|_p = 2^{-v_p(a)}$ with base 2, is to have $|a| \prod_p |a|_p = 1$ for all nonzero $a \in \mathbb{Z}$, where now the product is over all prime numbers p. Taking logarithms this identity becomes $\log |a| - \sum_p v_p(a) \log p = 0$, which mimics the identity $\deg f - \sum_{P \in \mathbb{C}} v_P(f) = 0$ for non-zero $f \in \mathbb{C}[X]$. Here $v_P(f)$ is the *order of vanishing* of f at the point P; a precise definition is given later. This strengthens the analogy between the ring $\mathbb{Z}$ and the polynomial ring $\mathbb{C}[X]$. This analogy can be pushed much further and is very productive in number theory.)

The absolute value $| \ |_p$ extends uniquely to an absolute value on $\mathbb{Q}$, which we call the *p-adic absolute value*. The completion of $\mathbb{Q}$ with respect to the p-adic absolute value is denoted by $\mathbb{Q}_p$, and called *the field of p-adic numbers*. The absolute value of $\mathbb{Q}_p$ is also denoted by $| \ |_p$, and referred to as *the p-adic absolute value*. The closure of $\mathbb{Z}$ in $\mathbb{Q}_p$ is denoted by $\mathbb{Z}_p$. By previous remarks $\mathbb{Z}_p$ is a subring of $\mathbb{Q}_p$, and is the completion of $\mathbb{Z}$ with respect to its p-adic norm. We call $\mathbb{Z}_p$ *the ring of p-adic integers*. Thus we have the following diagram where all maps are inclusions:

$$\begin{array}{ccc} \mathbb{Z}_p & \longrightarrow & \mathbb{Q}_p \\ \uparrow & & \uparrow \\ \mathbb{Z} & \longrightarrow & \mathbb{Q} \end{array}$$

Since $|a|_p \leq 1$ for $a \in \mathbb{Z}$, this remains true for $a \in \mathbb{Z}_p$. Here are some of the most basic facts about p-adic numbers:

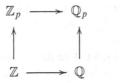

Proposition 2.8. *Given any sequence (a_n) of integers, $\lim_{n\to\infty}\sum_{i=0}^{n} a_i p^i$ exists in $\mathbb{Z}_p$, and for $a = \sum_{n=0}^{\infty} a_n p^n$ we have the equivalences*

$$p|a_0 \iff a \in p\mathbb{Z}_p \iff |a|_p < 1 \iff a \notin U(\mathbb{Z}_p).$$

Moreover:

(i) *the map $(a_n) \mapsto \sum_{n=0}^{\infty} a_n p^n : \{0, 1, \ldots, p-1\}^{\mathbb{N}} \to \mathbb{Z}_p$ is a bijection;*
(ii) *$\mathbb{Z}_p$ is a henselian local ring with maximal ideal $p\mathbb{Z}_p$;*
(iii) *for each integer $k \geq 1$ the ring morphism $\mathbb{Z} \to \mathbb{Z}_p/p^k\mathbb{Z}_p$ is surjective with kernel $p^k\mathbb{Z}$, and thus induces a ring isomorphism*

$$\mathbb{Z}/p^k\mathbb{Z} \to \mathbb{Z}_p/p^k\mathbb{Z}_p.$$

For $k = 1$ this isomorphism identifies $\mathbb{F}_p$ with the residue field of $\mathbb{Z}_p$.
(iv) *each nonzero $x \in \mathbb{Q}_p$ is of the form $x = p^k u$ with unique integer exponent k and $u \in U(\mathbb{Z}_p)$.*
(v) *$\mathbb{Z}_p = \{x \in \mathbb{Q}_p : |x|_p \leq 1\} = \{x \in \mathbb{Q}_p : |x|_p < p\}$.*
(vi) *$\mathbb{Z}_p$ is open and closed in $\mathbb{Q}_p$, and is compact.*

Proof. Let (a_n) be a sequence in $\mathbb{Z}$. By Lemma 2.4 the series $\sum_{n=0}^{\infty} a_n p^n$ converges, since $|a_n p^n|_p \leq p^{-n} \to 0$ as $n \to \infty$. Put $a := \sum_{n=0}^{\infty} a_n p^n$. If $p|a_0$, then $a_0 = pb$ with $b \in \mathbb{Z}$, so

$$a = a_0 + p\sum_{n=0}^{\infty} a_{n+1}p^n = p(b + \sum_{n=0}^{\infty} a_{n+1}p^n) \in p\mathbb{Z}_p.$$

The implications $a \in p\mathbb{Z}_p \Rightarrow |a|_p < 1$, and $|a|_p < 1 \Rightarrow a \notin U(\mathbb{Z}_p)$ follow from $|p|_p < 1$ and the fact that $|x|_p \leq 1$ for all $x \in \mathbb{Z}_p$. We obtain the implication $a \notin U(\mathbb{Z}_p) \Rightarrow p|a_0$ by proving its contrapositive: Suppose that $p \nmid a_0$. Take $b \in \mathbb{Z}$ with $a_0 b = 1 + pk$, $k \in \mathbb{Z}$. Then

$$ab = 1 + p\sum_{n=0}^{\infty} c_n p^n, \qquad c_0 := k + a_1 b,\ c_n = a_{n+1}b \text{ for } n > 0.$$

Since $|p\sum_{n=0}^{\infty} c_n p^n|_p \leq p^{-1} < 1$, it follows from Remark 2.5 that $ab \in U(\mathbb{Z}_p)$, hence $a \in U(\mathbb{Z}_p)$.

As to (i), to prove injectivity, let (a_n) and (b_n) be two distinct sequences in $\{0, 1, \ldots, p-1\}$, and $a = \sum_{n=0}^{\infty} a_n p^n$, $b = \sum_{n=0}^{\infty} b_n p^n$. Let m be the least n such that $a_n \neq b_n$. Then $a - b = p^m(\sum_{i=0}^{\infty}(a_{m+i} - b_{m+i})p^i) \neq 0$, since $p \nmid (a_m - b_m)$. For surjectivity, let A be the image of the map in (i). We claim that, given any $k \in \mathbb{Z}$ and real $\epsilon > 0$ there is $a \in A$ such that $|k - a|_p < \epsilon$. To see this, take n such that $p^{-n} \leq \epsilon$, and take $a_0, \ldots, a_{n-1} \in \{0, 1, \ldots, p-1\}$ such that $k \equiv a_0 + a_1 p + \cdots + a_{n-1}p^{n-1} \mod p^n$, so $a = a_0 + a_1 p + \cdots + a_{n-1}p^{n-1} + p^n b$ with $b \in \mathbb{Z}$, hence $|a - (a_0 + a_1 p + \cdots + a_{n-1}p^{n-1})|_p \leq p^{-n} < \epsilon$. This proves our claim, which says that $\mathbb{Z}$ is contained in the closure of A. It remains to show that A is closed in $\mathbb{Z}_p$. We equip the finite set $\{0, 1, \ldots, p-1\}$ with the discrete topology, and give $\{0, 1, \ldots, p-1\}^{\mathbb{N}}$ the corresponding product topology, making this set of sequences a compact hausdorff space (by Tychonov). It is easily checked that the map in (i) is continuous, so A is compact, hence A is closed in $\mathbb{Z}_p$.

Item (ii): By (i) and what we proved before (i), the set of nonunits in $\mathbb{Z}_p$ is the ideal $\{x \in \mathbb{Z}_p : |x|_p < 1\} = p\mathbb{Z}_p$. Hence $\mathbb{Z}_p$ is a local ring with maximal ideal $p\mathbb{Z}_p$. Since $\mathbb{Z}_p$ is complete with respect to $|\ |_p$, it is henselian by Hensel's Lemma.

Item (iii): Let k be a positive integer and $a \in \mathbb{Z}_p$. By (i) we have $a = a_0 + a_1 p + \cdots + a_{k-1}p^{k-1} + p^k b$ with $a_0, \ldots, a_{k-1} \in \{0, \ldots, p-1\}$ and $b \in \mathbb{Z}_p$, so $a + p^k\mathbb{Z}_p = (a_0 + a_1 p + \cdots + a_{k-1}p^{k-1}) + p^k\mathbb{Z}_p$. This gives surjectivity of the map in (iii). Suppose $\ell \in \mathbb{Z}$ lies in the kernel of this map, that is, $\ell = p^k b$ with $b \in \mathbb{Z}_p$. To show that $\ell \in p^k\mathbb{Z}$ we can replace ℓ by $\ell + p^k q$ where $q \in \mathbb{Z}$, and thus we can assume that $\ell = b_0 + b_1 p + \cdots + b_{k-1}p^{k-1}$ where the "base p-digits" b_i lie in $\{0, \ldots, p-1\}$. So $p^k b = b_0 + b_1 p + \cdots + b_{k-1}p^{k-1}$, and by the injectivity in (i) this yields $b_0 = \cdots = b_{k-1} = 0$.

Item (iv): By the above, each nonzero $a \in \mathbb{Z}_p$ has the form $p^n u$ with $u \in U(\mathbb{Z}_p)$, and we note that this determines n by $|a|_p = p^{-n}$. Using this fact, the set $K := \{a/p^m : a \in \mathbb{Z}_p, m \in \mathbb{N}\}$ is easily seen to be a subfield of $\mathbb{Q}_p$. Note that each nonzero $x \in K$ has the form $p^k u$ with $k \in \mathbb{Z}$ and $u \in U(\mathbb{Z}_p)$, and that then $|x|_p = p^{-k}$, so k and u are uniquely determined by x. It follows that if $x \in K$ and $|x|_p \leq 1$, then $x \in \mathbb{Z}_p$. It remains to show that $K = \mathbb{Q}_p$, and since $\mathbb{Q}_p$ is the closure of $\mathbb{Q}$, this will follow by showing that K is closed in $\mathbb{Q}_p$. Let (a_n) be a sequence in K converging to $a \in \mathbb{Q}_p$. Then the sequence $(|a_n|_p)$ is bounded, so we can take $k \in \mathbb{N}$ such that $|a_n|_p \leq p^k$ for all n, hence $|p^k a_n|_p \leq 1$ for all n, and thus $p^k a_n \in \mathbb{Z}_p$ for all n. Since $(p^k a_n)$ is a Cauchy sequence, it converges to an element $b \in \mathbb{Z}_p$, hence (a_n) converges to $b/p^k \in K$, that is, $a = b/p^k \in K$.

Item (v): The proof of (iv) shows $\mathbb{Z}_p = \{x \in \mathbb{Q}_p : |x|_p \leq 1\}$, and also that the p-adic norm of any $x \in \mathbb{Q}_p^\times$ is of the form p^k with $k \in \mathbb{Z}$. So if $x \in \mathbb{Q}_p$ and $|x|_p < p$, then $|x|_p \leq 1$.

Item (vi): that $\mathbb{Z}_p$ is open and closed in $\mathbb{Q}_p$ is immediate from (v), and that $\mathbb{Z}_p$ is compact follows from (i) and its proof: the map in (i) is a continuous bijection from the compact hausdorff space $\{0, 1, \ldots, p-1\}^{\mathbb{N}}$ onto $\mathbb{Z}_p$, so this map is even a homeomorphism. □

Note that the familiar base p representation of a natural number is extended in (i) to all p-adic integers: the natural numbers are exactly the p-adic integers whose base p representation $\sum a_n p^n$ as in (i) has $a_n = 0$ for all sufficiently large n. In connection with (vi), note that every $a \in \mathbb{Q}_p$ has a compact open neighborhood $a + \mathbb{Z}_p$.

Exercises. Find the base p representation of -1 as in (i). Prove:

1. $\mathbb{Z}_p$ is a valuation ring;
2. the $p^n\mathbb{Z}_p$ are exactly the nonzero ideals of $\mathbb{Z}_p$;
3. $\mathbb{Z}_p \cap \mathbb{Q} = \mathbb{Z}_{p\mathbb{Z}}$;
4. the only ring endomorphism of $\mathbb{Z}_p$ is the identity.

An Analogy Between the Real Field and the p-adic Field Involving Squares. The field $\mathbb{R}$ of real numbers is a completion of the field $\mathbb{Q}$ of rational numbers with respect to a certain absolute value, and so is the p-adic field $\mathbb{Q}_p$. The analogy goes further. In both cases the topology given by the absolute value makes the field locally compact, which is very powerful. Moreover, for $\mathbb{R}$ its topology is determined by its natural ordering, which in turn has a purely algebraic definition in terms of squares: for $x \in \mathbb{R}$ we have:

$$x \geq 0 \iff x = y^2 \text{ for some } y \in \mathbb{R}.$$

(This can be used to show that $\mathbb{R}$ is rigid in the sense that the only ring endomorphism of $\mathbb{R}$ is the identity.) Likewise, the topology on $\mathbb{Q}_p$ is the unique topology on $\mathbb{Q}_p$ making it a topological group with respect to addition such that $\{p^n\mathbb{Z}_p : n = 0, 1, 2, \dots\}$ is a neighborhood basis at 0. Also, the open subgroup $\mathbb{Z}_p$ has the following (first-order) definition in the field $\mathbb{Q}_p$: for $x \in \mathbb{Q}_p$ we have

$$x \in \mathbb{Z}_p \iff 1 + px^2 = y^2 \text{ for some } y \in \mathbb{Q}_p \qquad (p \neq 2),$$
$$x \in \mathbb{Z}_p \iff 1 + 8x^2 = y^2 \text{ for some } y \in \mathbb{Q}_p \qquad (p = 2).$$

We leave the proof of these equivalences to the reader. (For the forward direction, use that $\mathbb{Z}_p$ is henselian.) It now follows easily from an earlier exercise that $\mathbb{Q}_p$ is rigid in the sense that the only ring endomorphism of $\mathbb{Q}_p$ is the identity.

2.4 Lifting the Residue Field

Suppose R is a local ring with a subfield E, that is, E is a subring that happens to be a field. Then E is mapped isomorphically onto a subfield $\bar{E}$ of the residue field $\boldsymbol{k}$ by the residue map $x \mapsto \bar{x}$. (Why?) In case $\bar{E} = \boldsymbol{k}$ we call E a *lift* of $\boldsymbol{k}$. (Example: if $R = \mathbb{C}[[t]]$, then the subfield $\mathbb{C}$ is such a lift. Non-example: let p be a prime number; then $\mathbb{Z}/p^2\mathbb{Z}$ is a local ring but has no subfield.)

Suppose $\bar{E} \neq \boldsymbol{k}$. It would be nice if we could extend E to a lift of $\boldsymbol{k}$. In what follows we indicate how to do this. Let $y \in \boldsymbol{k} \setminus \bar{E}$ and take $x \in R$ such that $\bar{x} = y$.

Case 1. y is transcendental over $\bar{E}$. For $f(X) \in E[X] \setminus \{0\}$, we have $\bar{f}(X) \in \bar{E}[X] \setminus \{0\}$, hence $\overline{f(x)} = \bar{f}(y) \neq 0$, so $f(x) \in U(R)$. In particular, the subring $E[x]$ of R is mapped isomorphically onto the subring $\bar{E}[y]$ of $\boldsymbol{k}$ by the residue map. Thus $E[x]$ is a domain with fraction field $E(x)$ inside R, and $E(x)$ is mapped isomorphically onto the subfield $\bar{E}(y)$ of $\boldsymbol{k}$.

In this case we found a subfield $E(x)$ of R that strictly contains E.

Case 2. y is algebraic over E. Take the monic polynomial $f(X) \in E[X]$ such that its image $\bar{f}(X) \in \bar{E}[X]$ is the minimum polynomial of y over $\bar{E}$. Note that

$f(X) \in E[X]$ is irreducible, since $\bar{f}(X) \in \bar{E}[X]$ is. We have a surjective ring morphism

$$a \mapsto \bar{a} : E[x] \to \bar{E}[y],$$

where $E[x]$ is a subring of R and $\bar{E}[y]$ is a subfield of $\boldsymbol{k}$. If we happened to be so lucky that $f(x) = 0$, then this map would be injective (why?), and we would have extended E to a strictly larger subfield $E[x]$ of R. In general $f(x) \neq 0$, but if we can find an $\epsilon \in \mathfrak{m}$ such that $f(x + \epsilon) = 0$, then we can replace in the above argument x by $x + \epsilon$, and $E[x + \epsilon]$ would be a subfield of R that strictly contains E.

To find such an ϵ we assume at this point that *R is henselian and* char $\boldsymbol{k} = 0$. We have $\overline{f(x)} = \bar{f}(y) = 0$, that is, $f(x) \in \mathfrak{m}$. Since char $\boldsymbol{k} = 0$, the irreducible polynomial $\bar{f}(X) \in \bar{E}[X]$ is separable, so $\bar{f}'(y) \neq 0$, that is, $f'(x) \notin \mathfrak{m}$. The henselian assumption on R now yields an $\epsilon \in \mathfrak{m}$ such that $f(x + \epsilon) = 0$, as desired.

This discussion leads to the following important result.

Theorem 2.9. *Let R be a henselian local ring with* char $\boldsymbol{k} = 0$. *Then the residue field* $\boldsymbol{k}$ *can be lifted.*

Proof. The unique ring morphism $\mathbb{Z} \to R$ is injective, and as usual we identify $\mathbb{Z}$ with its image in R via this morphism. Each non-zero integer in R is a unit since its residue class is, so $\mathbb{Z}$ has a fraction field in R. Thus we have found a subfield of R. By Zorn's lemma there is a maximal subfield of R; by the earlier discussion, a maximal subfield of R is a lift of $\boldsymbol{k}$. □

In some cases this lift allows us to completely describe a local ring R in terms of its residue field. Here is such a case:

Proposition 2.10. *Let R be a local ring, with* $t \in R$ *and n such that:*

$$\text{char}\, \boldsymbol{k} = 0, \qquad \mathfrak{m} = tR, \qquad t^n \neq 0, \qquad t^{n+1} = 0 \tag{1}$$

Then we have a ring isomorphism $R \cong \boldsymbol{k}[T]/(T^{n+1})$.

A first step in proving this proposition is to show that a local ring R as in the proposition is henselian. This will follow from the next elementary result.

Lemma 2.11. *Let R be a local ring and let* $t \in R$ *be such that* $\mathfrak{m} = tR$. *Let* $A \subseteq R$ *be a set of representatives modulo* $\mathfrak{m}$, *that is, A maps bijectively onto* $\boldsymbol{k}$ *via the residue map. Then we can expand in powers of t with coefficients in A:*

(1) *For any n, each* $r \in R$ *is of the form*

$$r = a_0 + a_1 t + \cdots + a_n t^n + s t^{n+1}$$

with $a_0, \ldots, a_n \in A$ *and* $s \in R$.

(2) *If* $t^n \neq 0$, *then* $t^m \notin t^{m+1}R$ *for* $m \leq n$, *and the tuple* $(a_0, \ldots, a_n)$ *in* (1) *is uniquely determined by* n, r.
(3) *If* $t^n \neq 0$ *and* $t^{n+1} = 0$, *then the map*

$$(a_0, \ldots, a_n) \mapsto a_0 + a_1 t + \cdots + a_n t^n : A^{n+1} \to R$$

is a bijection.

Proof of Lemma. Let $r \in R$. We establish (1) by induction on n. We have $r \equiv a_0$ mod $\mathfrak{m}$ with $a_0 \in A$, that is, $r = a_0 + st$ with $s \in R$. This proves the case $n = 0$. Suppose $r = a_0 + a_1 t + \cdots + a_n t^n + st^{n+1}$ as in (1). Since $s = a_{n+1} + ut$ with $a_{n+1} \in A$ and $u \in R$, this gives

$$r = a_0 + a_1 t + \cdots + a_n t^n + a_{n+1} t^{n+1} + ut^{n+2}.$$

For (2), suppose $t^n \neq 0$, and $t^m = t^{m+1} r$ with $r \in R$. Then $t^m(1 - tr) = 0$, hence $t^m = 0$ since $1 - tr$ is a unit. This forces $m > n$. Next suppose

$$a_0 + a_1 t + \cdots + a_n t^n \equiv b_0 + b_1 t + \cdots + b_n t^n \mod t^{n+1},$$

where $a_i, b_i \in A$. Suppose $a_i \neq b_i$ for some $i \leq n$, and let m be the least i with this property. After subtracting the first m terms from each side we get $a_m t^m \equiv b_m t^m$ mod t^{m+1}, hence $(a_m - b_m)t^m \in t^{m+1}R$. But $a_m \not\equiv b_m$ mod $\mathfrak{m}$, so $a_m - b_m \in U(R)$, and thus $t^m \in t^{m+1}R$, a contradiction. Item (3) follows immediately from (1) and (2). □

Proof of Proposition 2.10. Let R, t and n be as in the hypothesis of the proposition. By (2) of the lemma above we have a strictly descending sequence of ideals

$$R = t^0 R \supset tR \supset \cdots \supset t^n R \supset t^{n+1} R = 0.$$

With this we can define an ultranorm $|\ |$ on R,

$$|r| = \begin{cases} 2^{-m} & \text{if } r \neq 0, m = \max\{i : r \in t^i R\}, \\ 0 & \text{if } r = 0 \end{cases}.$$

Note that $|r| \leq 1$ for all $r \in R$, and $|r| < 1$ if and only if $r \in \mathfrak{m}$. Since the norm takes only finitely many values, R is complete in this norm. Thus by Hensel's Lemma R is henselian. By Theorem 2.9 there is a ring embedding $j : \boldsymbol{k} \to R$ such that $\overline{j(x)} = x$ for all $x \in \boldsymbol{k}$, so $A := j(\boldsymbol{k})$ is a set of representatives modulo $\mathfrak{m}$. We extend j to a ring morphism $j_t : \boldsymbol{k}[T] \to R$ by sending T to t. By (1) of the above lemma the map j_t is surjective, and by (2) we have ker $j_t = (T^{n+1})$. Thus j_t induces a ring isomorphism $\boldsymbol{k}[T]/(T^{n+1}) \cong R$. □

Exercises. Prove the following:

1. Let $R := \boldsymbol{k}[T]/(T^{n+1})$, and let t be the image of T under the canonical map $\boldsymbol{k}[T] \to R$. Then R is a henselian local ring with maximal ideal tR.
2. Let p be a prime number, and $R := \mathbb{Z}/(p^{n+1})$ and let t be the image of p under the canonical map $\mathbb{Z} \to R$. Then R is a henselian local ring with maximal ideal tR.

2.5 *The Ax-Kochen Principle*

At this point we shall bring a little logic into the picture. In what follows L_t is the language of rings augmented by a new constant symbol t, that is $L_t = L \cup \{t\}$. Here is an easy consequence of Proposition 2.10:

Corollary 2.12. *For each sentence σ in the language L_t there is a sentence σ_n in L such that for any R, n and t as in Proposition 2.10,*

$$(R,t) \models \sigma \iff \boldsymbol{k} \models \sigma_n.$$

Note the resemblance between this result and 1.2 in the introduction. This corollary yields a "baby" Ax-Kochen principle, but for that we need one more lemma:

Lemma 2.13. *For each L-sentence τ there is an L-sentence τ^* such that, for any local ring R,*

$$R \models \tau^* \iff \boldsymbol{k} \models \tau.$$

This follows from the fact that the ideal $\mathfrak{m}$ is definable in R as the set of all non-units. Call a sentence τ^* as in this lemma a *lift* of τ.

The Ax-Kochen Principle says roughly that as the prime number p goes to infinity, the rings $\mathbb{Z}_p$ and $\mathbb{F}_p[[T]]$ get harder to tell apart. The simple version of this result below says the same about the rings $\mathbb{Z}/(p^n)$ and $\mathbb{F}_p[T]/(T^n)$ for any fixed $n \geq 1$. For example, for $n = 2$, the integers

$$a_0 + a_1 p, \qquad (a_0, a_1 \in \{0, 1, \dots, p-1\})$$

are in bijection with their images in $\mathbb{Z}/(p^2)$, and the polynomials

$$b_0 + b_1 T \in \mathbb{F}_p[T], \quad (b_0, b_1 \in \mathbb{F}_p)$$

are in bijection with their images in $\mathbb{F}_p[T]/(T^2)$. These two rings clearly resemble each other: both are henselian local rings with maximal ideal generated by an element whose square is 0, and both have exactly p^2 elements. But they are very different in one respect: $\mathbb{Z}/(p^2)$ has no subfield, but $\mathbb{F}_p[T]/(T^2)$ has (the image of)

$\mathbb{F}_p$ as subfield. The ring $\mathbb{Z}/(p^2)$ is an arithmetic object, while $\mathbb{F}_p[T]/(T^2)$ has more a geometric flavour.

Baby Ax-Kochen Principle 2.14. *Let σ be any sentence of L_t, and let $n \geq 1$. Then*

$$\big(\mathbb{Z}/(p^n), p_n\big) \models \sigma \iff \big(\mathbb{F}_p[T]/(T^n), t_n\big) \models \sigma$$

for all but finitely many primes p, where p_n and t_n are the image of p and T in $\mathbb{Z}/(p^n)$ and $\mathbb{F}_p[T]/(T^n)$, respectively.

Before we start the proof, we define for a prime number p the L-sentence ch_p to be $\forall x(p \cdot x \neq 1)$, where $p \cdot x$ denotes the term $x + \cdots + x$ where the logical variable x occurs exactly p times. So for any ring R we have: $R \models \mathrm{ch}_p \iff p \cdot 1 \notin U(R)$, and thus for any local ring R we have $R \models \mathrm{ch}_p \iff \mathrm{char}\, \boldsymbol{k} = p$.

Proof. Let Σ_n be a set of axioms in the language L_t whose models are exactly the pairs (R,t) such that R is a local ring with $\mathrm{char}\, \boldsymbol{k} = 0$, $\mathfrak{m} = tR$, $t^{n-1} \neq 0$, and $t^n = 0$. By the previous corollary we can take an L-sentence σ_{n-1} such that for every model (R,t) of Σ_n we have

$$(R,t) \models \sigma \iff \boldsymbol{k} \models \sigma_{n-1}.$$

Let σ^*_{n-1} be a lift of σ_{n-1} as in the lemma. Then for every model (R,t) of Σ_n we have

$$(R,t) \models \sigma \iff R \models \sigma^*_{n-1}.$$

Thus by the completeness theorem there is a formal proof of $\sigma \leftrightarrow \sigma^*_{n-1}$ from Σ_n. We can assume that Σ_n is the union of a finite set with an infinite set $\{\neg\mathrm{ch}_2, \neg\mathrm{ch}_3, \neg\mathrm{ch}_5, \ldots, \}$. Fix a formal proof of $\sigma \leftrightarrow \sigma^*_{n-1}$ from Σ_n, and take the natural number N so large that for $p > N$ the axiom $\neg\mathrm{ch}_p$ is not used in this proof. Then for $p > N$, the structures $\big(\mathbb{Z}/(p^n), p_n\big)$ and $\big(\mathbb{F}_p[T]/(T^n), t_n\big)$ satisfy all axioms of Σ_n used in this proof, hence

$$\begin{aligned}
\big(\mathbb{Z}/(p^n), p_n\big) \models \sigma &\iff \mathbb{Z}/(p^n) \models \sigma^*_{n-1} \\
&\iff \mathbb{F}_p \models \sigma_{n-1} \quad \text{by Lemma 2.13} \\
&\iff \mathbb{F}_p[T]/(T^n) \models \sigma^*_{n-1} \\
&\iff \big(\mathbb{F}_p[T]/(T^n), t_n\big) \models \sigma.
\end{aligned}$$

□

Remark 2.15. The proof as given shows that the equivalence holds for all $p > N(\sigma, n)$, where $(\sigma, n) \mapsto N(\sigma, n)$ is a computable function with values in $\mathbb{N}$.

Another (less constructive) argument to obtain this result is to show that for each non-principal ultrafilter $\mathcal{F}$ on the set of primes,

$$\prod_{\mathcal{F}} (\mathbb{Z}/(p^n), p_n) \cong \prod_{\mathcal{F}} (\mathbb{F}_p[T]/(T^n), T_n) \qquad \text{(as } L_t\text{-structures).}$$

The existence of the isomorphism follows from Proposition 2.10: both sides are local rings whose maximal ideal is generated by a single element (the interpretation of t) whose $(n-1)$th power is not zero, but whose nth power is 0, and both sides have residue field isomorphic to $\prod_{\mathcal{F}} \mathbb{F}_p$, which has characteristic 0.

To state other applications of the lifting theorem it is useful to introduce some terminology: Let R be a ring and $X = (X_1, \dots, X_n)$ a tuple of distinct indeterminates with $n \geq 1$. Given any polynomial $f(X) \in \mathbb{Z}[X]$, let the polynomial $f^R(X) \in R[X]$ be obtained from f by replacing each of its coefficients by its image under the ring morphism $\mathbb{Z} \to R$, and for $a \in R^n$, put $f(a) := f^R(a) \in R$. Given polynomials $f_1, \dots, f_k \in \mathbb{Z}[X]$ a *solution of the system* $f_1(X) = \cdots = f_k(X) = 0$ *in* R is an n-tuple $a \in R^n$ such that $f_1(a) = \cdots = f_k(a) = 0$. If R is a local ring, then a *lift* of a tuple $(x_1, \dots, x_n) \in \boldsymbol{k}^n$ is a tuple $(a_1, \dots, a_n) \in R^n$ such that $\bar{a}_i = x_i$ for $i = 1, \dots, n$. The next lemma is an immediate consequence of the lifting theorem.

Lemma 2.16. *Let R be a henselian local ring with* $\operatorname{char} \boldsymbol{k} = 0$*, and $f_1, \dots, f_k \in \mathbb{Z}[X]$. Then any solution of $f_1(X) = \cdots = f_k(X) = 0$ in $\boldsymbol{k}$ can be lifted to a solution in R.* □

The following corollary answered a question of S. Lang from the 1950s.

Corollary 2.17 (Greenleaf, Ax-Kochen). *Let $f_1, \dots, f_k \in \mathbb{Z}[X]$. Then for all but finitely many primes p, every solution of*

$$f_1(X) = \cdots = f_k(X) = 0$$

in $\mathbb{F}_p$ can be lifted to a solution in $\mathbb{Z}_p$.

Proof. Since $f_1, \dots, f_k$ are given by L-terms, we can construct an L-sentence σ_f that expresses in any local ring the liftability of solutions in the residue field, that is, for for every local ring R:

$$R \models \sigma_f \iff \text{for all } x \in R^n \text{ with } f_1(x), \dots, f_k(x) \in \mathfrak{m} \text{ there is } y \in R^n \text{ such that } f_1(y) = \cdots = f_k(y) = 0 \text{ and } y_i - x_i \in \mathfrak{m} \text{ for } i = 1, \dots, n.$$

Let HLR be a set of axioms in the language L whose models are exactly the henselian local rings, and put $\mathrm{HLR}(0) := \mathrm{HLR} \cup \{\neg\mathrm{ch}_2, \neg\mathrm{ch}_3, \neg\mathrm{ch}_5, \dots\}$. So the models of HLR(0) are exactly the henselian local rings with residue field of characteristic 0. By the lemma we have $\mathrm{HLR}(0) \models \sigma_f$, so by compactness there is $N \in \mathbb{N}$ such that

$$\mathrm{HLR} \cup \{\neg \mathrm{ch}_p : p \leq N\} \models \sigma_f.$$

Now observe that $\mathbb{Z}_p$ is a model of the left hand side if $p > N$. □

We cannot bound the exceptional primes in this corollary in terms of k, n and a bound on the degrees of the f_i: the polynomial $X^2 - p \in \mathbb{Z}[X]$ (with p a prime) in a single variable X is of degree 2, the equation $X^2 - p = 0$ has a solution in $\mathbb{F}_p$, but no solution in $\mathbb{Z}_p$.

Theorem 2.18 (Chevalley, Warning). *Let $q = p^e$, $e \in \mathbb{N}^{\geq 1}$, and suppose the polynomials $f_1, \ldots, f_k \in \mathbb{F}_q[X] \setminus \{0\}$ are such that $\sum \deg f_i < n$. Then*

$$\left|\{x \in \mathbb{F}_q^n : f_1(x) = \cdots = f_k(x) = 0\}\right| \equiv 0 \mod p;$$

in particular, if $f_1, \ldots, f_k$ have constant term 0, then there is a nonzero $x \in \mathbb{F}_q^n$ such that $f_1(x) = \cdots = f_k(x) = 0$.

See for example Serre's *Cours d'Arithmétique*, [49] or Greenberg's *Lectures on Forms in Many Variables*, [25] for a proof. In combination with the result of Greenleaf and Ax & Kochen, the theorem of Chevalley and Warning yields:

Corollary 2.19. *Let $f_1, \ldots, f_k \in \mathbb{Z}[X] \setminus \{0\}$ all have constant term 0, and suppose that $\sum \deg f_i < n$. Then for all sufficiently large primes p there exists a nonzero $x \in \mathbb{Z}_p^n$ such that $f_1(x) = \cdots = f_k(x) = 0$.*

Later in the course we shall prove:

Ax-Kochen Principle 2.20. *Let σ be any L-sentence. Then*

$$\mathbb{Z}_p \models \sigma \iff \mathbb{F}_p[[T]] \models \sigma$$

for all but finitely many primes p.

Here is an application that at the time (mid 1960s) created a stir:

2.21. *Given any positive integer d there is $N(d) \in \mathbb{N}$ such that for all prime numbers $p > N(d)$: if $f(X_1, \ldots, X_n) \in \mathbb{Z}_p[X_1, \ldots, X_n]$ is homogeneous of degree d and $n > d^2$, then there exists a nonzero $x \in \mathbb{Z}_p^n$ with $f(x) = 0$.* (Of course it suffices to consider the case $n = d^2 + 1$.)

This follows by applying the Ax-Kochen Principle to a theorem of Lang from the 1950s which says that if $f(X_1, \ldots, X_n) \in R[X_1, \ldots, X_n]$ is homogeneous of degree $d \geq 1$ and $n = d^2 + 1$, with $R = \mathbb{F}_p[[T]]$ and p any prime number, then there exists a nonzero $x \in R^n$ with $f(x) = 0$. (See Greenberg's book [25] for a proof of Lang's theorem, which is closely related to the theorem of Chevalley and Warning.) Here is some more history: it was known from the early twentieth century that a quadratic form (= homogeneous polynomial of degree 2) over a p-adic field in 5 variables always has a nontrivial zero in that field. Emil Artin had conjectured that this remained true for forms of degree d in d^2+1 variables. This conjecture was proved for $d=3$ in the 1950s. Ax and Kochen established the above asymptotic form of Artin's conjecture, and around the same time Terjanian gave a counterexample for $d=4$; see [51].

More on Completion. We still need to explain the term *complete local ring* that was used in motivating Hensel's Lemma. Let R be a ring and $I \subseteq R$ a proper ideal. (Here *proper* means that $I \neq R$.) Consider the decreasing sequence of ideals:

$$R = I^0 \supseteq I^1 \supseteq I^2 \supseteq \cdots .$$

Recall that I^n consists of the finite sums of products $a_1 \cdots a_n$ with each $a_i \in I$. Suppose $\bigcap_n I^n = 0$. (This is always true if R is local and noetherian.) Define $\operatorname{ord}_I : R \to \mathbb{N} \cup \{\infty\}$ by:

$$\operatorname{ord}_I(r) = \begin{cases} \max\{n : r \in I^n\} & \text{if } r \neq 0 \\ \infty & \text{if } r = 0 \end{cases}.$$

Note that then for all $r, s \in R$:

- $\operatorname{ord}_I(r+s) \geq \min\{\operatorname{ord}_I(r), \operatorname{ord}_I(s)\}$,
- $\operatorname{ord}_I(rs) \geq \operatorname{ord}_I(r) + \operatorname{ord}_I(s)$,
- $\operatorname{ord}_I(r) > 0 \iff r \in I$.

This defines an ultranorm by $|r| = 2^{-\operatorname{ord}_I(r)}$; note that for all $r \in R$:

$$|r| \leq 1, \qquad |r| < 1 \iff r \in I.$$

A ring R with proper ideal I is said to be *I-adically complete* if $\bigcap_n I^n = 0$ and R is complete with respect to this norm.

Lemma 2.22. *Let R be a ring with maximal ideal $\mathfrak{m}$ which is $\mathfrak{m}$-adically complete. Then R is a local ring (and thus henselian, by Hensel's lemma).*

Proof. Since $\mathfrak{m} = \{x \in R : |x| < 1\}$, Lemma 2.6 yields $R \setminus \mathfrak{m} = U(R)$. □

We define a *complete local ring* to be a local ring R that is $\mathfrak{m}$-adically complete.

Assume the proper ideal I in the ring R is finitely generated and $\bigcap_n I^n = 0$. Let $\hat{R}$ be the completion of R with respect to the I-adic norm. Then the closure of I in $\hat{R}$ is $I\hat{R}$, also denoted by $\hat{I}$. More generally, the closure of I^n in $\hat{R}$ is $I^n\hat{R}$, which equals $(\hat{I})^n$. Furthermore, the natural map

$$R/I^n \to \hat{R}/(\hat{I})^n$$

is a ring isomorphism. We leave the proofs of these claims as an exercise.

Consider now the ring $\mathbb{Z}$, with ideal $I = p\mathbb{Z}$, where p is a prime number. Although the p-adic norm is defined with base p instead of 2, this yields the same topology on $\mathbb{Z}$ as the I-adic norm. As rings we can identify $\mathbb{Z}_p$ with the I-adic completion $\hat{\mathbb{Z}}$ such that the p-adic norm is a fixed power of the I-adic norm: $|x|_p = |x|^{\log_2 p}$ for $x \in \mathbb{Z}_p$.

We finish this chapter with mentioning some terminology to be used occasionally in later chapters. An absolute value $|\cdot|$ on a field K is said to be *ultrametric* if it is an ultranorm, that is, $|a+b| \leq \max\{|a|, |b|\}$ for all $a, b \in K$. For example, the p-adic absolute value $|\cdot|_p$ on $\mathbb{Q}_p$ is ultrametric. (An older term for "ultrametric" is "non-archimedean".)

3 Valuation Theory

We begin this chapter with developing basic valuation theory, and end it with a quantifier elimination for algebraically closed valued fields. For the important rank 1 case the contents of the first three sections are mostly from Ostrowski [39]. Key ideas enabling us to go beyond this case are due to Hahn [26] and Krull [33].

3.1 Valuations

A *valuation* on an (additively written) abelian group A is a map

$$v : A \setminus \{0\} \to \Gamma$$

into a linearly ordered set Γ, such that $v(-x) = v(x)$ for nonzero $x \in A$ and $v(x+y) \geq \min\{v(x), v(y)\}$ for $x, y \in A$ such that $x, y, x+y \neq 0$.

By convention we extend v to all of A by setting $v(0) = \infty$, where ∞ is not in Γ. We extend $<$ to a linear order on $\Gamma_\infty = \Gamma \cup \{\infty\}$ by specifying $\gamma < \infty$ for all $\gamma \in \Gamma$. Then $v(-x) = v(x)$ and $v(x+y) \geq \min\{v(x), v(y)\}$ for all $x, y \in A$ without exception. For future use, note that if $x, y \in A$ and $v(x) < v(y)$, then $v(x+y) = v(x)$. (Use that $v(x) = v(x+y-y) \geq \min\{v(x+y), vy\}$.) More generally, if $x_1, \ldots, x_n \in A$, $n \geq 1$, and $v(x_1) < v(x_2), \ldots, v(x_n)$, then $v(x_1 + \cdots + x_n) = v(x_1)$. Another elementary fact that is often used is that

$$\{(x, y) \in A \times A : v(x-y) > v(x)\}$$

is an equivalence relation on $A \setminus \{0\}$. (The reader should check this.)

Our real interest is in valuations on *fields*, and then Γ is not just a linearly ordered set but has also a compatible group structure:

An *ordered abelian group* is an abelian (additively written) group Γ equipped with a translation invariant linear order $<$. Such Γ is called *archimedean* if, for any $a, b > 0$, there is $n \in \mathbb{N}$ such that $na > b$. (Examples: The additive groups of

$\mathbb{Z}, \mathbb{Q}, \mathbb{R}$ with the usual ordering are archimedean ordered abelian groups.) Note that ordered abelian groups are torsion-free.

The following result has historic roots in Eudoxos, Euclid, and Archimedes, but its present formulation is due to O. Hölder and became only possible after Dedekind's foundation of the real number system in the nineteenth century. It says that archimedean ordered abelian groups can be embedded into the ordered additive group of real numbers.

Lemma 3.1. *Let Γ be an archimedean ordered abelian group, and $\gamma \in \Gamma^{>0}$. Then there is a unique embedding $i : \Gamma \to \mathbb{R}$ of ordered groups such that $i(\gamma) = 1$. If there is no $\beta \in \Gamma$ with $0 < \beta < \gamma$, then $i(\Gamma) = \mathbb{Z}$.*

Proof. If Γ has γ as least positive element, then $\Gamma = \mathbb{Z}\gamma$, and the embedding is given by $i(k\gamma) = k$ for $k \in \mathbb{Z}$. If Γ has no least positive element, the embedding is given by $i(\beta) := \sup\{\frac{k}{n} : k \in \mathbb{Z},\ n \geq 1,\ n\beta < k\gamma\}$ for $\beta \in \Gamma$. The reader should fill in the details. □

Definition 3.2. A *valuation* on a domain A is a function $v : A \setminus \{0\} \to \Gamma$, where Γ is an ordered abelian group, such that for all $x, y \in A \setminus \{0\}$:

(V1) $v(x + y) \geq \min\{v(x), v(y)\}$, provided $x + y \neq 0$,
(V2) $v(xy) = v(x) + v(y)$,

Given a domain A we have a bijective correspondence between the set of ultrametric absolute values on A and the set of valuations $v : A \setminus \{0\} \to \mathbb{R}$, with the ultrametric absolute value $|\ |$ on A corresponding to the valuation $v : A \setminus \{0\} \to \mathbb{R}$ given by $v(a) := -\log|a|$.

Let $v : A \setminus \{0\} \to \Gamma$ be a valuation on the domain A. Note that then $v(1) = v(-1) = 0$, since v restricted to $U(A)$ is a group morphism. Hence $v(x) = v(-x)$ for all $x \in A \setminus \{0\}$, so v is a valuation on the additive group as defined earlier. We extend v uniquely to a valuation $v : K^{\times} \to \Gamma$ on the fraction field K of A, by

$$v(x/y) = v(x) - v(y), \qquad x, y \in A \setminus \{0\}.$$

Thus $v(K^{\times})$ is a subgroup of Γ. By convention we extend v to all of K by setting $v(0) = \infty$, where $\Gamma_{\infty} = \Gamma \cup \{\infty\}$ is linearly ordered as before, and where we extend $+$ to a binary operation on Γ_{∞} by $\gamma + \infty = \infty + \gamma = \infty$ for all $\gamma \in \Gamma_{\infty}$. Then (V1) and (V2) hold for all $x, y \in K$ without the proviso in (V1).

When we refer to a valuation $v : K^{\times} \to \Gamma$ on a field K, we shall assume from now on that $v(K^{\times}) = \Gamma$, unless specified otherwise. We call $\Gamma = v(K^{\times})$ the *value group of the valuation.*

Motivating Example. Let $K = \mathbb{C}(t)$. Each point $a \in \mathbb{C}$ gives rise to a valuation $v_a : K^{\times} \to \mathbb{Z}$ by:

$$v_a(f(t)) = k \qquad \text{if} \quad f(t) = (t - a)^k \frac{g(t)}{h(t)},$$

where $g, h \in \mathbb{C}[t]$ are such that $g(a), h(a) \neq 0$. Note that if $k > 0$, then $f(a) = 0$ and k is the *order of vanishing* of f at a; if $k = 0$, then $f(a) \in \mathbb{C}^\times$; and if $k < 0$, then f has a pole at a of order $-k$. Geometrically it is natural to consider $f \in K^\times$ as defining a meromorphic function on the Riemann sphere $\mathbb{C}_\infty := \mathbb{C} \cup \{\infty\}$, and to assign also to the point ∞ on this sphere a valuation $v_\infty : K^\times \to \mathbb{Z}$ with $v_\infty(a) = 0$ for $a \in \mathbb{C}^\times$ and $v_\infty(t^{-1}) = 1$, namely

$$v_\infty(f(t)) = \deg h(t) - \deg g(t) \quad \text{if} \quad f(t) = \frac{g(t)}{h(t)},\ g, h \in \mathbb{C}[t],\ h \neq 0.$$

These valuations are related by the fundamental identity

$$\sum_{a \in \mathbb{C}_\infty} v_a(f) = 0, \qquad f \in K^\times.$$

(This identity clearly holds for $f \in \mathbb{C}[t] \setminus \{0\}$, and the general case then follows easily.) This family of valuations is an important tool to understand the field $K = \mathbb{C}(t)$; the same role is played for other so-called *global* fields by a family of valuations satisfying a similar identity. In this course, however, we focus on fields with a single valuation, that is, we study *local* aspects of fields.

With this example in mind, given a valued field (K, v), we think of K as a field of (meromorphic) functions on some space in which v is associated with a certain point of the space, where $v(f) > 0$ indicates that the "function" f vanishes at that point, $v(f)$ measuring the order of vanishing. Likewise, $v(f) < 0$ means that f has a pole at that point (and $v(f^{-1})$ is the order of vanishing of f^{-1}). Indeed, we shall regard v itself as an abstract point at which the elements of K can be evaluated. This will become clear in the discussion after the next definition.

Definition 3.3. Let $v : K^\times \to \Gamma$ be a valuation on a field. We set:

(i) $\mathcal{O}_v := \{x \in K : v(x) \geq 0\}$, a subring of K,
(ii) $\mathfrak{m}_v := \{x \in K : v(x) > 0\}$, an ideal in $\mathcal{O}_v$,
(iii) $\boldsymbol{k}_v := \mathcal{O}_v/\mathfrak{m}_v$.

Note that $v(x) = 0 \iff v(x^{-1}) = 0$, for $x \in K^\times$, so $U(\mathcal{O}_v) = \mathcal{O}_v \setminus \mathfrak{m}_v$. Therefore $\mathcal{O}_v$ is a local ring with maximal ideal $\mathfrak{m}_v$, and $\boldsymbol{k}_v$ is the residue field.

When thinking of K as a field of functions as above, $\mathcal{O}_v$ would be the set of functions which have no pole at (the point corresponding to) v, and $\mathfrak{m}_v$ would be the set of functions that vanish at v. The residue map $\mathcal{O}_v \to \boldsymbol{k}_v$ is then to be thought of as evaluation at v. We extend the residue map to a map $K \to \boldsymbol{k}_v \cup \{\infty\}$ by sending all elements of $K \setminus \mathcal{O}_v$ to ∞.

Note that we have the following disjoint union in K,

$$K = \underbrace{\mathfrak{m}_v \,\dot\cup\, U(\mathcal{O}_v)}_{\mathcal{O}_v} \,\dot\cup\, (\mathfrak{m}_v \setminus \{0\})^{-1}.$$

For each $x \in K^\times$, either $x \in \mathcal{O}_v$ or $x^{-1} \in \mathcal{O}_v$, so $\mathcal{O}_v$ is a valuation ring in K.

We have, therefore, two maps:

$$\begin{array}{ccc} K & \xrightarrow{\ v\ } & \Gamma \cup \{\infty\} \\ \big\downarrow & & \\ \boldsymbol{k}_v \cup \{\infty\} & & \end{array}$$

The *raison d' être* of valuation theory is in analyzing K in terms of the (usually) simpler objects Γ and $\boldsymbol{k}_v$ via these two maps.

Terminology. We call v *trivial* if $\Gamma = \{0\}$, *discrete* if $\Gamma \cong \mathbb{Z}$ as ordered groups, and of *rank 1* if $\Gamma \neq 0$ and Γ is archimedean.

If v is discrete, say $\Gamma = \mathbb{Z}$, take $t \in K$ with $v(t) = 1$. Then $\mathfrak{m}_v = t\mathcal{O}_v$, and each $a \in K^\times$ can be written uniquely as $a = ut^k$ with $u \in \mathcal{O}_v^\times$ and $k \in \mathbb{Z}$. (Simply take $k = v(a)$, and $u = a/t^k$.) The only nontrivial ideals of $\mathcal{O}_v$ are those of the form $t^n\mathcal{O}_v$. In particular $\mathcal{O}_v$ is a principal ideal domain.

Examples.

(1) Let $K = \mathbb{C}(t)$, $v = v_0 : K^\times \to \mathbb{Z}$. Then $v(t) = 1$ and

$$\mathcal{O}_v = \left\{ \frac{f}{g} : f, g \in \mathbb{C}[t], g(0) \neq 0 \right\} = \mathbb{C}[t]_{t\mathbb{C}[t]}.$$

We have a surjective ring morphism

$$\mathcal{O}_v \to \mathbb{C} : \frac{f(t)}{g(t)} \mapsto \frac{f(0)}{g(0)}, \quad (f, g \in \mathbb{C}[t],\ g(0) \neq 0)$$

which has kernel $\mathfrak{m}_v$, and thus induces a field isomorphism

$$\boldsymbol{k}_v = \mathcal{O}_v/\mathfrak{m}_v \cong \mathbb{C}.$$

It is traditional to identify $\boldsymbol{k}_v$ with $\mathbb{C}$ via this isomorphism.

(2) Let $K = \mathbb{Q}_p$. Here we have the *p-adic valuation* $v = v_p : K^\times \to \mathbb{Z}$ defined by $v(x) = k$ if $x = p^k u$, $u \in U(\mathbb{Z}_p)$, which is related to the p-adic absolute value by $|x|_p = p^{-v(x)}$. We have $\mathcal{O}_v = \mathbb{Z}_p$, $v(p) = 1$, $\mathfrak{m}_v = p\mathbb{Z}_p$, and $\boldsymbol{k}_v = \mathbb{Z}_p/p\mathbb{Z}_p = \mathbb{F}_p$.

For each ordered abelian group Γ and field $\boldsymbol{k}$, there is a field $K = \boldsymbol{k}((t^\Gamma))$ with valuation v having Γ as the value group and $\boldsymbol{k}$ as the residue field. To construct $\boldsymbol{k}((t^\Gamma))$ we need some basic facts on well-ordered subsets of Γ.

Recall that a linearly ordered set A is said to be *well-ordered* if each nonempty subset of A has a least element, equivalently, there is no strictly decreasing sequence (α_n) in A. We leave the proof of the next two lemmas as an exercise.

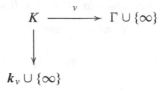

Lemma 3.4. *Let $A, B \subseteq \Gamma$ be well-ordered (by the ordering of Γ). Then $A \cup B$ is well-ordered, the set $A + B := \{\alpha + \beta : \alpha \in A,\ \beta \in B\}$ is well-ordered, and for each $\gamma \in \Gamma$ there are only finitely many $(\alpha, \beta) \in A \times B$ such that $\alpha + \beta = \gamma$.*

Lemma 3.5. *Let $A \subseteq \Gamma^{>0}$ be well-ordered. Then*

$$[A] := \{\alpha_1 + \cdots + \alpha_n : \alpha_1, \ldots, \alpha_n \in A\} \qquad (\textit{allowing } n = 0)$$

is also well-ordered, and for each $\gamma \in [A]$ there are only finitely many tuples $(n, \alpha_1, \ldots, \alpha_n)$ with $\alpha_1, \ldots, \alpha_n \in A$ such that $\gamma = \alpha_1 + \cdots + \alpha_n$.

We now define $K = \boldsymbol{k}((t^\Gamma))$ to be the set of all formal series $f(t) = \sum_{\gamma \in \Gamma} a_\gamma t^\gamma$, with coefficients $a_\gamma \in \boldsymbol{k}$, such that the support of f,

$$\operatorname{supp}(f) := \{\gamma \in \Gamma : a_\gamma \neq 0\},$$

is a well-ordered subset of Γ. Using the first lemma on well-ordered subsets of Γ we can define binary operations of addition and multiplication on the set $\boldsymbol{k}((t^\Gamma))$ as follows:

$$\sum a_\gamma t^\gamma + \sum b_\gamma t^\gamma = \sum (a_\gamma + b_\gamma) t^\gamma$$

$$\left(\sum a_\gamma t^\gamma\right)\left(\sum b_\gamma t^\gamma\right) = \sum_\gamma \Big(\sum_{\alpha+\beta=\gamma} a_\alpha b_\beta\Big) t^\gamma.$$

Note that K is a ring, even a domain, with $\boldsymbol{k}$ as a subfield via the identification $a \mapsto at^0$. To show K is a field we shall use the second lemma.

Define $v : K \setminus \{0\} \to \Gamma$ by

$$v\left(\sum a_\gamma t^\gamma\right) := \min\{\gamma : a_\gamma \neq 0\}.$$

Then v is a valuation on K, and if $v(f) > 0$, then by the second lemma $\sum_{n=0}^\infty f^n$ makes sense as an element of K: for any $\gamma \in \Gamma$ there are only finitely many n such that the coefficient of t^γ in f^n is not zero. Note that then

$$(1 - f) \sum_{n=0}^{\infty} f^n = 1.$$

It now follows in the usual way that K is a field: write an arbitrary $g \in K \setminus \{0\}$ as $g = ct^\gamma(1 - f)$, with $c \in k^\times$ and $v(f) > 0$. Then $g^{-1} = c^{-1} t^{-\gamma} \sum_n f^n$.

The valuation ring is

$$\mathcal{O}_v = \{f \in K : \operatorname{supp}(f) \subseteq \Gamma^{\geq 0}\}.$$

For $f = \sum a_\gamma t^\gamma$ in K, call a_0 the constant term of f. The map sending f to its constant term sends $\mathcal{O}_v$ onto $\boldsymbol{k}$, and this is a ring homomorphism. Its kernel is

$$\mathfrak{m}_v = \{f \in K : \operatorname{supp}(f) \subseteq \Gamma^{>0}\},$$

so this induces an isomorphism $\mathcal{O}_v/\mathfrak{m}_v \to \boldsymbol{k}$.

For $\Gamma = \mathbb{Z}$, the field $\boldsymbol{k}((t^{\mathbb{Z}}))$ is the usual field of Laurent series, i.e. $\boldsymbol{k}((t))$, with valuation ring $\boldsymbol{k}[\![t]\!]$. An early study [26] of the fields $\boldsymbol{k}((t^\Gamma))$ is due to Hahn (of "Hahn–Banach" fame and Gödel's thesis advisor), so these valued fields are often referred to as *Hahn fields*.

Topology Induced by a Valuation. An *ultrametric space* is a set X with a map $v : X^2 \to \Gamma_\infty$, where Γ_∞ is a linearly ordered set with largest element ∞, such that, for all $x, y, z \in X$ and $\gamma \in \Gamma_\infty$:

(i) $v(x, y) = \infty \iff x = y$,
(ii) $v(x, y) = v(y, x)$,
(iii) $v(x, y), v(y, z) \geq \gamma \implies v(x, z) \geq \gamma$.

Example. An abelian group A with valuation v is an ultrametric space with respect to $v(x, y) := v(x - y)$, and this is how we consider in particular a field with a valuation as an ultrametric space.

Let X be an ultrametric space as above. For a point $a \in X$ and $\gamma \in \Gamma_\infty$, we define two kinds of disks:

$$D_a(\gamma) = \{x \in X : v(x, a) > \gamma\}, \quad \text{an open disk,}$$
$$\bar{D}_a(\gamma) = \{x \in X : v(x, a) \geq \gamma\}, \quad \text{a closed disk.}$$

We refer to $D_a(\gamma)$ as the *open disk centered at a of valuation radius γ*, and to $\bar{D}_a(\gamma)$ as the *closed disk centered at a of valuation radius γ*. Note that we include here some degenerate cases: the open disks of valuation radius ∞ are empty, and the closed disks of valuation radius ∞ are the singletons $\{a\}$ with $a \in X$. The reader should verify that $D_a(\gamma) = D_b(\gamma)$ for each $b \in D_a(\gamma)$, and $\bar{D}_a(\gamma) = \bar{D}_b(\gamma)$ for each $b \in \bar{D}_a$, in stark contrast to disks in the euclidean plane, which have a unique center. It follows easily that if D, E are disks (of any kind) with nonempty intersection, then $D \subseteq E$ or $E \subseteq D$. The open disks form a basis for a topology on X, the *v-topology.* It is easy to see that all disks of valuation radius $\gamma \neq \infty$ are both open and closed in the v-topology, and that X with the v-topology is a hausdorff space. A field K with valuation v is a topological field with respect to the v-topology, that is, the field operations $+, -, \cdot : K^2 \to K$, and $x \mapsto x^{-1} : K^\times \to K^\times$ are continuous. Note that then $\mathcal{O}_v$ is the closed disk centered at 0 of valuation radius 0, and that the open disks of valuation radius 0 contained in $\mathcal{O}_v$ are exactly the cosets $a + \mathfrak{m}_v$ with $a \in \mathcal{O}$. In particular, $\mathcal{O}_v$ contains exactly $|\boldsymbol{k}_v|$ open disks of valuation radius 0. Indeed, any closed disk of valuation radius $\gamma \in v(K^\times)$ contains exactly $|\boldsymbol{k}_v|$ open disks of valuation radius γ: by a translation one can assume the closed disk contains 0, and then multiplication by an $a \in K$ with $v(a) = -\gamma$ makes the closed disk equal to $\mathcal{O}_v$.

The next result is due to Krull. The short proof below was found by Gravett.

Proposition 3.6. *Let $v : K^\times \to \Gamma$ be a valuation on the field K, and $\boldsymbol{k} := \boldsymbol{k}_v$. Then the cardinality of K is bounded in terms of the cardinalities of $\boldsymbol{k}$ and Γ,*

$$|K| \le |\boldsymbol{k}|^{|\Gamma|}.$$

Proof. Let $\gamma \in \Gamma$, and define equivalence relations on K by

$$x\gamma y \iff v(x-y) > \gamma, \qquad x\bar{\gamma}y \iff v(x-y) \ge \gamma.$$

Thus the γ-classes are the open disks of valuation radius γ, and the $\bar{\gamma}$-classes are the closed disks of valuation radius γ. Each γ-class D is contained in a unique $\bar{\gamma}$-class $\bar{D}$, and each $\bar{\gamma}$-class contains exactly $|\boldsymbol{k}|$-many γ-classes. Let $\mathcal{D}_\gamma$ be the set of all γ-classes. Then we have a map $f_\gamma : \mathcal{D}_\gamma \to \boldsymbol{k}$ such that for all $D, E \in \mathcal{D}_\gamma$, if $\bar{D} = \bar{E}$ and $f_\gamma(D) = f_\gamma(E)$, then $D = E$.

We now associate to each $x \in K$ the function $\gamma \mapsto x(\gamma) : \Gamma \to \boldsymbol{k}$ by $x(\gamma) = f_\gamma(D)$ where $D \in \mathcal{D}_\gamma$ contains x. Then the map $K \to \boldsymbol{k}^\Gamma$ that sends each $x \in K$ to the function $\gamma \mapsto x(\gamma)$ is injective: if $x, y \in K$ and $x \ne y$, then for $\gamma := v(x-y)$ we have $x(\gamma) \ne y(\gamma)$ since x and y are in the same closed disk of valuation radius γ but in different open disks of valuation radius γ. □

Even for Hahn fields $K = \boldsymbol{k}((t^\Gamma))$ the bound in the proposition is not always attained. For example, let Γ be the ordered additive group $\mathbb{R}$ of real numbers. Then all well-ordered subsets of Γ are countable, and so there are only $2^{\aleph_0}$ well-ordered subsets of Γ. Each of these subsets of Γ is the support of at most $|\boldsymbol{k}|^{\aleph_0}$ elements of K, and thus

$$|K| = |\boldsymbol{k}|^{\aleph_0} \cdot 2^{\aleph_0} = |\boldsymbol{k}|^{\aleph_0}.$$

For $\boldsymbol{k} = \mathbb{Q}$ and also for $\boldsymbol{k} = \mathbb{R}$, this gives $|K| = 2^{\aleph_0} < 2^{2^{\aleph_0}} = |\boldsymbol{k}|^{|\Gamma|}$.

Correspondence Between Valuation Rings and Valuations. A *valuation ring of a field* K is a subring A of K such that for each $x \in K^\times$, either $x \in A$ or $x^{-1} \in A$. Then A is indeed a valuation ring, with K as field of fractions.

Let A be a valuation ring of the field K. Then we have the disjoint union

$$K = \mathfrak{m}_A \cup U(A) \cup (\mathfrak{m}_A \setminus \{0\})^{-1}.$$

Consider the (abelian) quotient group $\Gamma_A = K^\times/U(A)$, written additively. The binary relation $\ge$ on Γ_A defined by

$$xU(A) \ge yU(A) \iff x/y \in A, \qquad (x, y \in K^\times)$$

makes Γ_A into an ordered abelian group, and the natural map

$$v_A : K^\times \to \Gamma_A, \quad v_A(x) = xU(A)$$

is a valuation. (Check this!) Note that $\mathcal{O}_{v_A} = A$.

Every valuation $v : K^\times \to \Gamma$ with valuation ring $\mathcal{O}_v = A$ is *equivalent* to v_A as follows: there is a unique isomorphism of ordered abelian groups $i : \Gamma_A \to \Gamma$ such that $i \circ v_A = v$. More generally, two valuations $v_1 : K^\times \to \Gamma_1$ and $v_2 : K^\times \to \Gamma_2$ are said to be *equivalent* if there is an isomorphism of ordered abelian groups $i : \Gamma_1 \to \Gamma_2$ such that $i \circ v_1 = v_2$; note that such an i is then uniquely determined. The reader should check that two valuations v_1 and v_2 on a field are equivalent iff they have the same valuation ring.

A valuation ring A is said to be *discrete* if the valuation v_A on its fraction field K is discrete. In that case there is a unique valuation $v : K^\times \to \mathbb{Z}$ such that $\mathcal{O}_v = A$; this v is called the *normalized valuation* of A. (For example, the value group of a discrete valuation v on a field could be $\frac{1}{2}\mathbb{Z}$, and then $2v$ would be the corresponding normalized valuation.) A common abbreviation for *discrete valuation ring* is DVR. Thus $\mathbb{Z}_p$ and $\boldsymbol{k}[[t]]$ are DVR's.

Definition 3.7. A *valued field* is a pair (K, A), where K is a field and A is a valuation ring of K.

If (K, A) is a valued field and E a subfield of K, then $A \cap E$ is a valuation ring of E. Note: $(E, A \cap E) \subseteq (K, A)$ (where "$\subseteq$" means "is a substructure of").

Exercise. Let (K, A) be an algebraically closed valued field. Show that the residue field $\boldsymbol{k}_A$ is algebraically closed and the value group Γ_A is divisible.

Let (K, A) be a valued field, and $L \supseteq K$ a field extension. Then a valuation ring B of L *lies over* A (or *dominates* A) if $A = B \cap K$, equivalently, $(K, A) \subseteq (L, B)$. Note that then $\mathfrak{m}_A = \mathfrak{m}_B \cap K$, so we have an induced embedding of residue fields, $A/\mathfrak{m}_A \to B/\mathfrak{m}_B$, by means of which $\boldsymbol{k}_A = A/\mathfrak{m}_A$ is identified with a subfield of $\boldsymbol{k}_B = B/\mathfrak{m}_B$. Also $U(B) \cap K = U(A)$, so we have an induced embedding $v_A(x) \mapsto v_B(x) : \Gamma_A \to \Gamma_B$ $(x \in K^\times)$, of ordered abelian groups, by means of which Γ_A is identified with an ordered subgroup of Γ_B.

3.2 Valuation Rings and Integral Closure

We begin with recalling some terminology and basic facts concerning rings and ideals. (See for example S. Lang's book "Algebra" for more details.)

Let $A \subseteq B$ be rings. For an ideal I of A we let

$$IB := \{a_1b_1 + \cdots + a_nb_n : a_1, \ldots, a_n \in I, b_1, \ldots, b_n \in B\}$$

be the ideal of B generated by I. If $\mathfrak{p}$ and $\mathfrak{q}$ are prime ideals of A and B respectively, we say that $\mathfrak{q}$ *lies over* $\mathfrak{p}$ if $\mathfrak{q} \cap A = \mathfrak{p}$. In that case we have a ring embedding $a + \mathfrak{p} \mapsto a + \mathfrak{q} : A/\mathfrak{p} \to B/\mathfrak{q}$.

We call an element $b \in B$ *integral* over A if b is a zero of a monic polynomial over A, that is,

$$b^n + a_1 b^{n-1} + \cdots + a_n = 0$$

for suitable $n \geq 1$ and $a_1, \ldots, a_n \in A$. (Then $A[b] = A + Ab + \cdots + Ab^{n-1}$, so $A[b]$ is a finitely generated A-module. The converse is also true: if $b \in B$ and $A[b]$ is a finitely generated A-module, then b is integral over A.) The elements of B that are integral over A form a subring of B, called the *integral closure* of A in B. Call A *integrally closed in* B if each $b \in B$ that is integral over A lies already in A. Call B *integral over* A if every $b \in B$ is integral over A.

Given rings $A \subseteq B \subseteq C$, the ring C is integral over A iff C is integral over B and B is integral over A.

Exercise. Let $A \subseteq B$ be domains, and suppose B is integral over A. Show: A is a field iff B is a field.

As before, let $A \subseteq B$ be rings, and suppose B is integral over A. Let $\mathfrak{p}$ be a prime ideal of A. Then there is a prime ideal $\mathfrak{q}$ of B that lies over $\mathfrak{p}$. Moreover, for each such $\mathfrak{q}$, the ideal $\mathfrak{p}$ is maximal iff $\mathfrak{q}$ is maximal. (Explanation: For such $\mathfrak{q}$ we have a natural ring embedding $A/\mathfrak{p} \to B/\mathfrak{q}$, and after identifying $A/\mathfrak{p}$ with its image in $B/\mathfrak{q}$, this last domain is integral over the domain $A/\mathfrak{p}$. Hence by the exercise above, $A/\mathfrak{p}$ is a field iff $B/\mathfrak{q}$ is a field.)

We say that a domain A is *integrally closed* if it is integrally closed in its field of fractions.

Given a domain A with fraction field K, and a prime ideal $\mathfrak{p}$ of A we let

$$A_{\mathfrak{p}} := \left\{ \frac{x}{y} \in K : x, y \in A,\ y \notin \mathfrak{p} \right\}$$

be the localization of A with respect to $\mathfrak{p}$. It is easy to check that $A_{\mathfrak{p}}$ is a local domain with maximal ideal $\mathfrak{p}A_{\mathfrak{p}}$, and that this maximal ideal lies over $\mathfrak{p}$.

Proposition 3.8. *Valuation rings are integrally closed.*

Proof. Suppose A is a valuation ring of the field K, and let $v = v_A$ be the canonical valuation. To show A is integrally closed, let $x \in K^\times$ be such that $x^n + a_{n-1}x^{n-1} + \cdots + a_0 = 0$ with $n \geq 1$ and all $a_i \in A$. If $x \notin A$, then $v(x) < 0$, hence

$$v(x^n) = nv(x) < iv(x) + v(a_i) = v(a_i x^i), \qquad i = 0, \ldots, n-1,$$

hence $v(x^n + \cdots + a_0) = v(x^n) < 0$. But $v(x^n + \cdots + a_0) = \infty$, a contradiction. □

Corollary 3.9. *Let* (K, A) *be an algebraically closed valued field. Then* A *is henselian.*

Proof. Consider a polynomial $f(Y) = Y^n + Y^{n-1} + a_2 Y^{n-2} + \cdots + a_n$ with $n \geq 2$ and $a_2, \ldots, a_n \in \mathfrak{m}$. By Lemma 2.2 it is enough to show that f has a zero in $U(A)$. By the proposition above, $f = \prod_{i=1}^{n}(Y + \alpha_i)$ with all $\alpha_i \in A$. Now $\sum_i \alpha_i = 1$, so at least one α_i must be outside $\mathfrak{m}$, and thus be in $U(A)$. □

Definition 3.10. Given local rings A, B, we say B *dominates* A (or that B *lies over* A) if $A \subseteq B$ and $\mathfrak{m}_A \subseteq \mathfrak{m}_B$ (hence $\mathfrak{m}_B \cap A = \mathfrak{m}_A$).

One should check that this agrees with our definition of domination for valuation rings in the previous subsection. We now state the main results to be proved in this subsection.

Proposition 3.11. *Let A be a local subring of the field K. Consider the class of all local subrings of K that dominate A, partially ordered by domination. Any maximal element of this class is a valuation ring of K.*

In particular, given any local subring A of a field K, there is a valuation ring of K lying above A.

Proposition 3.12. *Let A be a local subring of a field K. The integral closure of A in K is the intersection of all valuation rings of K that dominate A.*

Let $\mathrm{Frac}(A)$ be the fraction field of a domain A, and let $\mathrm{Aut}(L|K)$ be the group of automorphisms of a field L that are the identity on a subfield K of L.

Proposition 3.13. *Let A be a local domain integrally closed in $K = \mathrm{Frac}(A)$, and let L be a normal field extension of K. Let B be the integral closure of A in L. Then, given any maximal ideals $\mathfrak{n}$ and $\mathfrak{n}'$ of B there exists $\sigma \in \mathrm{Aut}(L|K)$ such that* $\sigma(\mathfrak{n}) = \mathfrak{n}'$. (Note: all maximal ideals in B lie over $\mathfrak{m}_A$.)

Proposition 3.14. *Let A be a valuation ring of the field K, let $L \supseteq K$ be an algebraic field extension, and let B be the integral closure of A in L. Then every valuation ring of L dominating A is of the form $B_{\mathfrak{n}}$ for some maximal ideal $\mathfrak{n} \subseteq B$.*

Corollary 3.15. *With the same assumption as in the proposition, we have a bijection $\mathfrak{n} \mapsto B_{\mathfrak{n}}$ from the set of maximal ideals in B onto the set of valuation rings of L dominating A. If in addition L is a normal field extension of K, then for any valuation rings V, V' of L dominating A there exists $\sigma \in \mathrm{Aut}(L|K)$ such that* $\sigma(V) = V'$.

Proof. Let $\mathfrak{n}$ be a maximal ideal of B. We first show that the localization $B_{\mathfrak{n}}$ is a valuation ring of L dominating A. Note that $\mathfrak{n}$ lies over $\mathfrak{m}_A$, hence $B_{\mathfrak{n}}$ dominates A. The set of local subrings of L dominating $B_{\mathfrak{n}}$ is partially ordered by domination. Take a maximal element V of this partially ordered set. Then V is a valuation ring of L by Proposition 3.11. As V dominates A, Proposition 3.14 gives $V = B_{\mathfrak{n}'}$ with $\mathfrak{n}'$ a maximal ideal of B. Then $\mathfrak{n}' = \mathfrak{n}$, so $B_{\mathfrak{n}} = V$ is a valuation ring of L. The "bijective correspondence" part of the Corollary now follows from Proposition 3.14, and the second part follows from this correspondence and Proposition 3.13. □

Corollary 3.16. *Let A be a valuation ring of K, and A^{ac} a valuation ring of K^{ac} dominating A. Then any valued field embedding $(K, A) \to (L, B)$ with algebraically closed L extends to an embedding $(K^{\mathrm{ac}}, A^{\mathrm{ac}}) \to (L, B)$.*

Proof. Let a valued field embedding $(K, A) \to (L, B)$ be given with algebraically closed L. Take a field embedding $j : K^{\mathrm{ac}} \to L$ that extends the given field embedding $K \to L$. Then $j^{-1}(B)$ is a valuation ring of K^{ac} lying over A. Hence there is $\sigma \in \mathrm{Aut}(K^{\mathrm{ac}}|K)$ such that $j^{-1}(B) = \sigma(A^{\mathrm{ac}})$. Then $j\sigma$ is a valued field embedding of $(K^{\mathrm{ac}}, A^{\mathrm{ac}})$ into (L, B). □

It remains to prove Propositions 3.11–3.14. For the first two we use the next lemma:

Lemma 3.17. *Let A be a local subring of the field K, and suppose $x \in K^{\times}$ is such that $1 \in \mathfrak{m}A[x^{-1}] + x^{-1}A[x^{-1}]$. Then x is integral over A.*

Proof. We have $1 = a_n x^{-n} + \cdots + a_1 x^{-1} + a_0$ with $a_1, \dots, a_n \in A$ and $a_0 \in \mathfrak{m}$. Multiplying both sides by x^n yields:

$$x^n(1 - a_0) + (\text{terms of lower degree in } x) = 0.$$

Since $1 - a_0$ is a unit in A, it follows that x is integral over A. □

Proof of Proposition 3.11. Let V be a maximal element in the class of local subrings of K that dominate A, and let $x \in K^{\times}$. Put $\mathfrak{m} := \mathfrak{m}_V$.

Consider first the case that x is integral over V. Then $V[x]$ is integral over V, so we can take a maximal ideal $\mathfrak{n}$ of $V[x]$ that lies over $\mathfrak{m}$, and then $V[x]_{\mathfrak{n}}$ is a local subring of K dominating V. By V's maximality this gives $x \in V$.

Next, suppose x is not integral over V. Then by the above lemma we have $1 \notin \mathfrak{m}V[x^{-1}] + x^{-1}V[x^{-1}]$, so we have a maximal ideal $\mathfrak{n}$ of $V[x^{-1}]$ such that $\mathfrak{n} \supseteq \mathfrak{m}$, and thus $\mathfrak{n} \cap V = \mathfrak{m}$. The local subring $V[x^{-1}]_{\mathfrak{n}}$ of K dominates V, so the maximality of V yields $x^{-1} \in V$. □

Proof of Proposition 3.12. Any $x \in K$ integral over A lies in every valuation ring of K containing A as a subring. Next, suppose $x \in K^{\times}$ is not integral over A. By the lemma above, $\mathfrak{m}A[x^{-1}] + x^{-1}A[x^{-1}]$ is a proper ideal of $A[x^{-1}]$. So we can take a maximal ideal $\mathfrak{n} \supseteq \mathfrak{m}$ of $A[x^{-1}]$ that contains x^{-1}. This yields a local subring $A[x^{-1}]_{\mathfrak{n}}$ of K that dominates A. Let V be a maximal element of the class of local subrings of K that dominate $A[x^{-1}]_{\mathfrak{n}}$. Then V is a valuation ring (by Proposition 3.11), V dominates A, and $x^{-1} \in \mathfrak{m}_V$, so $x \notin V$. □

In the next proof we use the Chinese Remainder Theorem: *Let A be an abelian (additive) group with subgroups $B_1, \dots, B_n$ $(n \geq 1)$ such that $B_i + B_j = A$ for all i, j with $1 \leq i < j \leq n$, and let $a_1, \dots, a_n \in A$. Then there exists $x \in A$ such that $x - a_i \in B_i$ for $i = 1, \dots, n$.*

Proof of Proposition 3.13. As usual, $[L : K]$ denotes the dimension of L as a vector space over K, with $[L : K] := \infty$ if this dimension is infinite. The general case follows from the case $[L : K] < \infty$ (how?), so we assume $[L : K] < \infty$ below.

Let $\mathfrak{n}, \mathfrak{n}'$ be maximal ideals in B, lying over $\mathfrak{m}$, and assume towards a contradiction that $\sigma(\mathfrak{n}) \neq \mathfrak{n}'$ for all $\sigma \in G := \mathrm{Aut}(L|K)$. Thus the two sets of maximal ideals of B,

$$\{\sigma(\mathfrak{n}) : \sigma \in G\}, \qquad \{\sigma(\mathfrak{n}') : \sigma \in G\}$$

are disjoint; since $[L : K] < \infty$, these two sets are also finite. By the Chinese Remainder Theorem there is $x \in B$ such that

$$x \equiv 0 \mod \sigma(\mathfrak{n}), \qquad x \equiv 1 \mod \sigma(\mathfrak{n}')$$

for all $\sigma \in G$. Hence $\sigma(x) \in \mathfrak{n} \setminus \mathfrak{n}'$ for all $\sigma \in G$. Recall that

$$\mathrm{N}_K^L(x) := \Big(\prod_{\sigma \in G} \sigma(x)\Big)^{\ell}$$

lies in K, where $\ell = 1$ if char $K = 0$, and $\ell = p^e$ for some $e \in \mathbb{N}$ if char $K = p > 0$. Each $\sigma(x)$ is integral over A, so $\mathrm{N}_K^L(x) \in A$. Also $\mathrm{N}_K^L(x) \in \mathfrak{n} \setminus \mathfrak{n}'$ because $\mathfrak{n}$ and $\mathfrak{n}'$ are prime ideals, hence $\mathrm{N}_K^L(x) \in A \cap \mathfrak{n} = \mathfrak{m} \subseteq \mathfrak{n}'$, a contradiction. □

Proof of Proposition 3.14. Let V be a valuation ring of L lying over A. Valuation rings are integrally closed, so $B \subseteq V$. We claim that $V = B_{\mathfrak{n}}$ where $\mathfrak{n} := \mathfrak{m}_V \cap B$. Clearly $B \setminus \mathfrak{n} \subseteq V \setminus \mathfrak{m}_V$, hence $B_{\mathfrak{n}} \subseteq V$. For the other inclusion, let $x \in V$, $x \neq 0$. We have a relation $a_n x^n + \cdots + a_0 = 0$, where $a_0, \ldots, a_n \in A$ and $a_n \neq 0$. Take $s \in \{0, \ldots, n\}$ maximal such that $v_A(a_s) = \min_i v_A(a_i)$. Put $b_i := a_i/a_s$. Dividing by $a_s x^s$ yields

$$\underbrace{(b_n x^{n-s} + \cdots + b_{s+1}x + 1)}_{y} + x^{-1}\underbrace{(b_{s-1} + \cdots + b_0/x^{s-1})}_{z} = 0.$$

So $y = -(z/x)$, and thus $-xy = z$. Since $b_i \in \mathfrak{m}_A$ for $s < i \leq n$, we have $y \in U(V)$, and hence $y \notin \mathfrak{n}$. Thus to get $x \in B_{\mathfrak{n}}$ it suffices to show that $y, z \in B$. This will follow if we show that y, z lie in every valuation ring of L dominating A. If such a ring contains x, it also contains y, hence contains $z = -yx$. If such a ring contains x^{-1}, it contains z, hence also $y = -zx^{-1}$. □

Proposition 3.18. *Let A be a local ring, $f(X) \in A[X]$ a monic polynomial of degree $d \geq 1$, and $A[x] := A[X]/(f)$ with $x := X + (f)$. Let $\overline{f}(X) \in \boldsymbol{k}[X]$ factor in $\boldsymbol{k}[X]$ as*

$$\overline{f} = \prod_{i=1}^{n} \overline{f_i}^{e_i}$$

where each $e_i \geq 1$, each $f_i \in A[X]$ is monic, and $\overline{f_1}, \ldots, \overline{f_n}$ are irreducible in $\boldsymbol{k}[X]$ and distinct. Put $\mathfrak{m}_i := (\mathfrak{m}, f_i(x))A[x]$. Then $\mathfrak{m}_1, \ldots, \mathfrak{m}_n$ are exactly the distinct maximal ideals of $A[x]$, and $A[x]/\mathfrak{m}_i \cong \boldsymbol{k}[X]/(\overline{f_i})$ (as rings) for each i.

Proof. Let $1 \leq i \leq n$, and consider the canonical morphisms,

$$A[x] = A[X]/(f) \xrightarrow{\pi} \boldsymbol{k}[X]/(\overline{f}) \xrightarrow{\pi_i} \boldsymbol{k}[X]/(\overline{f_i}).$$

Now $\ker \pi = \mathfrak{m}A[x]$, and $\ker \pi_i$ is generated as an ideal by $\overline{f_i} + (\overline{f})$. Thus

$$\ker(\pi_i \pi) = (\mathfrak{m}, f_i(x)) = \mathfrak{m}_i,$$

since $f_i(x)$ is the preimage of $\overline{f_i} + (\overline{f})$ in $A[x]$. Because $\boldsymbol{k}[X]/(\overline{f_i})$ is a field, $(\mathfrak{m}, f_i(x))$ is maximal in $A[x]$. If $1 \leq j \leq n$, $i \neq j$, then $\overline{f_i} \not\equiv 0 \mod \overline{f_j}$ in $\boldsymbol{k}[X]$, and so $\mathfrak{m}_i \neq \mathfrak{m}_j$. It remains to show that the $\mathfrak{m}_i$'s are the only maximal ideals in $A[x]$. First note that the ring $A[x]$ is integral over A, so each maximal ideal in $A[x]$ lies over $\mathfrak{m}$. The polynomial $f - \prod f_i^{e_i}$ is in $\mathfrak{m}[X]$, and $f(x) = 0$, so $\prod f_i(x)^{e_i} \in \mathfrak{m}A[x]$. Thus each maximal ideal of $A[x]$ contains $\prod f_i(x)^{e_i}$, hence contains some $f_i(x)$, and thus is one of the $\mathfrak{m}_i$'s. □

3.3 Extensions of Valued Fields

Our main goal in this chapter is to show that the theory of algebraically closed *valued* fields admits elimination of quantifiers. This will use Corollary 3.16, in combination with some results about extending a valuation on a field K to $K(x)$ where x is transcendental over K. In this section we treat this last issue, and establish other basic facts on valued field extensions. The first result below is a key inequality for valued field extensions of finite degree. The proof is easy but fundamental, since the same arguments apply in other situations.

Given a field extension $K \subseteq L$ we let $[L : K]$ be its *degree*, that is, the dimension of L as a vector space over K; to keep things simple we set $[L : K] = \infty$ if this dimension is infinite. Likewise, given an extension $\Gamma \subseteq \Gamma'$ of abelian groups we let $[\Gamma' : \Gamma]$ be its index, which by convention is ∞ if Γ'/Γ is infinite.

Proposition 3.19. *Let $(K, A) \subseteq (L, B)$ be valued fields, with corresponding inclusions $\boldsymbol{k}_A \subseteq \boldsymbol{k}_B$ and $\Gamma_A \subseteq \Gamma_B$ between their residue fields and value groups. Then:*

$$[L : K] \geq [\boldsymbol{k}_B : \boldsymbol{k}_A] \cdot [\Gamma_B : \Gamma_A].$$

Proof. Let $b_1, \ldots, b_p \in B$ be such that $\overline{b}_1, \ldots, \overline{b}_p$ are linearly independent over $\boldsymbol{k}_A$, $p \geq 1$. Likewise, let $c_1, \ldots, c_q \in L^\times$ be such that $v(c_1), \ldots, v(c_q)$ lie in distinct cosets of Γ_A, $q \geq 1$. It is enough to show that then $[L : K] \geq pq$; this will follow from the K-linear independence of the family $(b_i c_j)$, which in turn follows from the identity:

$$v(\sum a_{ij} b_i c_j) = \min_{i,j} v(a_{ij} b_i c_j) = \min_{i,j} \{v(a_{ij}) + v(c_j)\} \qquad (\text{all } a_{ij} \in K) \quad (2)$$

First we show that $v(a_1b_1 + \cdots + a_pb_p) = \min_i v(a_i)$, where $a_1, \ldots, a_p \in K$. We can assume that some $a_i \neq 0$, and then dividing by an a_i of minimum valuation, we can reduce to the case that all a_i are in A and $v(a_i) = 0$ for some i. We must show that then $v(a_1b_1 + \cdots + a_pb_p) = 0$. We have:

$$\overline{a_1b_1 + \cdots + a_pb_p} = \overline{a}_1\overline{b}_1 + \cdots + \overline{a}_p\overline{b}_p \neq 0,$$

since $\overline{b}_1, \ldots, \overline{b}_p$ are linearly independent over $\boldsymbol{k}_A$ and $\overline{a}_i \neq 0$ for some i. This gives what we want. Next, to prove (2) we can assume that for each j there is i such that $a_{ij} \neq 0$. Then for any j,

$$\begin{aligned} v\Big(\sum_i a_{ij}b_ic_j\Big) &= v\Big(\big(\sum_i a_{ij}b_i\big)c_j\Big) \\ &= v\Big(\sum_i a_{ij}b_i\Big) + v(c_j) \\ &= \min_i v(a_{ij}) + v(c_j) \\ &\in \Gamma_A + v(c_j). \end{aligned}$$

For $j_1 \neq j_2$ we have $\Gamma_A + v(c_{j_1}) \neq \Gamma_A + v(c_{j_2})$, so

$$v\Big(\sum_i a_{ij_1}b_ic_{j_1}\Big) \neq v\Big(\sum_i a_{ij_2}b_ic_{j_2}\Big),$$

hence,

$$\begin{aligned} v\Big(\sum a_{ij}b_ic_j\Big) &= \min_j\{\min_i v(a_{ij}) + v(c_j)\} \\ &= \min_{i,j}\{v(a_{ij}) + v(c_j)\} \end{aligned}$$

□

For algebraic extensions this yields:

Corollary 3.20. *Let $(K, A) \subseteq (L, B)$ be as in the proposition above.*

If $[L : K] = n$, then $[\boldsymbol{k}_B : \boldsymbol{k}_A] \leq n$ (so $\boldsymbol{k}_B$ is algebraic over $\boldsymbol{k}_A$), and $[\Gamma_B : \Gamma_A] \leq n$ (so $m\Gamma_B \subseteq \Gamma_A$ for some $m \in \{1, \ldots, n\}$).

If L is an algebraic closure of K, then $\boldsymbol{k}_B$ is an algebraic closure of $\boldsymbol{k}_A$, and Γ_B is a divisible hull of Γ_A.

This is used in the proof of the next result.

Lemma 3.21. *Let (K, A) be a valued field. Let β lie in an algebraic closure of $\boldsymbol{k} := \boldsymbol{k}_A$, let $P \in A[x]$ be monic such that $\overline{P} \in \boldsymbol{k}_A[x]$ is the minimum polynomial of β over $\boldsymbol{k}_A$, and let b be a zero of P in an algebraic closure of K. Then $v := v_A$*

extends uniquely to a valuation $v' : K(b)^\times \to \Gamma_A$ *on* $K(b)$. *The residue field* $\boldsymbol{k}' \supseteq \boldsymbol{k}$ *of* v' *is isomorphic over* $\boldsymbol{k}$ *to* $\boldsymbol{k}(\beta)$, *and* $[K(b) : K] = [\boldsymbol{k}' : \boldsymbol{k}]$.

Proof. Let $v' : K(b)^\times \to \Gamma' \supseteq \Gamma_A$ be a valuation on $K(b)$ that extends v, with residue field $\boldsymbol{k}' \supseteq \boldsymbol{k}$. Then $v'(b) \geq 0$ since b is integral over A. Now $\overline{P}(\overline{b}) = 0$ in $\boldsymbol{k}'$, which gives a $\boldsymbol{k}$-isomorphism $\boldsymbol{k}(\overline{b}) \to \boldsymbol{k}(\beta)$ sending $\overline{b}$ to β. With $n := \deg P = \deg \overline{P}$, it follows that $1, \overline{b}, \ldots, \overline{b}^{n-1}$ are linearly independent over $\boldsymbol{k}$, and so by the proof of Proposition 3.19,

$$v'(\sum_{i=0}^{n-1} a_i b^i) \;=\; \min_i v(a_i) \in \Gamma_A$$

whenever $a_0, \ldots, a_{n-1} \in K$ are not all zero. In particular, $1, b, \ldots, b^{n-1}$ are linearly independent over K, and thus P is the minimum polynomial of b over K. The equality above then gives $\Gamma' = \Gamma_A$, and the uniqueness of v'. Moreover,

$$[K(b) : K] \;=\; \deg P \;=\; \deg \overline{P} \;=\; [\boldsymbol{k}(\overline{b}) : \boldsymbol{k}] \;\leq\; [\boldsymbol{k}' : \boldsymbol{k}],$$

and $[K(b) : K] \geq [\boldsymbol{k}' : \boldsymbol{k}]$ by Corollary 3.20, and so $\boldsymbol{k}' = \boldsymbol{k}(\overline{b})$. □

For simple transcendental extensions it is the *proof* of Proposition 3.19 that really matters, as the next two lemmas will show.

Lemma 3.22. *Let* (K, A) *be a valued field, and let* $L = K(x)$ *be a field extension with* x *transcendental over* K. *There is a unique valuation ring* B *of* L *lying over* A, *such that* $x \in B$, *and* $\bar{x}$ *is transcendental over* $\boldsymbol{k}_A$. *Moreover, this* B *satisfies* $\boldsymbol{k}_B = \boldsymbol{k}_A(\bar{x})$ *and* $\Gamma_B = \Gamma_A$.

Proof. Let B be a valuation ring as in the lemma. Then $\bar{1}, \bar{x}, \bar{x}^2, \ldots$ are linearly independent over $\boldsymbol{k}_A$, so by the proof of Proposition 3.19,

$$v_B(f_0 + f_1 x + \cdots + f_n x^n) \;=\; \min_i v_A(f_i) \qquad (f_i \in K).$$

Thus v_B is uniquely determined by v_A on $K[x]$, hence on L. In particular, there is at most one B as in the lemma.
(Existence.) Define $v : K[x] \setminus \{0\} \to \Gamma_A$ by:

$$v(f_0 + f_1 x + \cdots + f_n x^n) \;=\; \min_i v_A(f_i) \qquad (f_i \in K).$$

Claim. v is a valuation on $K[x]$ (hence extends to a valuation on L).

Condition (V1) in Definition 3.2 is clearly satisfied. To check (V2), let $f, g \in K[x] \setminus \{0\}$, so $f = \sum_i f_i x^i$, $g = \sum_j g_j x^j$ $(f_i, g_j \in K)$. Take i_0 minimal such that $v(f_{i_0}) = \min_i v(f_i)$, and take j_0 minimal such that $v(g_{j_0}) = \min_j v(g_j)$. Then

$$fg = \sum_n \Big(\sum_{i+j=n} f_i g_j \Big) x^n,$$

and for each n

$$v\Big(\sum_{i+j=n} f_i g_j \Big) \geq \min_{i+j=n} \{v(f_i) + v(g_j)\} \geq v(f) + v(g).$$

Now consider

$$\sum_{i+j=i_0+j_0} f_i g_j = f_{i_0} g_{j_0} + \underbrace{\text{terms } f_i g_j \text{ with } i < i_0 \text{ or } j < j_0}_{\text{each has valuation } > v(f_{i_0}) + v(g_{j_0})}$$

Therefore:

$$v\Big(\sum_{i+j=i_0+j_0} f_i g_j \Big) = v(f_{i_0}) + v(g_{j_0}) = v(f) + v(g).$$

This finishes the proof of the claim. Put $B := \mathcal{O}_v$. We still need to check:

Claim. $x \in B$, and $\bar{x}$ is transcendental over $\boldsymbol{k}_A$.

To see this, note that $v(x) = 0$, so $x \in B$. Let $a_1, \dots, a_n \in A$ be such that $\bar{a}_1, \dots, \bar{a}_n$ are not all zero. Then $v(x^n + a_1 x^{n-1} + \cdots + a_n) = 0$, hence

$$\bar{x}^n + \bar{a}_1 \bar{x}^{n-1} + \cdots + \bar{a}_n \neq 0.$$

This proves that $\bar{x}$ is transcendental over $\boldsymbol{k}_A$.

Claim. $\boldsymbol{k}_B = \boldsymbol{k}_A(\bar{x})$.

Let $b \in B$ be such that $v(b) = 0$. Then $b = \frac{f(x)}{g(x)}$, with $f, g \in K[x] \setminus \{0\}$, so $v(f) = v(g)$. Dividing all coefficients of f and g by a coefficient of g of minimal valuation, we may assume $v(g) = 0$, so $v(f) = 0$. Then $\bar{f}(\bar{x}) \neq 0$, $\bar{g}(\bar{x}) \neq 0$. Hence $bg(x) = f(x)$ yields $\bar{b}\bar{g}(\bar{x}) = \bar{f}(\bar{x})$, so

$$\bar{b} = \frac{\bar{f}(\bar{x})}{\bar{g}(\bar{x})} \in \boldsymbol{k}_A(\bar{x}),$$

as desired. Finally, it is clear from the above that $\Gamma_B = \Gamma_A$. □

Lemma 3.23. *Let $v : K^\times \to \Gamma$ be a valuation on a field K, and let $L = K(x)$ be a field extension with x transcendental over K. Let δ, in some ordered abelian group extending Γ, satisfy $n\delta \notin \Gamma$ for all $n \geq 1$. Then v extends uniquely to a valuation $w : L^\times \to \Gamma + \mathbb{Z}\delta$ such that $w(x) = \delta$. Moreover, $\boldsymbol{k}_v = \boldsymbol{k}_w$.*

Proof. Let $w : L^\times \to \Gamma + \mathbb{Z}\delta$ be a valuation such that $w(x) = \delta$. Then

$$w(1) = 0,\ w(x) = \delta,\ w(x^2) = 2\delta,\ \ldots$$

lie in different cosets of Γ, so by the proof of Proposition 3.19,

$$w(f_0 + f_1 x + \cdots + f_n x^n) \ = \ \min_i \{v(f_i) + i\delta\} \qquad (\text{all } f_i \in K). \tag{3}$$

Thus w is completely determined by v.
(*Existence.*) Define $w : K[x] \setminus \{0\} \to \Gamma + \mathbb{Z}\delta$ by (3). Again it is clear that (V1) is satisfied. To verify (V2) we mimic the proof of the previous lemma: take i_0 minimal such that $w(f_{i_0}) = \min_i\{v(f_i) + i\delta\}$, and choose j_0 minimal such that $w(g_{j_0}) = \min_j\{v(g_j) + j\delta\}$. Put $n_0 := i_0 + j_0$. First note that

$$w((\sum_{i+j=n} f_i g_j)x^n) \ \geq \ \min_{i+j=n}\{v(f_i) + v(g_j) + n\delta\} \ \geq \ w(f) + w(g).$$

Next,

$$\sum_{n \leq n_0}(\sum_{i+j=n} f_i g_j)x^n \ = \ f_{i_0} g_{j_0} x^{n_0} + \underbrace{\text{terms } f_i g_j x^n \text{ with } i < i_0 \text{ or } j < j_0}_{\text{valuation} > v(f_{i_0}) + v(g_{j_0}) + n_0\delta}.$$

Therefore:

$$w((\sum_{i+j=n_0} f_i g_j)x^{n_0}) \ = \ v(f_{i_0}) + v(g_{j_0}) + n_0\delta \ = \ w(f) + w(g).$$

So w is a valuation on $K[x]$, hence yields a valuation on L extending v and satisfying $w(x) = \delta$.

It is an instructive exercise to prove $\boldsymbol{k}_v = \boldsymbol{k}_w$. □

Note that if Γ is a *divisible* ordered abelian group, δ is in some ordered abelian group extension, and $\delta \notin \Gamma$, then $n\delta \notin \Gamma$ for all $n \geq 1$.

The Zariski-Abhyankar Inequality. The proof of Proposition 3.19 yields also a useful inequality for valued field extensions of finite transcendence degree. We discuss this now, but this inequality, due in various forms to Zariski and Abhyankar, will not be needed in these notes. We include it here for its intrinsic interest. For what follows the reader is assumed to be familiar with algebraic independence and transcendence degree.

Let $(K, A) \subseteq (L, B)$ be a valued field extension, let $(b_i)_{i \in I}$ be a family of elements of B such that the family $(\overline{b}_i)$ in $\boldsymbol{k}_B$ is algebraically independent over $\boldsymbol{k}_A$, and let $(c_j)_{j \in J}$ be a family of elements of $L^\times$ such that the family $\big(v(c_j)\big)$ in

$\mathbb{Q}\Gamma_B$ is $\mathbb{Q}$-linearly independent over $\mathbb{Q}\Gamma_A$. Assume $I \cap J = \emptyset$, and define $d_k \in L$ for $k \in I \cup J$ by $d_i = b_i$ for $i \in I$ and $d_j = c_j$ for $j \in J$.

Lemma 3.24. *The family* (d_k) *in* L *is algebraically independent over* K.

Proof. The assumption on (b_i) gives the linear independence over $\boldsymbol{k}_A$ of suitably indexed products of powers of the $\overline{b}_i$. Likewise, the assumption on (c_j) means that the valuations of suitably indexed products of powers of the c_j lie in distinct cosets of Γ_A. The reader should formulate these claims precisely, verify them, and observe that the lemma then follows from the proof of Proposition 3.19. □

Let dim V denote the dimension of a vector space V over $\mathbb{Q}$, and let $\operatorname{trdeg}(L|K)$ be the transcendence degree of a field extension $L|K$.

Corollary 3.25. *Let* $(K, A) \subseteq (L, B)$ *be valued fields, with* L *of finite transcendence degree over* K. *Then* $\boldsymbol{k}_B$ *has finite transcendence degree over* $\boldsymbol{k}_A$ *and* $\mathbb{Q}\Gamma_B/\mathbb{Q}\Gamma_A$ *has finite dimension as a vector space over* $\mathbb{Q}$, *with*

$$\operatorname{trdeg}(\boldsymbol{k}_B|\boldsymbol{k}_A) + \dim \mathbb{Q}\Gamma_B/\mathbb{Q}\Gamma_A \le \operatorname{trdeg}(L|K).$$

We refer to [1] and [53] for further elaborations and applications.

3.4 Algebraically Closed Valued Fields

In this section we establish quantifier elimination (QE) for algebraically closed valued fields. More precisely, we restrict to the case where the valuation is nontrivial. Since quantifier elimination is rather sensitive as to the choice of primitives, we shall be explicit about this.

Let (K, A) be an algebraically closed valued field with $A \neq K$, and consider it as a structure for the language of rings with a unary relation symbol U (interpreted as the set A). Then the binary relation "$v_A(x) \le v_A(y)$" on K is definable in (K, A) by the formula $\exists z\big(U(z) \wedge xz = y\big)$. However, one can show that this relation is not quantifier-free definable in (K, A), even if we allow names for the elements of K to be used. Thus QE for algebraically closed valued fields needs a different language. It turns out that the binary relation above is the only *obstruction* to QE in the language above.

Definition 3.26. A *valuation divisibility* on a domain R is a binary relation $|$ on R such that for all $x, y, z \in R$

(VD1) not $0 \mid 1$;
(VD2) $x \mid y$ and $y \mid z \implies x \mid z$;
(VD3) $x \mid y$ and $x \mid z \implies x \mid y + z$;
(VD4) $x \mid y \iff xz \mid yz$ for $z \neq 0$;
(VD5) $x \mid y$ or $y \mid x$.

Lemma 3.27. *Valuation divisibilities and valuations are related as follows:*

(i) *Each valuation ring A of a field K gives rise to a valuation divisibility $|_A$ on K by*

$$x \,|_A\, y \iff v_A(x) \leq v_A(y) \quad (\iff ax = y \text{ for some } a \in A\,).$$

(ii) *Given a field K, the map*

$$A \mapsto |_A : \{\,\text{valuation rings of } K\,\} \longrightarrow \{\,\text{valuation divisibilities on } K\,\}$$

is a bijection.

(iii) *For each valuation divisibility $|$ on a domain R there is a unique valuation divisibility $|'$ on $K = \mathrm{Frac}(R)$ such that $(R, |) \subseteq (K, |')$; it is given by*

$$\frac{a_1}{b} \,|'\, \frac{a_2}{b} \iff a_1 \mid a_2 \qquad (a_1, a_2, b \in R,\ b \neq 0).$$

(iv) *The valuation divisibilities on $\mathbb{Z}$ are exactly the $|_p$ (p a prime number) and $|_t$, where $|_t$ is the trivial valuation divisibility:*

$$a \,|_p\, b \iff v_p(a) \leq v_p(b),$$

$$a \,|_t\, b \text{ for all } a, b \text{ with } a \neq 0.$$

Definition 3.28. Let $L_{\mathrm{val}} := \{\,0, 1, +, -, \cdot, |\,\}$ be the language of rings with a binary relation symbol $|$, and let $\mathrm{ACF}_{\mathrm{val}}$ be the theory in this language whose models are the structures $(K, |\,)$ such that $K \models \mathrm{ACF}$, and $|$ is a *nontrivial* valuation divisibility on K. (Here a valuation divisibility on a field K is said to be *trivial* if the corresponding valuation ring is K, equivalently, the associated valuation is trivial.)

We shall obtain quantifier elimination (QE) for $\mathrm{ACF}_{\mathrm{val}}$ using the following test, which is an easy variant of possibly more familiar back-and-forth criteria.

QE-Test. Given a theory T, the following are equivalent:

(1) T has QE;
(2) given any models $\mathcal{M}, \mathcal{N}$ of T where $\mathcal{N}$ is $|\mathcal{M}|^+$-saturated, and given any embedding $i : \mathcal{A} \to \mathcal{N}$ of a proper substructure $\mathcal{A}$ of $\mathcal{M}$ into $\mathcal{N}$, we can extend i to an embedding $j : \mathcal{A}' \to \mathcal{N}$ of some substructure $\mathcal{A}'$ of $\mathcal{M}$ that properly extends $\mathcal{A}$.

Thus the algebraic counterpart of T having QE is that isomorphisms between substructures of models of T can be extended.
Any substructure of a model of $\mathrm{ACF}_{\mathrm{val}}$ is clearly a domain with a valuation divisibility on it. Conversely, every domain with a valuation divisibility on it is a substructure of a model of $\mathrm{ACF}_{\mathrm{val}}$: first extend the domain with its valuation divisibility to its fraction field; then extend further to the algebraic closure of the

fraction field; in case a valuation is trivial, adjoin a transcendental to make it nontrivial.

Let $(R, |)$ be a domain with valuation divisibility $|$ on R, and let

$$i : (R, |) \to (F, |^*)$$

be an embedding into an algebraically closed field F with valuation divisibility $|^*$ on F. By item (iii) of the lemma above, and using its notation, we can extend i uniquely to an embedding $i' : (K, |') \to (F, |^*)$. Let K^{ac} be an algebraic closure of K, and $|^{\mathrm{ac}}$ a valuation divisibility on K^{ac} such that $(K, |') \subseteq (K^{\mathrm{ac}}, |^{\mathrm{ac}})$. Then by Corollary 3.16 we can extend i' to an embedding $i^{\mathrm{ac}} : (K^{\mathrm{ac}}, |^{\mathrm{ac}}) \to (F, |^*)$.

Theorem 3.29. $\mathrm{ACF}_{\mathrm{val}}$ *has* QE.

Proof. By the remarks preceding the theorem, and the QE-test it suffices to prove the following:
Let (E, A), (F, B) be nontrivially valued algebraically closed fields such that (F, B) is $|E|^+$-saturated. Let K be a proper algebraically closed subfield of E, and $i : (K, A \cap K) \longrightarrow (F, B)$ a valued field embedding. Then there is $x \in E \setminus K$ such that i extends to a valued field embedding $j : (K(x), A \cap K(x)) \longrightarrow (F, B)$.

$$\begin{array}{ccc}
(E, A) & & (F, B) \\
\Big| & & \Big| \\
(K(x), A \cap K(x)) & \xrightarrow[j]{\sim} & ((iK)(y), B \cap (iK)(y)) \\
\Big| & & \Big| \\
(K, A \cap K) & \xrightarrow[i]{\sim} & (iK, B \cap iK)
\end{array}$$

To find such x and j we shall distinguish three cases. To simplify notation, denote both v_A and v_B by v, and let $\overline{z}$ be the residue class of z in $\boldsymbol{k}_A$ for $z \in A$, and also the residue class of z in $\boldsymbol{k}_B$ for $z \in B$.

Case 1: $\boldsymbol{k}_{A\cap K} \neq \boldsymbol{k}_A$. Then we take $x \in A$ such that $\overline{x} \notin \boldsymbol{k}_{A\cap K}$. Since $\boldsymbol{k}_{A\cap K}$ is algebraically closed, $\overline{x}$ is transcendental over $\boldsymbol{k}_{A\cap K}$. Also $x \notin K$, so x is transcendental over K. By the saturation assumption on (F, B) we can find $y \in B$ such that $\overline{y} \notin \boldsymbol{k}_{B\cap iK}$. So $\overline{y}$ is transcendental over $\boldsymbol{k}_{B\cap iK}$ and y is transcendental over iK. Extend i to a field isomorphism $j : K(x) \longrightarrow (iK)(y)$ by $j(x) = y$. By the uniqueness part of Lemma 3.22, j is also an isomorphism

$$\big(K(x), A \cap K(x)\big) \xrightarrow{\sim} \big((iK)(y), B \cap (iK)(y)\big) \subseteq (F, B).$$

Case 2: $v(K^\times) \neq v(E^\times)$. Take any $\gamma \in v(E^\times) \setminus v(K^\times)$. Note that $v(K^\times)$ and $v(iK^\times)$ are divisible. Since (F, B) is κ^+-saturated where $\kappa = |K|$, so is $v(F^\times)$

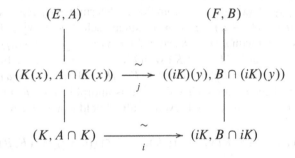

as an ordered abelian group. Also $v(F^\times) \neq \{0\}$. It follows that we can take $\delta \in v(F^\times) \setminus v(iK^\times)$ such that for all $a \in K^\times$,

$$\gamma < v(a) \iff \delta < v(ia).$$

Thus the isomorphism of ordered abelian groups

$$va \mapsto v(ia) : v(K^\times) \longrightarrow v(iK^\times), \qquad (a \in K^\times)$$

extends to an isomorphism of ordered abelian groups

$$va + k\gamma \mapsto v(ia) + k\delta : v(K^\times) + \mathbb{Z}\gamma \longrightarrow v(iK^\times) + \mathbb{Z}\delta.$$

Take $x \in E^\times$ and $y \in F^\times$ such that $v(x) = \gamma$, and $v(y) = \delta$. Since $x \notin K$, x is transcendental over K; likewise, y is transcendental over iK. By the uniqueness part of Lemma 3.23, the field isomorphism $j : K(x) \longrightarrow (iK)(y)$ extending i and sending x to y is even a valued field isomorphism

$$\big(K(x), A \cap K(x)\big) \xrightarrow{\sim} \big((iK)(y), B \cap (iK)(y)\big) \subseteq (F, B).$$

Case 3: $\boldsymbol{k}_{A\cap K} = \boldsymbol{k}_A$ and $v(K^\times) = v(E^\times) =: \Gamma$. Take any $x \in E \setminus K$. The valuation $v|_{K(x)} : K(x) \to \Gamma_\infty$ is completely determined by $v|_K : K \to \Gamma_\infty$ and by the map $a \mapsto v(x - a) : K \longrightarrow \Gamma$, since each $f \in K[x] \setminus \{0\}$ factors as $f = c \prod_i (x - a_i)$ with $c, a_i \in K$, so $v(f) = v(c) + \sum_i v(x - a_i)$. To simplify notation, let us identify $(K, A \cap K)$ with $(iK, B \cap iK)$ via i (so both now have value group Γ). It is enough to find $y \in F \setminus iK$ such that $v(y - a) = v(x - a)$ for all $a \in K$, because for such y the field isomorphism $j : K(x) \longrightarrow (iK)(y)$ extending i and sending x to y is even a valued field isomorphism

$$(K(x), A \cap K(x)) \xrightarrow{\sim} ((iK)(y), B \cap (iK)(y)) \subseteq (F, B).$$

Such an element y exists by saturation and the following general lemma.

□

Lemma 3.30. *Let $(K, A) \subseteq (L, B)$ be a valued field extension such that $\boldsymbol{k}_A = \boldsymbol{k}_B$, and let $v = v_B$. Let $a_1, \ldots, a_n \in K$, $n \geq 1$, and let $x \in L \setminus K$ be such that $v(x - a_i) \in v(K^\times)$ for $i = 1, \ldots, n$. Then there exists $a \in K$ such that $v(x - a_i) = v(a - a_i)$ for $i = 1, \ldots, n$.*

Proof. Any $a \in K$ such that $v(a - x) > v(a - a_i)$ for $i = 1, \ldots, n$ has the desired property. We may assume $v(x - a_1) \geq v(x - a_i)$ for $i = 2, \ldots, n$. Since $v(x - a_1) \in v(K^\times)$ we can take $b \in K$ such that $v(x - a_1) = vb$. So $v(\frac{x-a_1}{b}) = 0$, and since $\boldsymbol{k}_A = \boldsymbol{k}_B$, $\frac{x-a_1}{b} = c + \varepsilon$ with $c \in K$, $v(c) = 0$, $v(\varepsilon) > 0$. Then $a = a_1 + bc$ works because $x - a = b\varepsilon$ and $v(b\varepsilon) > v(x - a_i)$. □

By Theorem 3.29, the theory of nontrivially valued algebraically closed fields construed as structures (K, A) is model complete. This model completeness is what

Robinson [43] actually proved; one can obtain our QE from it with a few extra lines of valuation theory. It may be worth mentioning a converse to Theorem 3.29: if $|$ is any nontrivial valuation divisibility on a field K and the L_{val}-theory of $(K, |)$ admits QE, then $(K, |) \models \text{ACF}_{\text{val}}$; see [37].

Let (K, A) be a nontrivially valued algebraically closed field, with valuation

$$v : K^{\times} \to \Gamma, \quad v = v_A, \ \Gamma = \Gamma_A.$$

Call a set $X \subseteq K$ *boolean* if it is in the boolean algebra of subsets of K generated by the disks in K. Thus K is the largest boolean subset of K, and the empty subset of K is the smallest. Obviously, if $X \subseteq K$ is boolean, then X is definable in (K, A). A consequence of Theorem 3.29 is the converse, Corollary 3.32 below. This also requires the next lemma of independent interest.

On a notational issue, we have been using and are going to use the letter x in different ways, but the context ("let $x \in K$", or "let x be an indeterminate") should always make it clear what is intended. The same holds for our diverse uses of the letters X and t.

Lemma 3.31. *Let x be an indeterminate and $f(x) \in K[x]$, $f \notin K$. Then there is a partition of K into finitely many boolean sets X, on each of which $v\big(f(t)\big)$ is a very simple function, namely, there are $m \geq 1$, $a \in K$, and $\alpha \in \Gamma = v(K^{\times})$ such that $v\big(f(t)\big) = mv(t-a) + \alpha$ for all $t \in X$.*

Proof. We have a factorization $f(x) = c(x - a_1)\cdots(x - a_n)$ with $c \in L^{\times}$ and $n \geq 1, a_1, \ldots, a_n \in K$. Let t range over K. Then

$$v\big(f(t)\big) = v(c) + v(t - a_1) + \cdots + v(t - a_n).$$

This gives the desired result (with $m = n$) if $a_1 = \cdots = a_n$. Suppose we have summands $v(t - a)$ and $v(t - b)$ with $a = a_i$, $b = a_j$, $a \neq b$. Then we observe:

(i) on the boolean set given by $v(t - a) < v(a - b)$ we have $v(t - a) = v(t - b)$,
(ii) on the boolean set given by $v(t - a) = v(a - b)$ we have $v(t - a) = v(a - b)$,
(iii) on the boolean set given by $v(t - a) > v(a - b)$ we have $v(t - b) = v(a - b)$.

Using this fact we can shorten the above sum for $v\big(f(t)\big)$ on each of these three boolean sets (which are disjoint and cover K). We repeat this process until we eventually arrive at a partition of K as required. □

The proof shows that we can take the a in Lemma 3.31 among the zeros of f in K. There are further refinements of the lemma, but the above suffices for us.

Corollary 3.32. *If $X \subseteq K$ is definable in (K, A), then X is boolean.*

Proof. Let $f(x), g(x) \in K[x]$. By Theorem 3.29 it is enough to show that

$$\{t \in K : v\big(f(t)\big) \leq v\big(g(t)\big)\}$$

is boolean. We leave it to the reader to check that this set is boolean by applying Lemma 3.31 to f and g, using also the decomposition argument in its proof. □

We can improve this result a bit by defining a *swiss cheese* to be a set

$$D \setminus (E_1 \cup \cdots \cup E_n) \subseteq K$$

where $D \subseteq K$ is a disk or $D = K$, and where $E_1, \ldots, E_n \subseteq K$ are disks; since for any disks B, C in K we have $B \cap C = \emptyset$, or $B \subseteq C$, or $C \subseteq B$, it follows that in such a swiss cheese we can actually take $E_1, \ldots, E_n$ to be disjoint and contained in D. (We allow $n = 0$, so every disk is a swiss cheese, and so is K.)

Proposition 3.33. *If $X \subseteq K$ is definable in (K, A), then X is a finite disjoint union of swiss cheeses.*

Proof. Note that the intersection of two swiss cheeses is a swiss cheese, and that the complement of a swiss cheese is a finite disjoint union of swiss cheeses. □

This material on "one-variable" definable sets is taken from [29], and is closely related to [24], I.1.2.

3.5 *The Complete Extensions of* $\mathbf{ACF}_{\mathbf{val}}$

For a valued field (K, A) with residue field $\boldsymbol{k} = \boldsymbol{k}_A$ the pair $(\mathrm{char}(K), \mathrm{char}(\boldsymbol{k}))$ can take the following values:

$(0, 0)$: "equicharacteristic 0"; an example is $\boldsymbol{k}((t^\Gamma))$ with $\mathrm{char}(\boldsymbol{k}) = 0$ and the usual valuation.

$(0, p)$, p a prime number: "mixed characteristic p"; an example is $\mathbb{Q}[p]$ with the p-adic valuation, whose residue field is $\mathbb{F}_p$.

(p, p), p a prime number: "equicharacteristic p"; an example is $\boldsymbol{k}((t^\Gamma))$ with $\mathrm{char}(\boldsymbol{k}) = p$ and the usual valuation.

It is easy to check that $(p, 0)$ with p prime, and (p, q) with p, q distinct primes, are impossible as values of $(\mathrm{char}(K), \mathrm{char}(\boldsymbol{k}))$. Let us call $(\mathrm{char}(K), \mathrm{char}(\boldsymbol{k}))$ the *characteristic* of the valued field (K, A).

Let (a, b) be a possible characteristic of a valued field, and let $\mathrm{ACF}_{\mathrm{val}}(a, b)$ be obtained from $\mathrm{ACF}_{\mathrm{val}}$ by adding axioms specifying that the characteristic of the underlying field is a and that the characteristic of the residue field is b.

Corollary 3.34. $\mathrm{ACF}_{\mathrm{val}}(a, b)$ *is complete.*

Proof. Construing valued fields as fields with a valuation divisibility, those of characteristic $(0, 0)$ all have (an isomorphic copy of) $(\mathbb{Z}, |_{\mathrm{t}})$ as substructure, those of characteristic $(0, p)$ have $(\mathbb{Z}, |_p)$ and those of characteristic (p, p) have $(\mathbb{F}_p, |_{\mathrm{t}})$ as substructure, where $|_{\mathrm{t}}$ is the trivial valuation divisibility on $\mathbb{F}_p$. To see this, use the classification of valuation divisibilities on $\mathbb{Z}$, and the fact that finite fields only admit the trivial valuation divisibility. □

Corollary 3.34 is from [43].

Let (K, A) be a nontrivially valued algebraically closed field. The results above suggests some natural model-theoretic questions.

(i) Find natural models for the completions of ACF_{val}.
(ii) What structure does (K, A) induce on $\boldsymbol{k}_A$ and Γ_A?
(iii) What is the definable closure in (K, A) of a given subset of K?
(iv) Find some natural expansion of ACF_{val} by definable sorts that admits elimination of imaginaries.

Here are some answers to and comments on these questions:

(i) If $\boldsymbol{k}$ is algebraically closed and Γ is a divisible ordered abelian group, then the power series field $\boldsymbol{k}((t^{\Gamma}))$ is algebraically closed, in particular, $\mathbb{C}((t^{\mathbb{Q}}))$ with its usual valuation divisibility is a model of $\text{ACF}_{\text{val}}(0, 0)$, and $\mathbb{F}_p^{\text{ac}}((t^{\mathbb{Q}}))$ with its usual valuation divisibility is a model of $\text{ACF}_{\text{val}}(p, p)$. (Here p is a prime number and $\mathbb{F}_p^{\text{ac}}$ is the algebraic closure of $\mathbb{F}_p$.) These are among the facts established in the next chapter.

Without proof, we mention the following about the mixed characteristic case. Let p be a prime number and fix an algebraic closure $\mathbb{Q}_p^{\text{ac}}$ of the field $\mathbb{Q}_p$ of p-adic numbers. The p-adic absolute value $|\cdot|_p$ on $\mathbb{Q}_p$ extends uniquely to an ultrametric absolute value on $\mathbb{Q}_p^{\text{ac}}$. The completion of $\mathbb{Q}_p^{\text{ac}}$ with respect to this ultrametric absolute value is denoted by $\mathbb{C}_p$. Then $\mathbb{C}_p$ is so-to-say the p-adic analogue of the field $\mathbb{C}$ of complex numbers: like $\mathbb{C}$ it is an algebraically closed field (even isomorphic as a field to $\mathbb{C}$), but its absolute value is ultrametric and extends the absolute value $|\cdot|_p$ on $\mathbb{Q}_p$. Thus $\mathbb{C}_p$ with the valuation divisibility corresponding to its absolute value is a model of $\text{ACF}_{\text{val}}(0, p)$.

(ii) Let $\pi : A \to \boldsymbol{k} = \boldsymbol{k}_A$ be the residue map, and let $X \subseteq A^n$ be definable in (K, A). Then the set

$$\pi(X) := \{(\pi(a_1), \ldots, \pi(a_n)) : (a_1, \ldots, a_n) \in X\} \subseteq \boldsymbol{k}^n$$

is definable in the field $\boldsymbol{k}$. Let $v = v_A : K^{\times} \to \Gamma = \Gamma_A$ be the valuation, and let $Y \subseteq (K^{\times})^n$ be definable in (K, A). Then the set

$$v(Y) := \{(v(a_1), \ldots, v(a_n)) : (a_1, \ldots, a_n) \in Y\} \subseteq \Gamma^n$$

is definable in the divisible ordered abelian group Γ. This is a special case of Corollary 5.25 below, but only for (K, A) of equicharacteristic 0. One way to get this result for all models of ACF_{val} is to consider valued fields as 3-sorted structures, with the residue field and value group as extra sorts, and to prove an analogue of Theorem 3.29 in this setting.

(iii) A complete answer to this question involves the notion of the henselization of a valued field, which we define in the next chapter. The model-theoretic algebraic closure of a set $S \subseteq K$ in (K, A) is the field-theoretic algebraic closure in K of the subfield of K generated by S.

(iv) We do not have elimination of imaginaries just with the residue field and the value group as extra sorts, and the maps π and v as in (ii); see [29]. An explicit list of extra sorts adequate for eliminating imaginaries has been given by Haskell, Hrushovski, and Macpherson in [27]. Their work requires deeper tools from model theory than we use in these notes; see also the book [28].

Exercise on Puiseux Series Fields. Let $\boldsymbol{k}$ be a field, and define the field of *Puiseux series* over $\boldsymbol{k}$ to be the subfield of $\boldsymbol{k}((t^{\mathbb{Q}}))$ given by

$$\mathrm{P}(\boldsymbol{k}) := \bigcup_{d=1}^{\infty} \boldsymbol{k}((t^{\frac{1}{d}\mathbb{Z}})).$$

Note that $\mathrm{P}(\boldsymbol{k})$ contains the field $\boldsymbol{k}((t))$ of Laurent series over $\boldsymbol{k}$ as a subfield. Show that if d is a positive integer and $\alpha = t^{\frac{1}{d}} \in P(\boldsymbol{k})$, then

$$\boldsymbol{k}((t^{\frac{1}{d}\mathbb{Z}})) = \boldsymbol{k}((t)) + \boldsymbol{k}((t))\alpha + \cdots + \boldsymbol{k}((t))\alpha^{d-1}.$$

Use this to show that $\boldsymbol{k}((t^{\frac{1}{d}\mathbb{Z}}))$ is an algebraic extension of $\boldsymbol{k}((t))$ of degree d. Thus $\mathrm{P}(\boldsymbol{k})$ is an algebraic extension of $\boldsymbol{k}((t))$.

Show that the closure of $\mathrm{P}(\boldsymbol{k})$ in $\boldsymbol{k}((t^{\mathbb{Q}}))$ (with respect to the valuation topology) is the subfield

$$\{\, f \in \boldsymbol{k}((t^{\mathbb{Q}})) \mid \operatorname{supp} f \cap (-\infty, n) \text{ is finite for each } n \,\}$$

of $\boldsymbol{k}((t^{\mathbb{Q}}))$. In particular, $\mathrm{P}(\boldsymbol{k})$ is not dense in $\boldsymbol{k}((t^{\mathbb{Q}}))$, although it has the same residue field $\boldsymbol{k}$ and the same value group $\mathbb{Q}$ as $\boldsymbol{k}((t^{\mathbb{Q}}))$.

Comment on exercise: It will be shown later that if $\boldsymbol{k}$ is algebraically closed of characteristic 0, then $\mathrm{P}(\boldsymbol{k})$ is algebraically closed, and thus an algebraic closure of $\boldsymbol{k}((t))$. On the other hand, if $\boldsymbol{k}$ is algebraically closed of characteristic $p \neq 0$, then $\mathrm{P}(\boldsymbol{k})$ is not algebraically closed, since the equation $x^p - x = t^{-1}$ over $\mathrm{P}(\boldsymbol{k})$ has a root

$$x = t^{-1/p} + t^{-1/p^2} + t^{-1/p^3} + \cdots$$

in $\boldsymbol{k}((t^{\mathbb{Q}}))$ that does not lie in $\mathrm{P}(\boldsymbol{k})$.

4 Immediate Extensions

In this chapter it will be convenient to construe a valued field as a 2-sorted structure $(K, \Gamma; v)$ where $v : K^{\times} \to \Gamma = v(K^{\times})$ is a valuation on the field K. In particular, a Hahn field $\boldsymbol{k}((t^{\Gamma}))$ is considered as the valued field $(\boldsymbol{k}((t^{\Gamma})), \Gamma; v)$ with v as in

Sect. 3.1. Puiseux series fields $\mathrm{P}(\boldsymbol{k})$ are considered as valued fields $(\mathrm{P}(\boldsymbol{k}), \mathbb{Q}; v)$ in the same way.

Consider a valued field extension $(K, \Gamma; v) \subseteq (K', \Gamma'; v')$. We identify $\boldsymbol{k}_v$ with a subfield of $\boldsymbol{k}_{v'}$ in the usual way. This extension is said to be *immediate* if $\boldsymbol{k}_v = \boldsymbol{k}_{v'}$ and $\Gamma = \Gamma'$. The basic facts on immediate extensions, due to Ostrowski, Krull, Kaplansky (first half of the twentieth century), have their own intrinsic interest, and are also crucial in the later work by Ax–Kochen and Ershov. These facts are related to the notion of *pseudoconvergence*. To explain this, consider for example the valued field $\mathrm{P}(\boldsymbol{k})$ of Puiseux series over a field $\boldsymbol{k}$ and its immediate valued field extension $\boldsymbol{k}((t^{\mathbb{Q}}))$ (with valuation v). The series

$$a = 1 + t^{1/2} + t^{2/3} + t^{3/4} + \ldots = \sum_n t^{n/(n+1)}$$

in $\boldsymbol{k}((t^{\mathbb{Q}}))$ does not lie in $\mathrm{P}(\boldsymbol{k})$. While a is approximated in a certain sense by the sequence $\{a_n\}$ in $\mathrm{P}(\boldsymbol{k})$, where $a_n := \sum_{i=0}^{n} t^{i/(i+1)}$, a is not the limit of this sequence in the valuation topology of $\boldsymbol{k}((t^{\mathbb{Q}}))$: $v(a-a_n) \not\to \infty$ as $n \to \infty$, in fact, $v(a-a_n) < 1$ for all n. On the other hand, $v(a - a_n)$ is strictly increasing as a function of n, and this suggests a notion of (pseudo)limit that turns out to be very useful. The property of pseudoconvergence is due to Ostrowski [39] in the rank one case. With Krull's [33] it became reasonable to study this property for any rank, and this was done by Kaplansky. Accordingly, most of the results in this chapter are from [31], while many of the ideas originate with Ostrowski.

In this chapter $(K, \Gamma; v)$ *is a valued field, and* $\mathcal{O} := \mathcal{O}_v$, $\mathfrak{m} := \mathfrak{m}_v$, $\boldsymbol{k} := \boldsymbol{k}_v$.

4.1 Pseudoconvergence

A *well-indexed sequence in* K is a sequence $\{a_\rho\}$ in K whose terms a_ρ are indexed by the elements ρ of an infinite well-ordered set without a last element. Let $\{a_\rho\}$ be a *well-indexed sequence in* K, and $a \in K$. Then $\{a_\rho\}$ is said to *pseudoconverge to* a (notation: $a_\rho \rightsquigarrow a$), if $\{v(a - a_\rho)\}$ is eventually strictly increasing, that is, for some index ρ_0 we have $v(a - a_\sigma) > v(a - a_\rho)$ whenever $\sigma > \rho > \rho_0$. We also say in that case that a *is a pseudolimit of* $\{a_\rho\}$. Note that if $a_\rho \rightsquigarrow a$, then $a_\rho + b \rightsquigarrow a + b$ for each $b \in K$, and $a_\rho b \rightsquigarrow ab$ for each $b \in K^\times$. We also have the equivalence

$$a_\rho \rightsquigarrow 0 \iff \{v(a_\rho)\} \text{ is eventually strictly increasing.}$$

Lemma 4.1. *Let* $\{a_\rho\}$ *be a well-indexed sequence in* K *such that* $a_\rho \rightsquigarrow a$ *where* $a \in K$. *With* $\gamma_\rho := v(a - a_\rho)$, *we have:*

(i) *for all* $b \in K$: $\quad a_\rho \rightsquigarrow b \iff v(a - b) > \gamma_\rho$ *eventually;*
(ii) *either* $v(a_\rho) < v(a)$ *eventually, or* $v(a_\rho) = v(a)$ *eventually;*
(iii) $\{v(a_\rho)\}$ *is either eventually strictly increasing, or eventually constant.*

Proof. We leave (i) to the reader. Let ρ_0 be as in the definition of "$a_\rho \rightsquigarrow a$". Suppose $v(a) \le v(a_\rho)$ for a certain $\rho > \rho_0$. Then we claim that $v(a) = v(a_\sigma)$ for all $\sigma > \rho$. This is because for $\sigma > \rho$ we have $v(a - a_\sigma) > v(a - a_\rho) \ge v(a)$, so $v(a) = v(a_\sigma)$. This proves (ii). Now (iii) follows from (ii) by noting that if $v(a_\rho) < v(a)$ eventually, then $v(a - a_\rho) = v(a_\rho)$ eventually, so $\{v(a_\rho)\}$ is eventually strictly increasing. □

If $\{a_\rho\}$ is a well-indexed sequence in K and $a \in K'$ where $(K', \Gamma'; v')$ is a valued field extension of $(K, \Gamma; v)$, then "$a_\rho \rightsquigarrow a$" is of course to be interpreted in this valued field extension, that is, by considering the sequence $\{a_\rho\}$ in K as a sequence in K'.

Lemma 4.2. *Suppose* $(K', \Gamma; v')$ *is an immediate valued field extension of* $(K, \Gamma; v)$, *and let* $a' \in K' \setminus K$. *Then there is well-indexed sequence* $\{a_\rho\}$ *in* K *such that* $a_\rho \rightsquigarrow a'$ *and* $\{a_\rho\}$ *has no pseudolimit in* K.

Proof. We claim that the subset $\{v'(a' - x) : x \in K\}$ of Γ has no largest element. To see this, let $x \in K$; we shall find $y \in K$ such that $v'(a' - y) > v'(a' - x)$. Take $b \in K$ such that $v'(a' - x) = v(b)$. Then $v'(\frac{a'-x}{b}) = 0$, so $\frac{a'-x}{b} = c + \epsilon$ with $c \in K$, $v*c) = 0$, and $v'(\epsilon) > 0$. Then $a' - x = bc + b\epsilon$, and thus $y = x + bc$ has the desired property. It follows that we can take a well-indexed sequence $\{a_\rho\}$ in K such that the sequence $\{v'(a' - a_\rho)\}$ is strictly increasing and cofinal in $\{v'(a' - x) : x \in K\}$. Thus $a_\rho \rightsquigarrow a'$. If $a_\rho \rightsquigarrow a \in K$, then $v'(a' - a) > v(a' - a_\rho)$ for all ρ by Lemma 4.1, part (i), hence $v'(a' - a) > v(a' - x)$ for all $x \in K$, a contradiction. □

An illustration of this lemma comes from the beginning of this chapter: we have the immediate valued field extension $\mathrm{P}(\boldsymbol{k}) \subseteq \boldsymbol{k}((t^{\mathbb{Q}}))$, the sequence $\{a_n\}$ in $\mathrm{P}(\boldsymbol{k})$ has no pseudolimit in $\mathrm{P}(\boldsymbol{k})$ but pseudoconverges to $a \in \boldsymbol{k}((t^{\mathbb{Q}}))$. Note also that the pseudolimits of $\{a_n\}$ in $\boldsymbol{k}((t^{\mathbb{Q}}))$ are exactly the series $a + b$ with $v(b) \ge 1$, by part (i) of Lemma 4.1.

To capture within $(K, \Gamma; v)$ that a well-indexed sequence $\{a_\rho\}$ in K has a pseudolimit in some valued field extension we define a *pseudocauchy sequence in* K (more precisely, in $(K, \Gamma; v)$) to be a well-indexed sequence $\{a_\rho\}$ in K such that for some index ρ_0 we have

$$\tau > \sigma > \rho > \rho_0 \implies v(a_\tau - a_\sigma) > v(a_\sigma - a_\rho).$$

We also write "pc-sequence" for "pseudocauchy sequence".

Lemma 4.3. *Let* $\{a_\rho\}$ *be a well-indexed sequence in* K. *Then* $\{a_\rho\}$ *is a pc-sequence in* K *if and only if* $\{a_\rho\}$ *has a pseudolimit in some valued field extension of* $(K, \Gamma; v)$. *In that case,* $\{a_\rho\}$ *has even a pseudolimit in some elementary extension of* $(K, \Gamma; v)$.

Proof. Suppose $\{a_\rho\}$ is a pc-sequence in K, and let ρ_0 be as in the definition of "pseudocauchy sequence". We consider the partial type in the variable x consisting of the formulas

$$v(x - a_\sigma) > v(x - a_\rho), \qquad (\sigma > \rho > \rho_0).$$

Every finite subset of this partial type is realized by some a_τ. By compactness we can realize this partial type by a suitable $a \in K'$ for some elementary extension $(K', \Gamma'; v')$ of $(K, \Gamma; v)$, and then $a_\rho \rightsquigarrow a$.

For the converse, suppose $a_\rho \rightsquigarrow a$ where $a \in K'$ and $(K', \Gamma'; v')$ is a valued field extension of $(K, \Gamma; v)$. Let ρ_0 be as in the definition of pseudolimit, and let $\sigma > \rho > \rho_0$. Then $a_\sigma - a_\rho = (a_\sigma - a) - (a_\rho - a)$, so $v(a_\sigma - a_\rho) = v(a - a_\rho)$. So if in addition $\tau > \sigma$, then $v(a_\tau - a_\sigma) = v(a - a_\sigma) > v(a - a_\rho) = v(a_\sigma - a_\rho)$. □

The "elementary extension" part (Macintyre) goes through for any expansion of (K, v, Γ). Lemma 4.3 and part (iii) of Lemma 4.1 yield:

Corollary 4.4. *If $\{a_\rho\}$ is a pc-sequence in K, then $\{v(a_\rho)\}$ is either eventually strictly increasing, or eventually constant.*

We are going to show that polynomials behave well on pc-sequences, and in this connection it is useful to fix some notation. For a polynomial $f(x) \in K[x]$ of degree $\leq n$ we have a unique Taylor expansion for $f(x+y)$ in the ring $K[x, y]$ of polynomials over K in the distinct indeterminates x, y:

$$f(x+y) = \sum_{i=0}^{n} f_{(i)}(x) \cdot y^i,$$

where each $f_{(i)}(x) \in K[x]$. For convenience we also set $f_{(i)} = 0$ for $i > n$. Note that $f_{(0)} = f$ and $f_{(1)} = f'$. (If $\mathrm{char}(K) = 0$, then $f_{(i)}(x) = f^{(i)}(x)/i!$ where $f^{(i)}$ is the usual ith formal derivative of f.) It is convenient to record here the following identity, although it is needed only much later in this section:

Lemma 4.5. $f_{(i)(j)} = \binom{i+j}{i} f_{(i+j)}, \qquad (i, j \in \mathbb{N}).$

We also need an elementary fact on ordered abelian groups Γ whose proof we leave to the reader as an easy exercise.

Lemma 4.6. *For each i in a finite nonempty set I, let $\beta_i \in \Gamma$, and $n_i \in \mathbb{N}^{>0}$, and let $\lambda_i : \Gamma \to \Gamma$ be the linear function given by $\lambda_i(\gamma) = \beta_i + n_i\gamma$. Assume that $n_i \neq n_j$ for distinct $i, j \in I$. Let $\rho \mapsto \gamma_\rho$ be a strictly increasing function from an infinite linearly ordered set without largest element into Γ. Then there is an $i_0 \in I$ such that if $i \in I$ and $i \neq i_0$, then $\lambda_{i_0}(\gamma_\rho) < \lambda_i(\gamma_\rho)$, eventually.*

We can now prove an important continuity property of polynomials:

Proposition 4.7. *Let $\{a_\rho\}$ be a well-indexed sequence in K such that $a_\rho \rightsquigarrow a$, where $a \in K$, and let $f(x) \in K[x]$ be a nonconstant polynomial (that is, $f \notin K$). Then $f(a_\rho) \rightsquigarrow f(a)$.*

Proof. Let f be of degree $\leq n$, so we have the identity

$$f(x+y) = f(x) + f_{(1)}(x)y + \cdots + f_{(n)}(x)y^n$$

in the polynomial ring $K[x,y]$. Substituting a for x and $a_\rho - a$ for y yields

$$f(a_\rho) - f(a) = f_{(1)}(a)(a_\rho - a) + \cdots + f_{(n)}(a)(a_\rho - a)^n = \sum_{i=1}^{n} f_{(i)}(a)(a_\rho - a)^i.$$

Since $f(x) = f(a) + \sum_{i=1}^{n} f_{(i)}(a)(x-a)^i$ and f is not constant, there exists $i \in \{1,\dots,n\}$ such that $f_{(i)}(a) \neq 0$. Now $v\big(f_{(i)}(a)(a_\rho - a)^i\big) = \beta_i + i\gamma_\rho$ where $\beta_i := v\big(f_{(i)}(a)\big)$ and $\gamma_\rho := v(a_\rho - a)$. Since $\{\gamma_\rho\}$ is eventually strictly increasing, there is by the previous lemma an $i_0 \in \{1,\dots,n\}$ such that for every $i \in \{1,\dots,n\}$ with $i \neq i_0$ we have

$$\beta_{i_0} + i_0\gamma_\rho < \beta_i + i\gamma_\rho, \quad \text{eventually.}$$

Then $v\big(f(a_\rho) - f(a)\big) = \beta_{i_0} + i_0\gamma_\rho$ eventually, in particular, $\{v\big(f(a_\rho) - f(a)\big)\}$ is eventually strictly increasing, that is, $f(a_\rho) \rightsquigarrow f(a)$. □

Corollary 4.8. *Suppose $\{a_\rho\}$ is a pc-sequence in K, and $f(x) \in K[x]$ is nonconstant. Then $\{f(a_\rho)\}$ is a pc-sequence.*

Proof. By Lemma 4.3 we have $a_\rho \rightsquigarrow a$ with a in a valued field extension. Then $f(a_\rho) \rightsquigarrow f(a)$ in this extension, and thus $\{f(a_\rho)\}$ is a pc-sequence. □

Let $\{a_\rho\}$ be a pc-sequence in K, and let $f(x) \in K[x]$ be nonconstant. Then by the two corollaries above there are two possibilities: either

$$\{v\big(f(a_\rho)\big)\} \text{ is eventually strictly increasing,} \quad (\text{equivalently, } f(a_\rho) \rightsquigarrow 0),$$

or

$$\{v\big(f(a_\rho)\big)\} \text{ is eventually constant,} \quad (\text{equivalently, } f(a_\rho) \not\rightsquigarrow 0).$$

We say that $\{a_\rho\}$ is of *algebraic type over* K if the first possibility is realized for some nonconstant $f \in K[x]$, and then such an f of least degree is called a *minimal polynomial of* $\{a_\rho\}$ *over* K.

We say that $\{a_\rho\}$ is of *transcendental type over* K if the second possibility is realized for all nonconstant $f \in K[x]$. (Note that then $\{v\big(f(a_\rho)\big)\}$ is eventually constant for any $f \in K[x]$, including the case $f \in K$.) A pc-sequence in K of transcendental type determines an essentially unique immediate extension:

Theorem 4.9. *Let $\{a_\rho\}$ be a pc-sequence in K of transcendental type over K. Then $\{a_\rho\}$ has no pseudolimit in K. The valuation v on K extends uniquely to a valuation $v : K(x)^\times \to \Gamma$ (x transcendental over K) such that*

$$v(f) = \text{eventual value of } v\big(f(a_\rho)\big)$$

for each $f \in K[x]$. With this valuation $K(x)$ is an immediate valued field extension of $(K,\Gamma;v)$ in which $a_\rho \rightsquigarrow x$.

Conversely, if $a_\rho \rightsquigarrow a$ in a valued field extension of $(K, \Gamma; v)$, then there is a valued field isomorphism $K(x) \to K(a)$ over K that sends x to a.

Proof. If $a_\rho \rightsquigarrow a$ with $a \in K$, then the polynomial $x - a$ witnesses that $\{a_\rho\}$ is of algebraic type over K. So $\{a_\rho\}$ has no pseudolimit in K.

It is clear that by defining $v(f)$ for $f \in K[x]$ as in the above display, we have a valuation on $K[x]$, and thus on $K(x)$. The value group of this valuation is still Γ, and one checks easily that $a_\rho \rightsquigarrow x$. To verify that the residue field of $K(x)$ is the residue field of K we first note that because the value groups are equal, each $a \in K(x)$ with $v(a) = 0$ has the form $a = f/g$ where $f, g \in K[x]$ with $v(f) = v(g) = 0$. So it is enough to consider a nonconstant $f \in K[x]$ with $v(f) = 0$, and find $b \in K$ with $v(f - b) > 0$. We have $0 = v(f) = v\big(f(a_\rho)\big)$ eventually, and $\{v\big(f - f(a_\rho)\big)\}$ is eventually strictly increasing, so $v\big(f - f(a_\rho)\big) > 0$, eventually. Thus $b = f(a_\rho)$ for big enough ρ will do the job.

Finally, suppose $a_\rho \rightsquigarrow a$ with a in a valued field extension whose valuation we also indicate by v for simplicity. For nonconstant $f \in K[x]$ we have $f(a_\rho) \rightsquigarrow f(a)$, and thus $v\big(f(a)\big) = v\big(f(a_\rho)\big)$ eventually; in particular, $f(a) \neq 0$ and $v(f(a)) = v(f) \in \Gamma$. Thus a is transcendental over K and the field isomorphism $K(x) \to K(a)$ over K that sends x to a is even a valued field isomorphism. □

Here is the analogue for pc-sequences of *algebraic* type over K:

Theorem 4.10. *Let $\{a_\rho\}$ be a pc-sequence in K of algebraic type over K, without pseudolimit in K, and let $\mu(x)$ be a minimal polynomial of $\{a_\rho\}$ over K. Then μ is irreducible in $K[x]$ and $\deg \mu \geq 2$. Let a be a zero of μ in an extension field of K. Then v extends uniquely to a valuation $v : K(a)^\times \to \Gamma$ such that*

$$v\big(f(a)\big) = \textit{eventual value of } v\big(f(a_\rho)\big)$$

for each nonzero polynomial $f \in K[x]$ of degree $< \deg \mu$. With this valuation $K(a)$ is an immediate valued field extension of $(K, \Gamma; v)$, and $a_\rho \rightsquigarrow a$.

Conversely, if $\mu(b) = 0$ and $a_\rho \rightsquigarrow b$ in a valued field extension of $(K, \Gamma; v)$, then there is a valued field isomorphism $K(a) \to K(b)$ over K sending a to b.

Proof. Much of the proof duplicates the proof for the case of transcendental type. A difference is in how we obtain the multiplicative law for $v : K(a)^\times \to \Gamma$ as defined above. Let $s, t \in K(a)^\times$, and write $s = f(a), t = g(a)$ with nonzero $f, g \in K[x]$ of degree $< \deg \mu$. Then $fg = q\mu + r$ with $q, r \in K[x]$ and $\deg r < \deg \mu$, so $st = r(a)$, and thus eventually

$$v(s) = v\big(f(a_\rho)\big), \quad v(t) = v\big(g(a_\rho)\big), \quad v(st) = v\big(r(a_\rho)\big).$$

Also $v(s) + v(t) = v\big(f(a_\rho)g(a_\rho)\big) = v\big(q(a_\rho)\mu(a_\rho) + r(a_\rho)\big)$, eventually. Since $\{v\big(q(a_\rho)\mu(a_\rho)\big)\}$ is either eventually strictly increasing, or eventually ∞, this forces $v(s) + v(t) = v\big(r(a_\rho)\big)$, eventually, so $v(s) + v(t) = v(st)$. □

The minimal polynomial $\mu(x)$ in this theorem, even if we assume it to be monic, is in general highly non-unique. For example, adding to its constant term an $\varepsilon \in K$ such that eventually $v(\varepsilon) > v\big(\mu(a) - \mu(a_\rho)\big)$, does not change its status as minimal polynomial of $\{a_\rho\}$. That is why we have to specify a particular minimal polynomial in this theorem to achieve the uniqueness up to isomorphism of $K(a)$.

4.2 Maximal Valued Fields

A *maximal valued field* is by definition a valued field that has no immediate proper valued field extension. If $\boldsymbol{k}$ is algebraically closed and Γ is divisible, then by Corollary 3.20 the algebraic closure of K equipped with any valuation that extends v is necessarily an immediate extension of $(K, \Gamma; v)$. Therefore:

Corollary 4.11. *If* $(K, \Gamma; v)$ *is maximal,* $\boldsymbol{k}$ *is algebraically closed and* Γ *is divisible, then* K *is algebraically closed.*

Theorems 4.9 and 4.10 in combination with Lemma 4.2 yield:

Corollary 4.12. $(K, \Gamma; v)$ *is maximal if and only if each pc-sequence in* K *has a pseudolimit in* K.

We use this equivalence to get the next result.

Corollary 4.13. *The Hahn field* $\boldsymbol{k}((t^\Gamma))$ *is maximal. If* $\boldsymbol{k}$ *is algebraically closed and* Γ *is divisible, then* $\boldsymbol{k}((t^\Gamma))$ *is algebraically closed.*

Proof. The first assertion implies the second one, so it is enough to consider an arbitrary pc-sequence $\{a_\rho\}$ in $\boldsymbol{k}((t^\Gamma))$ and show that it has a pseudolimit in $\boldsymbol{k}((t^\Gamma))$. Take an index ρ_0 such that

$$\tau > \sigma > \rho > \rho_0 \Longrightarrow v(a_\tau - a_\sigma) > v(a_\sigma - a_\rho).$$

Below we restrict ρ to be $> \rho_0$. Note that we have $\gamma_\rho \in \Gamma$ with $v(a_\sigma - a_\rho) = \gamma_\rho$ for all $\sigma > \rho$, and that $\{\gamma_\rho\}$ is strictly increasing. We split up a_ρ as follows:

$$a_\rho = \sum_{\gamma < \gamma_\rho} c_{\gamma\rho} t^\gamma + \sum_{\gamma \geq \gamma_\rho} c_{\gamma\rho} t^\gamma, \qquad (\text{all } c_{\gamma\rho} \in \boldsymbol{k}).$$

Then $c_{\gamma\rho} = c_{\gamma\sigma}$ for $\gamma < \gamma_\rho$ and $\sigma > \rho$. For any $\gamma \in \Gamma$, either $\gamma < \gamma_\rho$ for *some* ρ, or $\gamma > \gamma_\rho$ for *all* ρ. Thus we can define the series $a := \sum_\gamma c_\gamma t^\gamma \in \boldsymbol{k}((t^\Gamma))$ by setting $c_\gamma := c_{\gamma\rho}$ whenever $\gamma < \gamma_\rho$, and $c_\gamma := 0$ if $\gamma > \gamma_\rho$ for all ρ. Then $a_\rho \rightsquigarrow a$, as is easily verified. □

Under certain conditions on residue field and value group, a maximal valued field is isomorphic to a Hahn field, see Corollary 4.32. First a consequence of Krull's cardinality bound $|K| \leq |\boldsymbol{k}|^{|\Gamma|}$ (Proposition 3.6) and Zorn:

Corollary 4.14. *$(K, \Gamma; v)$ has an immediate maximal valued field extension.*

This raises of course the question whether such an extension is unique up to isomorphism. In the equicharacteristic 0 case the answer is yes (Corollary 4.29), but first we take care of a loose end in Theorem 4.10:

Lemma 4.15. *Let a in an immediate valued field extension of $(K, \Gamma; v)$ be algebraic over K, and $a \notin K$. Then there is a pc-sequence $\{a_\rho\}$ in K of algebraic type over K that has no pseudolimit in K, such that $a_\rho \rightsquigarrow a$.*

Proof. By Lemma 4.2 we have a pc-sequence $\{a_\rho\}$ in K without pseudolimit in K, such that $a_\rho \rightsquigarrow a$. Let $f(x)$ be the minimum polynomial of a over K. Then by the Taylor identity preceding Lemma 4.5 we have

$$\begin{aligned} f(a_\rho) &= f(a_\rho) - f(a) = (a_\rho - a) \cdot g(a_\rho), \quad g(x) \in K(a)[x], \text{ so} \\ v\big(f(a_\rho)\big) &= v(a_\rho - a) + v\big(g(a_\rho)\big). \end{aligned}$$

Since $\{v(a_\rho - a)\}$ is eventually strictly increasing and $\{v(g(a_\rho))\}$ is eventually strictly increasing or eventually constant, $\{v\big(f(a_\rho)\big)\}$ is eventually strictly increasing, so $\{a_\rho\}$ is of algebraic type over K. □

A valued field is *algebraically maximal* if it has no immediate proper algebraic valued field extension. (Thus *maximal* $\Longrightarrow$ *algebraically maximal*.) Obviously, every algebraically closed valued field is algebraically maximal.

Theorem 4.10 and Lemma 4.15 yield:

Corollary 4.16. *$(K, \Gamma; v)$ is algebraically maximal if and only if each pc-sequence in K of algebraic type over K has a pseudolimit in K.*

Just from the definition of *algebraically maximal* (and Zorn) it follows that $(K, \Gamma; v)$ has an immediate valued field extension that is algebraically maximal and algebraic over K. A natural question is whether such an extension is determined up to isomorphism over $(K, \Gamma; v)$. The answer is *yes* in equicharacteristic 0 and in some other cases, see Theorem 4.27 and its consequences.

First we show that algebraic maximality implies henselianity. A valued field is said to be *henselian* if its valuation ring is henselian.

Lemma 4.17. *Let $f \in \mathcal{O}[x]$ and $a \in \mathcal{O}$ be such that $v\big(f(a)\big) > 0$, $v\big(f'(a)\big) = 0$, and f has no zero in $a + \mathfrak{m}$. Then there is a pc-sequence $\{a_\rho\}$ in K such that $f(a_\rho) \rightsquigarrow 0$, and $\{a_\rho\}$ has no pseudolimit in K.*

Proof. We build such a sequence by the Newton approximation at the beginning of Sect. 2.2, starting with $a_0 = a$. Indeed, this method yields $b \in \mathcal{O}$ such that $v(b - a) = v\big(f(a)\big) > 0$ and $v\big(f(b)\big) \geq 2v\big(f(a)\big) > 0$. In particular, $a \equiv b \mod \mathfrak{m}$, hence $f'(a) \equiv f'(b) \mod \mathfrak{m}$, so $v\big(f'(b)\big) = 0$. We now take $a_1 = b$, and continue with b as new input. More precisely, let λ be an ordinal > 0 and $\{a_\rho\}$ a sequence in $a + \mathfrak{m}$ indexed by the ordinals $\rho < \lambda$ such that $a_0 = a$, and

$v(a_\sigma - a_\rho) = v\big(f(a_\rho)\big)$ and $v\big(f(a_\sigma)\big) \geq 2v\big(f(a_\rho)\big)$ whenever $\lambda > \sigma > \rho$. (For $\lambda = 2$ we have such a sequence with $a_0 = a$ and $a_1 = b$.) If $\lambda = \mu + 1$ is a successor ordinal, then we construct the next term $a_\lambda \in a + \mathfrak{m}$ by Newton approximation so that $v(a_\lambda - a_\mu) = v\big(f(a_\mu)\big)$ and $v\big(f(a_\lambda)\big) \geq 2v\big(f(a_\mu)\big)$.

Suppose now that λ is a limit ordinal. Then $\{a_\rho\}$ is clearly a pc-sequence and $f(a_\rho) \rightsquigarrow 0$. If $\{a_\rho\}$ has no pseudolimit in K, then we are done. If $\{a_\rho\}$ has a pseudolimit in K, let a_λ be such a pseudolimit. Then $f(a_\rho) \rightsquigarrow f(a_\lambda)$; since $\{v\big(f(a_\rho)\big)\}$ is strictly increasing, this yields $v\big(f(a_\lambda)\big) \geq v\big(f(a_{\rho+1})\big) \geq 2v\big(f(a_\rho)\big)$ for each index $\rho < \lambda$. It is also clear that $v(a_\lambda - a_\rho) = v(a_{\rho+1} - a_\rho) = v\big(f(a_\rho)\big)$ for each $\rho < \lambda$, in particular, $a_\lambda \in a + \mathfrak{m}$. Thus we have extended our sequence by one more term. This building process must come to an end. □

In combination with Theorem 4.10 the previous lemma yields:

Corollary 4.18. *Each algebraically maximal valued field is henselian (and thus each maximal valued field is henselian).*

For example, the Puiseux series field $\mathrm{P}(\boldsymbol{k})$ is henselian, since it is the directed union of its maximal valued subfields $\boldsymbol{k}((t^{\frac{1}{d}\mathbb{Z}}))$ $(d = 1, 2, 3, \ldots)$.
In the equicharacteristic 0 case, *algebraically maximal* is equivalent to *henselian*; see Corollary 4.22. For $(K, \Gamma; v)$ to be henselian is a condition on polynomials over its valuation ring $\mathcal{O}$. It is convenient to have an equivalent condition for polynomials over K. Let $f(x) \in K[x]$ be of degree $\leq n$, and let $a \in K$, so

$$f(a + x) \;=\; \sum_{i=0}^{n} f_{(i)}(a)x^i \;=\; f(a) + f'(a)x + \sum_{i=2}^{n} f_{(i)}(a)x^i.$$

Definition 4.19. We say f, a is in *Hensel configuration* if $f'(a) \neq 0$, and either $f(a) = 0$ or there is $\gamma \in \Gamma$ such that

$$v\big(f(a)\big) \;=\; v\big(f'(a)\big) + \gamma \;<\; v\big(f_{(i)}(a)\big) + i \cdot \gamma$$

for $2 \leq i \leq n$. (Such γ is unique: $\gamma = v\big(f(a)\big) - v\big(f'(a)\big)$.)

Suppose f, a is in Hensel configuration and $f(a) \neq 0$. Take γ as in the definition, take $c \in K$ with $vc = \gamma$ and put $g(x) := f(cx)/f(a)$ and $\alpha := a/c$. Then $g(\alpha) = 1$, $v\big(g'(\alpha)\big) = 0$, and $v\big(g_{(i)}(\alpha)\big) > 0$ for $i > 1$, as is easily verified. Thus

$$h(x) \;:=\; g(\alpha + x) \;=\; 1 + g'(\alpha)x + \sum_{i=2}^{n} g_{(i)}(\alpha)x^i$$

lies in $\mathcal{O}[x]$. The set $\{u \in \mathcal{O} : v\big(1 + g'(\alpha)u\big) > 0\}$ is a congruence class modulo $\mathfrak{m}$. For u in this set we have $v\big(h(u)\big) = v\big(g(\alpha + u)\big) > 0$,

$$h'(u) = g'(\alpha + u) = g'(\alpha) + \sum_{j=1}^{n-1} g_{(1)(j)}(\alpha)u^j,$$

and $v\big(g_{(1)(j)}(\alpha)\big) \geq v\big(g_{(1+j)}(\alpha)\big) > 0$ for $j > 0$, so $v\big(h'(u)\big) = 0$.

Suppose now that $(K, \Gamma; v)$ is henselian; this allows us to pick u as above such that $g(\alpha + u) = 0$. By Lemma 2.1 this u is unique in the sense that there is no $u' \in \mathcal{O}$ such that $g(\alpha + u') = 0$ and $u' \neq u$. Now $b := a + cu$ satisfies $f(b) = 0$ and $v(a - b) = \gamma$. Summarizing:

Lemma 4.20. *If $(K, \Gamma; v)$ is henselian and f, a is in Hensel configuration, then there is a unique $b \in K$ such that $f(b) = 0$ and $v(a - b) \geq v\big(f(a)\big) - v\big(f'(a)\big)$; this b satisfies $v(a - b) = v\big(f(a)\big) - v\big(f'(a)\big)$.*

Suppose $(K, \Gamma; v)$ is of mixed characteristic $(0, p)$. Then we say that $(K, \Gamma; v)$ is *finitely ramified* if the set $\{\gamma \in \Gamma : 0 \leq \gamma < vp\}$ is finite. For example, $\mathbb{Q}_p$ with its p-adic valuation is finitely ramified.

Proposition 4.21. *Suppose $(K, \Gamma; v)$ is of equicharacteristic 0, or finitely ramified of mixed characteristic. Let $\{a_\rho\}$ be a pc-sequence in K and let $f(x) \in K[x]$ be such that $f(a_\rho) \rightsquigarrow 0$ and $f_{(j)}(a_\rho) \not\rightsquigarrow 0$ for all $j \geq 1$. Then*

(i) *f, a_ρ is in Hensel configuration, eventually;*
(ii) *in any henselian valued field extension of $(K, \Gamma; v)$ there is a unique b such that $a_\rho \rightsquigarrow b$ and $f(b) = 0$.*

Proof. Let a be a pseudolimit of $\{a_\rho\}$ in some valued field extension of $(K, \Gamma; v)$ whose valuation we also denote by v, and put $\gamma_\rho := v(a - a_\rho)$. The proof of Proposition 4.7 yields a unique $j_0 \geq 1$ such that for each $k \geq 1$ with $k \neq j_0$,

$$v\big(f(a_\rho) - f(a)\big) \;=\; v\big(f_{(j_0)}(a)\big) + j_0\gamma_\rho \;<\; v\big(f_{(k)}(a)\big) + k \cdot \gamma_\rho, \quad \text{eventually.}$$

Now $f(a_\rho) \rightsquigarrow 0$, so for $k \geq 1$, $k \neq j_0$:

$$v\big(f(a_\rho)\big) \;=\; v\big(f_{(j_0)}(a)\big) + j_0 \cdot \gamma_\rho < v\big(f_{(k)}(a)\big) + k \cdot \gamma_\rho, \quad \text{eventually.}$$

We claim that $j_0 = 1$. Let $k > 1$; our claim that $j_0 = 1$ will follow by deriving

$$v\big(f'(a)\big) + \gamma_\rho \;<\; v\big(f_{(k)}(a)\big) + k\gamma_\rho, \quad \text{eventually.}$$

The proof of Proposition 4.7 applied to f' instead of f also yields

$$v\big(f'(a_\rho) - f'(a)\big) \;\leq\; v\big(f_{(1)(j)}(a)\big) + j \cdot \gamma_\rho, \quad \text{eventually}$$

for all $j \geq 1$. Since $v\big(f'(a_\rho)\big) = v\big(f'(a)\big)$ eventually, this yields

$$v\big(f'(a)\big) \;\leq\; v\big(f_{(1)(j)}(a)\big) + j \cdot \gamma_\rho, \quad \text{eventually}$$

for all $j \geq 1$. Using $f_{(1)(j)} = (1+j) f_{(1+j)}$ (Lemma 4.5), this gives for $j \geq 1$:

$$v\big(f'(a)\big) \leq v(1+j) + v\big(f_{(1+j)}(a)\big) + j \cdot \gamma_\rho, \quad \text{eventually.}$$

For $j = k-1$, this yields

$$v\big(f'(a)\big) \leq vk + v\big(f_{(k)}(a)\big) + (k-1) \cdot \gamma_\rho, \quad \text{eventually.}$$

The assumption on $(K, \Gamma; v)$ then yields:

$$v\big(f'(a)\big) < v\big(f_{(k)}(a)\big) + (k-1) \cdot \gamma_\rho, \quad \text{eventually, hence}$$
$$v\big(f'(a)\big) + \gamma_\rho < v\big(f_{(k)}(a)\big) + k \cdot \gamma_\rho, \quad \text{eventually.}$$

Thus $j_0 = 1$, as claimed. Since for each $k \geq 1$ we have $v\big(f_{(k)}(a_\rho)\big) = v\big(f_{(k)}(a)\big)$, eventually, the above equalities and inequalities also show that f, a_ρ is in Hensel configuration, eventually. This proves (i).

For (ii), let $(K', \Gamma'; v')$ be a henselian valued field extension of $(K, \Gamma; v)$. After deleting an initial segment of the sequence $\{a_\rho\}$ we can assume that $\{\gamma_\rho\}$ is strictly increasing, that $v(a_\sigma - a_\rho) = \gamma_\rho$ whenever $\sigma > \rho$, and that $f'(a_\rho) \neq 0$ for all ρ. Likewise, by (i) and Lemma 4.20 we can assume that for every ρ there is a unique $b_\rho \in K'$ such that $f(b_\rho) = 0$ and $v(a_\rho - b_\rho) = v\big(f(a_\rho)\big) - v\big(f'(a_\rho)\big)$. The proof of (i) shows that we can also assume that $v\big(f(a_\rho)\big) - v\big(f'(a_\rho)\big) = \gamma_\rho$ for all ρ, so $v(a_\rho - b_\rho) = \gamma_\rho$ for all ρ. The uniqueness of b_ρ shows that $b_\sigma = b_\rho$ whenever $\sigma > \rho$. Thus all b_ρ are equal to a single b, which has the desired properties. □

Corollary 4.22. *Let $(K, \Gamma; v)$ be of equicharacteristic 0, or finitely ramified of mixed characteristic. Then $(K, \Gamma; v)$ is henselian if and only if it is algebraically maximal.*

Proof. Assume $(K, \Gamma; v)$ is henselian, and let $\{a_\rho\}$ be a pc-sequence in K of algebraic type over K. Take a monic minimal polynomial $f(x)$ of $\{a_\rho\}$ over K. Then $f(a_\rho) \rightsquigarrow 0$ and $f_{(j)}(a_\rho) \not\rightsquigarrow 0$ for each $j \geq 1$, so f, a_ρ is in Hensel configuration, eventually, and thus by Lemma 4.20 we have $b \in K$ such that $f(b) = 0$. Since $f(x)$ is irreducible, this gives $f(x) = x - b$, so $a_\rho \rightsquigarrow b$. □

This leads to a partial converse to some earlier results:

Corollary 4.23. *Let $(K, \Gamma; v)$ be of equicharacteristic 0. Then K is algebraically closed if and only if $(K, \Gamma; v)$ is henselian, $\boldsymbol{k}$ is algebraically closed, and Γ is divisible.*

Puiseux series fields $\mathrm{P}(\boldsymbol{k})$ are henselian with value group $\mathbb{Q}$, hence:

Application 4.24. *Suppose $\boldsymbol{k}$ is algebraically closed of characteristic 0. Then the Puiseux series field $\mathrm{P}(\boldsymbol{k})$ is algebraically closed.*

Exercise. Let $\boldsymbol{k} \subseteq \boldsymbol{k}'$ be a field extension of finite degree n, and let $b_1, \ldots, b_n$ be a basis of $\boldsymbol{k}'$ over $\boldsymbol{k}$. Show that $\boldsymbol{k}'((t^\Gamma))$ is also of finite degree n over its subfield $\boldsymbol{k}((t^\Gamma))$ with basis $b_1, \ldots, b_n$.

Show that if $\boldsymbol{k}$ has characteristic 0 with algebraic closure $\boldsymbol{k}^{\mathrm{ac}}$, then the algebraic closure of $\boldsymbol{k}((t))$ in the algebraically closed field extension $\boldsymbol{k}^{\mathrm{ac}}((t^{\mathbb{Q}}))$ is the union of the Puiseux series fields $\mathrm{P}(\boldsymbol{k}')$ where $\boldsymbol{k}'$ ranges over the subfields of $\boldsymbol{k}^{\mathrm{ac}}$ that contain $\boldsymbol{k}$ and are of finite degree over $\boldsymbol{k}$.

4.3 Henselization

The following notion is fundamental in valuation theory and beyond.

Definition 4.25. A *henselization* of $(K, \Gamma; v)$ is a henselian valued field extension $(K^{\mathrm{h}}, \Gamma^{\mathrm{h}}; v^{\mathrm{h}})$ of $(K, \Gamma; v)$ such that any valued field embedding

$$(K, \Gamma; v) \to (K', \Gamma'; v')$$

into a henselian valued field $(K', \Gamma'; v')$ extends uniquely to an embedding

$$(K^{\mathrm{h}}, \Gamma^{\mathrm{h}}; v^{\mathrm{h}}) \to (K', \Gamma'; v').$$

As usual with such definitions, existence guarantees uniqueness: if $(K_1, \Gamma_1; v_1)$ and $(K_2, \Gamma_2; v_2)$ are henselizations of $(K, \Gamma; v)$, then the unique embedding $(K_1, \Gamma_1; v_1) \to (K_2, \Gamma_2; v_2)$ over $(K, \Gamma; v)$ is an isomorphism. Thus there is no harm in referring to *the* henselization of $(K, \Gamma; v)$ if there is one. Of course, if $(K, \Gamma; v)$ is henselian, it is its own henselization. Also, if $(K, \Gamma; v)$ has a henselization and $(K', \Gamma'; v')$ is any henselian valued field extension of $(K, \Gamma; v)$, then the *henselization of* $(K, \Gamma; v)$ *in* $(K', \Gamma'; v')$ is by definition the henselization $(K^{\mathrm{h}}, \Gamma^{\mathrm{h}}; v^{\mathrm{h}})$ of $(K, \Gamma; v)$ such that

$$(K, \Gamma; v) \subseteq (K^{\mathrm{h}}, \Gamma^{\mathrm{h}}; v^{\mathrm{h}}) \subseteq (K', \Gamma'; v').$$

Corollary 4.26. *Any henselization of* $(K, \Gamma; v)$ *is an immediate valued field extension of* $(K, \Gamma; v)$ *and algebraic over* K.

Proof. We already know that $(K, \Gamma; v)$ has an immediate algebraically maximal valued field extension that is algebraic over K. Any henselization of $(K, \Gamma; v)$ must embed over $(K, \Gamma; v)$ into such an extension. □

Each valued field has a henselization, but for the Ax–Kochen–Ershov story as we tell it here it is enough to consider the equicharacteristic 0 case and the finitely ramified mixed characteristic case. In those cases the henselization has a particularly nice characterization:

Theorem 4.27. *Let* $(K, \Gamma; v)$ *be of equicharacteristic* 0*, or finitely ramified of mixed characteristic. Let* $(K_1, \Gamma; v_1)$ *be an immediate henselian valued field extension of* $(K, \Gamma; v)$ *such that* K_1 *is algebraic over* K*. Then* $(K_1, \Gamma; v_1)$ *is a henselization of* $(K, \Gamma; v)$*.*

Proof. Call a field L with $K \subseteq L \subseteq K_1$ *nice* if each valued field embedding from $(K, \Gamma; v)$ into a henselian valued field $(K', \Gamma'; v')$ extends uniquely to an embedding from $(L, \Gamma; v_L)$ into $(K', \Gamma'; v')$, where $v_L := v_1|_{L^\times}$. Consider a nice field L such that $L \neq K_1$. With Zorn in mind, it clearly suffices to show that then there is a nice field L' with $L \subseteq L' \subseteq K_1$ and $L \neq L'$.

Lemma 4.15 gives a pc-sequence $\{a_\rho\}$ in L of algebraic type over L without pseudolimit in L. Take a minimal polynomial $\mu(x)$ of $\{a_\rho\}$ over L. Then $\mu(a_\rho) \rightsquigarrow 0$ and $\mu_{(j)}(a_\rho) \not\rightsquigarrow 0$ for all $j \geq 1$. Item (ii) of Proposition 4.21 yields $a \in K_1$ such that $a_\rho \rightsquigarrow a$ and $\mu(a) = 0$. Let a valued field embedding i from $(K, \Gamma; v)$ into a henselian valued field $(K', \Gamma'; v')$ be given, and let i_L be the unique extension of i to an embedding from $(L, \Gamma; v_L)$ into $(K', \Gamma'; v')$. To simplify notation, identify $(L, \Gamma; v_L)$ with a valued subfield of $(K', \Gamma'; v')$ via i_L. Item (ii) of Proposition 4.21 also gives a unique $b \in K'$ such that $a_\rho \rightsquigarrow b$ and $\mu(b) = 0$. Now apply Theorem 4.10 to conclude that $L(a)$ is nice. □

Since $(K, \Gamma; v)$ has an immediate algebraically maximal valued field extension that is algebraic over K, this theorem shows in particular:

Corollary 4.28. *If* $(K, \Gamma; v)$ *is of equicharacteristic* 0 *or finitely ramified of mixed characteristic, then* $(K, \Gamma; v)$ *has a henselization, and any immediate algebraically maximal valued field extension of* $(K, \Gamma; v)$ *that is algebraic over* K *is isomorphic over* $(K, \Gamma; v)$ *to this henselization by a unique isomorphism.*

One can define and construct for any local ring R its henselization R^{h}: this is a henselian local ring extension of R whose maximal ideal contains the maximal ideal of R. For more on this, see for example [38]. What we defined to be the henselization of a valued field is a bit of a misnomer: this valued field henselization is actually the fraction field of the henselization of its valuation ring in the "local ring" sense.

4.4 Uniqueness of Immediate Maximal Extensions

The results above have some very nice consequences:

Corollary 4.29. *Let* $(K, \Gamma; v)$ *be of equicharacteristic* 0*, or finitely ramified of mixed characteristic. Then any two immediate maximal valued field extensions of* $(K, \Gamma; v)$ *are isomorphic over* $(K, \Gamma; v)$*.*

Proof. Let $(K_1, \Gamma; v_1)$ and $(K_2, \Gamma; v_2)$ be immediate maximal valued field extensions of $(K, \Gamma; v)$. Below, each subfield of K_i ($i = 1, 2$) is viewed as valued field by taking as valuation the restriction of v_i to the subfield. Consider fields L_1, L_2 with

$K \subseteq L_1 \subseteq K_1$ and $K \subseteq L_2 \subseteq K_2$, and suppose we have a valued field isomorphism $L_1 \cong L_2$. Note that $L_1 = K_1$ iff $L_2 = K_2$.

Suppose $L_1 \neq K_1$ and $L_2 \neq K_2$. It suffices to show that then we can extend the isomorphism $L_1 \cong L_2$ to a valued field isomorphism $L'_1 \cong L'_2$ where L'_1 and L'_2 are fields with $L_i \subseteq L'_i \subseteq K_i$ and $L_i \neq L'_i$ for $i = 1, 2$. If L_1 is not henselian, then we can take for L'_i the henselization of L_i in K_i, by Theorem 4.27. Suppose L_1 is henselian. Take $b \in K_1 \setminus L_1$, and take a pc-sequence $\{a_\rho\}$ in L_1 such that $a_\rho \rightsquigarrow b$ and $\{a_\rho\}$ has no pseudolimit in L_1. Since L_1 is algebraically maximal by Corollary 4.22, $\{a_\rho\}$ is of transcendental type over L_1. The image of $\{a_\rho\}$ under our isomorphism $L_1 \cong L_2$ has a pseudolimit $c \in K_2$, and then Theorem 4.9 allows us to extend this isomorphism to an isomorphism $L_1(b) \cong L_2(c)$. □

The next variant is just what we need for the proof of the Ax-Kochen-Ershov theorem in the next section.

Lemma 4.30. *Let $(K, \Gamma; v)$ be of equicharacteristic 0, or finitely ramified of mixed characteristic, and let $(K^*, \Gamma; v^*)$ be an immediate maximal valued field extension of $(K, \Gamma; v)$. Then we can embed $(K^*, \Gamma; v^*)$ over $(K, \Gamma; v)$ into any $|\Gamma|^+$-saturated henselian valued field extension $(K', \Gamma'; v')$ of $(K, \Gamma; v)$.*

Proof. Exercise. □

Are there useful conditions on $\boldsymbol{k}$ and Γ implying that if $(K, \Gamma; v)$ is maximal, then $(K, \Gamma; v)$ is isomorphic to the Hahn field $\boldsymbol{k}((t^\Gamma))$? The answer is yes, and here liftings and cross-sections come into play. A *lifting* of the residue field of $(K, \Gamma; v)$ is a field embedding $i : \boldsymbol{k} \to K$ such that $\overline{i(a)} = a$ for all $a \in \boldsymbol{k}$. A *cross-section* of the valuation $v : K^\times \to \Gamma$ is a group morphism $s : \Gamma \to K^\times$ such that $v(s(\gamma)) = \gamma$ for all $\gamma \in \Gamma$. For example, the valuation of the Hahn field $\boldsymbol{k}((t^\Gamma))$ has the cross-section $\gamma \mapsto t^\gamma$. Also, if $\Gamma = \mathbb{Z}$ and $v(\pi) = 1$, $\pi \in K$, then $k \mapsto \pi^k : \mathbb{Z} \to K^\times$ is a cross-section; thus $\mathbb{Q}[p]$ has cross section $k \mapsto p^k$.

To see how a lifting of the residue field and a cross-section of the valuation can help in answering the question above, we first single out the valued subfield $\boldsymbol{k}(t^\Gamma)$ of $\boldsymbol{k}((t^\Gamma))$: its underlying field is the subfield of $\boldsymbol{k}((t^\Gamma))$ generated over $\boldsymbol{k}$ by the multiplicative group t^Γ. Note that $\boldsymbol{k}((t^\Gamma))$ is an immediate extension of $\boldsymbol{k}(t^\Gamma)$. Suppose now that we have both a lifting $i : \boldsymbol{k} \to K$ of the residue field of $(K, \Gamma; v)$ and a cross-section $s : \Gamma \to K^\times$ of v. Jointly i and s yield a unique valued field embedding from $\boldsymbol{k}(t^\Gamma)$ into $(K, \Gamma; v)$ that extends i and maps t^γ to $s(\gamma)$ for each $\gamma \in \Gamma$. (Exercise.) It is clear that $(K, \Gamma; v)$ is an immediate extension of the image of this embedding. Making now the additional assumption that $(K, \Gamma; v)$ is maximal and char$(\boldsymbol{k}) = 0$, we conclude from Corollary 4.29 that $(K, \Gamma; v)$ is isomorphic to the Hahn field $\boldsymbol{k}((t^\Gamma))$ under an isomorphism that maps $i(a)$ to a for each $a \in \boldsymbol{k}$ and $s(\gamma)$ to t^γ for each $\gamma \in \Gamma$.

Lemma 4.31. *Suppose $(K, \Gamma; v)$ is henselian, char$(\boldsymbol{k}) = 0$ and the abelian groups $\boldsymbol{k}^\times$ and Γ are divisible. Then $K^\times$ is divisible, and the valuation v has a cross-section.*

Proof. Let $a \in K^\times$, and $n \geq 1$; we shall find $b \in K$ so that $a = b^n$. First take $d \in K$ with $n \cdot v(d) = v(a)$, so $a = ud^n$ with $vu = 0$. Now the residue class $\bar{u}$ is an nth power in $\boldsymbol{k}^\times$, so $u = c^n$ with $c \in K$ by an exercise at the end of Sect. 2.1. Hence $a = (cd)^n$. The second statement follows from the first: As $K^\times$ is divisible, so is $U(\mathcal{O})$ and thus the exact sequence

$$1 \to U(\mathcal{O}) \hookrightarrow K^\times \xrightarrow{v} \Gamma \to 0$$

splits. Therefore v has a cross-section. □

The first two assumptions in this lemma imply also that there is a lifting of the residue field of $(K, \Gamma; v)$, by Theorem 2.9. Therefore:

Corollary 4.32. *Suppose $(K, \Gamma; v)$ is maximal, char$(\boldsymbol{k}) = 0$ and the abelian groups $\boldsymbol{k}^\times$ and Γ are divisible. Then $(K, \Gamma; v)$ is isomorphic to the Hahn field $\boldsymbol{k}((t^\Gamma))$ via an isomorphism that induces the identity on $\boldsymbol{k}$ and on Γ.*

Corollary 4.33. *Suppose $(K, \Gamma; v)$ is maximal, char$(\boldsymbol{k}) = 0$, $\boldsymbol{k}$ is algebraically closed, and Γ is divisible. Then $(K, \Gamma; v)$ is isomorphic to the Hahn field $\boldsymbol{k}((t^\Gamma))$ via an isomorphism that induces the identity on $\boldsymbol{k}$ and on Γ.*

Another situation where a maximal valued field is isomorphic to a Hahn field is as follows. If $(K, \Gamma; v)$ is maximal, char$(\boldsymbol{k}) = 0$, and Γ is free as an abelian group, then v has a cross-section and so $(K, \Gamma; v)$ is isomorphic to the Hahn field $\boldsymbol{k}((t^\Gamma))$ via an isomorphism that induces the identity on $\boldsymbol{k}$ and on Γ.

5 The Theorem of Ax–Kochen and Ershov

We prove here the original AKE-theorem, where AKE abbreviates *Ax–Kochen and Ershov* (really two groups who proved the theorem independently, one from the USA: Ax and Kochen, the other from Russia: Ershov). Later in this chapter we prove a stronger version, the Equivalence Theorem, which allows us to deal with definable sets. All this is restricted to the equicharacteristic zero case, though with consequences beyond this, as we saw. In Chap. 7 we consider the case of unramified mixed characteristic with perfect residue field. Recall from the Introduction that we use the symbols $\equiv$ and $\preccurlyeq$ for the relations of elementary equivalence and being an elementary submodel.

AKE-Theorem 5.1. *Let (K, A) and (K', A') be two henselian valued fields with residue fields $\boldsymbol{k}$ and $\boldsymbol{k}'$, and value groups Γ and Γ'(viewed as ordered abelian groups). Suppose that char$(\boldsymbol{k}) = 0$. Then*

$$(K, A) \equiv (K', A') \iff \boldsymbol{k} \equiv \boldsymbol{k}' \text{ and } \Gamma \equiv \Gamma'.$$

The forward direction $\Rightarrow$ should be clear (and does not need the henselian or residue characteristic zero assumption). It will be convenient to prove first a variant of theorem where we have also a lifting of the residue field and a cross-section of the valuation as part of the structure. So we begin with some additional facts on cross-sections.

5.1 Existence of Cross-Sections in Elementary Extensions

In this section we fix a valued field $(K, \Gamma; v)$. We shall write $\times$-*section* to abbreviate the word *cross-section*.

Definition 5.2. A *partial* $\times$-*section* of v is a group morphism $s : \Delta \longrightarrow K^{\times}$ from a subgroup Δ of Γ back into $K^{\times}$ such that $v(s(\delta)) = \delta$ for all $\delta \in \Delta$.

A subgroup Δ of Γ is said to be *pure in* Γ if Γ/Δ is torsion free, that is, whenever $n\gamma \in \Delta$ with $n \geq 1$ and $\gamma \in \Gamma$, then $\gamma \in \Delta$.

Lemma 5.3. *Let $s : \Delta \longrightarrow K^{\times}$ be a partial $\times$-section of v such that Δ is pure in Γ. Then there is an elementary extension $(K_1, \Gamma_1; v_1)$ of $(K, \Gamma; v)$ with a partial $\times$-section $s_1 : \Gamma \longrightarrow K_1^{\times}$ of v_1.*

Proof. Let Δ' be a subgroup of Γ that contains Δ such that Δ'/Δ is finitely generated as an abelian group. Since Δ'/Δ is torsion-free, it is free as abelian group, so we can take nonzero $\gamma_1, \dots, \gamma_k \in \Delta'$ such that

$$\Delta' = \Delta \oplus \mathbb{Z}\gamma_1 \oplus \cdots \oplus \mathbb{Z}\gamma_k \quad \text{(internal direct sum of subgroups of } \Gamma\text{)}.$$

Take $a_1, \dots, a_k \in K^{\times}$ such that $v(a_1) = \gamma_1, \dots, v(a_k) = \gamma_k$. Extend s to a group morphism $s' : \Delta' \longrightarrow K^{\times}$ by $s'(\gamma_i) = a_i$, $i = 1, \dots, k$. It is easy to check that s' is a partial $\times$-section of v.

Since Δ' was arbitrary, a compactness argument gives an elementary extension $(K_1, \Gamma_1; v_1)$ of $(K, \Gamma; v)$ with a partial $\times$-section $s_1 : \Gamma \longrightarrow K_1^{\times}$ of v_1. □

Since $\Gamma \preccurlyeq \Gamma_1$ as ordered abelian groups, Γ is pure in Γ_1, so the purity assumption on s is inherited by s_1. Thus we can *iterate* the lemma:

Proposition 5.4. *Let $s : \Delta \longrightarrow K^{\times}$ be a partial $\times$-section of v such that Δ is pure in Γ. Then there is an elementary extension $(K', \Gamma'; v')$ of $(K, \Gamma; v)$ with a $\times$-section $s' : \Gamma' \longrightarrow K'^{\times}$ of v' that extends s.*

Proof. Use the previous lemma to build an elementary chain

$$(K, \Gamma; v) = (K_0, \Gamma_0; v_0) \preccurlyeq (K_1, \Gamma_1; v_1) \preccurlyeq (K_2, \Gamma_2; v_2) \preccurlyeq \dots$$

with a partial $\times$-section $s_n : \Delta_n \longrightarrow K_n^{\times}$ of v_n for each n, such that $\Delta_0 = \Delta$, $\Delta_{n+1} = \Gamma_n$ and s_{n+1} extends s_n, for each n. Then $(K', \Gamma'; v') = \bigcup_n (K_n, \Gamma_n; v_n)$ works with $s' = \bigcup_n s_n$ as $\times$-section. □

By taking $\Delta = \{0\}$ in this proposition we get:

Corollary 5.5. *$(K, \Gamma; v)$ has an elementary extension with a cross-section.*

Exercise. Suppose K is algebraically closed and $s : \Delta \to K^\times$ is a partial $\times$-section of v. Show that s extends to a $\times$-section of v.

5.2 Extending the Value Group

We need a few more basic results on how the value group can change when we extend a valued field. Let $(K, \Gamma; v)$ be a valued field, and let

$$\mathbb{Q}\Gamma = \{\gamma/n : \gamma \in \Gamma, n > 0\}$$

be the divisible hull of Γ.

Lemma 5.6. *Let p be a prime number, and x an element in a field extension of K such that $x^p = a \in K^\times$ but $v(a) \notin p\Gamma$. Then $X^p - a$ is the minimum polynomial of x over K, and v extends uniquely to a valuation $w : K(x)^\times \to \Delta$ with $\Delta \subseteq \mathbb{Q}\Gamma$ (as ordered groups). Moreover, $\boldsymbol{k}_w = \boldsymbol{k}_v$, and $[\Delta : \Gamma] = p$, with*

$$\Delta = \bigcup_{i=0}^{p-1} (\Gamma + iw(x)) \qquad (\textit{disjoint union}).$$

Proof. Let $w : K(x)^\times \to \Delta$ with $\Delta \subseteq \mathbb{Q}\Gamma$ be a valuation extending v. (By earlier results we know that there is such an extension.) Since $v(a) \notin p\Gamma$, the elements

$$w(x^0) = 0,\ w(x^1) = \frac{v(a)}{p},\ \ldots,\ w(x^{p-1}) = \frac{(p-1)v(a)}{p}$$

of Δ are in distinct cosets of Γ, so $1, x, \ldots, x^{p-1}$ are K-linearly independent, and thus $X^p - a$ is the minimum polynomial of x over K. Also, for an arbitrary nonzero element $b = b_0 + b_1 x + \cdots + b_{p-1}x^{p-1}$ of $K(x)$ with $b_0, \ldots, b_{p-1} \in K$ not all zero, we have

$$w(b) = \min\{v(b_i) + \frac{iv(a)}{p} : 0 \le i \le p-1\},$$

showing the uniqueness of w. This also proves the claims made by the lemma about Δ. Since $p = [K(x) : K] \ge [\Delta : \Gamma] \cdot [\boldsymbol{k}_w : \boldsymbol{k}_v]$, this yields $[\boldsymbol{k}_w : \boldsymbol{k}_v] = 1$, that is, $\boldsymbol{k}_w = \boldsymbol{k}_v$. □

Lemma 5.7. *Let $(K', \Gamma'; v')$ be a valued field extension of $(K, \Gamma; v)$, let $x \in K'$, and put $v'(x - K) := \{v'(x - a) : a \in K\}$. If $v'(x - K)$ has no largest element, then $v'(x - K) \subseteq \Gamma$. If $v'(x - K)$ has a largest element $v'(x - a)$, $a \in K$, then $v'(x - K) \subseteq \Gamma \cup \{v'(x - a)\}$.*

Proof. For $b, c \in K$ we have: $\quad v'(x-b) < v'(x-c) \implies v'(x-b) = v(b-c)$. $\square$

Corollary 5.8. *Let $(K, \Gamma; v)$ be algebraically maximal and have the valued field extension $(K', \Gamma'; v')$, and let $x \in K' \setminus K$. Then one of the following holds:*

(i) *x is a pseudolimit of a pc-sequence in K of transcendental type over K;*
(ii) *there exists $a \in K$ such that $v'(x-a) \notin \Gamma$;*
(iii) *there exist $a, b \in K$, $b \neq 0$, such that $v'(x-a) = v(b)$ and the residue class of $(x-a)/b$ does not lie in $\boldsymbol{k}$.*

Proof. Suppose $v'(x-K)$ has no largest element. As at the end of the proof of Lemma 4.2, this yields a pc-sequence in K that pseudoconverges to x but has no pseudolimit in K. By the algebraic maximality of $(K, \Gamma; v)$ this pc-sequence must be of transcendental type over K.

Suppose next that $v'(x-K)$ has a largest element $v'(x-a)$, $a \in K$, and that $v'(x-a) \in \Gamma$. Then $v'(x-a) = v(b)$ with $b \in K$, $b \neq 0$. If the residue class of $(x-a)/b$ were in $\boldsymbol{k}$, then $(x-a)/b = u + \epsilon$ with $u \in K$, $v(u) = 0$, and $v'(\epsilon) > 0$, so $v'\big(x - (a+bu)\big) > v(b) = v'(x-a)$, a contradiction. Thus the residue class of $(x-a)/b$ cannot lie in $\boldsymbol{k}$. $\square$

Corollary 5.9. *Let $(K', \Gamma'; v')$ be a valued field extension of $(K, \Gamma; v)$, and let $x \in K'$, $x \notin K$. If $\Gamma = v(K^\times)$ is countable, so is $v'\big(K(x)^\times\big)$.*

Proof. By passing to an algebraic closure of K' and extending the valuation v' to this algebraic closure we can assume that K' is algebraically closed. Replacing K by its algebraic closure inside K' we further reduce to the case that K is algebraically closed. Suppose first that $v'(x-K) \subseteq \Gamma$. Then, given any non-zero element $f(x) \in K[x]$ we have a factorization $f(x) = c(x-a_1)\cdots(x-a_m)$ with $c, a_1, \ldots, a_m \in K$, $c \neq 0$, so

$$v'\big(f(x)\big) = v(c) + v'(x-a_1) + \cdots + v'(x-a_m) \in \Gamma,$$

hence $v'(K(x)^\times) \subseteq \Gamma$. Next, suppose that $a \in K$ is such that $v'(x-a) \notin \Gamma$. Then $v'(x-K) \subseteq \Gamma \cup \{v'(x-a)\}$ by Lemma 5.7, and thus by the same argument $v'(K(x)^\times) \subseteq \Gamma + \mathbb{Z} \cdot v'(x-a)$. $\square$

5.3 AKE-Theorem with Lifting and Cross-Section

It will be convenient to consider 3-sorted structures

$$\mathcal{K} = (K, \boldsymbol{k}, \Gamma; \pi, v)$$

where K and $\boldsymbol{k}$ are fields, Γ is an ordered abelian group, $v : K^\times \to \Gamma$ is a valuation, and $\pi : \mathcal{O}_v \to \boldsymbol{k}$ is a surjective ring morphism. Note that then π has kernel $\mathfrak{m}_v$, and thus induces a field isomorphism $\boldsymbol{k}_v \cong \boldsymbol{k}$ between the residue field $\boldsymbol{k}_v$ and $\boldsymbol{k}$ such that the diagram

$$\begin{array}{ccc} & \mathcal{O}_v & \\ \swarrow & & \searrow^{\pi} \\ \mathbf{k}_v & \xrightarrow{\ \sim\ } & \mathbf{k} \end{array}$$

commutes. We call K the underlying field of $\mathcal{K}$, Γ its value group, and $\boldsymbol{k}$ its residue field (even though the latter is only isomorphic to the residue field $\boldsymbol{k}_v$ of $\mathcal{O}_v$). We shall refer to $\mathcal{K}$ as a valued field, since it represents a way to construe the (one-sorted) valued field $(K, \mathcal{O}_v)$ as a three-sorted model-theoretic structure where the residue field and the value group are more explicitly present.

For technical reasons it is useful to expand our valued fields with a lifting of the residue field and with a $\times$-section, so we consider 3-sorted structures

$$\mathcal{K} = (K, \boldsymbol{k}, \Gamma; \pi, v, i, s)$$

where $(K, \boldsymbol{k}, \Gamma; \pi, v)$ is a valued field as before, $i : \boldsymbol{k} \to \mathcal{O}_v$ is a ring morphism such that $\pi(ir) = r$ for all $r \in \boldsymbol{k}$, and $s : \Gamma \to K^\times$ is a $\times$-section of v. Such a structure $\mathcal{K}$ will be called an *lc-valued field*, with the letters l and c in "lc" as reminders of the lifting and the cross-section.

Remark 5.10. Each henselian valued field (K, A) of equicharacteristic 0 has an elementary extension (K', A') that can be expanded to an lc-valued field

$$(K', \boldsymbol{k}', \Gamma'; \pi', v', i', s')$$

where $v' := v_{A'} : K'^\times \to \Gamma' := \Gamma_{A'}$ and $\pi' : A' \to \boldsymbol{k}' := \boldsymbol{k}_{A'}$ is the residue map.

This follows from our lifting theorem for henselian local rings with residue field of characteristic 0, and Corollary 5.5.

In the rest of this section $\mathcal{K}$ and $\mathcal{K}'$ are lc-valued fields,

$$\mathcal{K} = (K, \boldsymbol{k}, \Gamma; \pi, v, i, s), \qquad \mathcal{K}' = (K', \boldsymbol{k}', \Gamma'; \pi', v', i', s').$$

An *embedding* $\mathcal{K} \to \mathcal{K}'$ is a triple (f, f_r, f_v) with field embeddings $f : K \to K'$ and $f_r : \boldsymbol{k} \to \boldsymbol{k}'$, and an ordered group embedding $f_v : \Gamma \to \Gamma'$, such that

$$f_r\big(\pi(a)\big) = \pi'\big(f(a)\big) \text{ for } a \in \mathcal{O}_v, \quad f\big(i(r)\big) = i'\big(f_r(r)\big) \text{ for } r \in \boldsymbol{k},$$
$$f_v\big(v(a)\big) = v'\big(f(a)\big) \text{ for } a \in K^\times, \quad f\big(s(\gamma)\big) = s'\big(f_v(\gamma)\big) \text{ for } \gamma \in \Gamma.$$

If $K \subseteq K'$, $\boldsymbol{k} \subseteq \boldsymbol{k}'$, and $\Gamma \subseteq \Gamma'$ (as sets), and the corresponding inclusion maps $f : K \to K'$, $f_r : \boldsymbol{k} \to \boldsymbol{k}'$, and $f_v : \Gamma \to \Gamma'$ yield an embedding

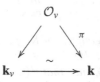

$$(f, f_r, f_v) : \mathcal{K} \to \mathcal{K}',$$

then we call $\mathcal{K}$ an *lc-valued subfield* of $\mathcal{K}'$; notation: $\mathcal{K} \subseteq \mathcal{K}'$.

Consider an embedding $(f, f_r, f_v) : \mathcal{K} \to \mathcal{K}'$. Then f_r and f_v are completely determined by f. If f is a bijection, then f_r and f_v are bijections, and we call (f, f_r, f_v) an *isomorphism from* $\mathcal{K}$ *onto* $\mathcal{K}'$; note that then $(f^{-1}, f_r^{-1}, f_v^{-1})$ is an isomorphism from $\mathcal{K}'$ onto $\mathcal{K}$.

In view of Remark 5.10 the AKE-theorem 5.1 is a consequence of the following stronger result.

Theorem 5.11. *Suppose* $\mathcal{K}$ *and* $\mathcal{K}'$ *are henselian of equicharacteristic* 0. *Then*

$$\mathcal{K} \equiv \mathcal{K}' \iff \boldsymbol{k} \equiv \boldsymbol{k}' \text{ and } \Gamma \equiv \Gamma'.$$

Proof. Since the forward direction $\Rightarrow$ is clear, we assume below that $\boldsymbol{k} \equiv \boldsymbol{k}'$ and $\Gamma \equiv \Gamma'$, and our task is to derive $\mathcal{K} \equiv \mathcal{K}'$. The case that $\Gamma = \{0\}$ is trivial, so we assume from now on that $\Gamma \neq \{0\}$ and thus $\Gamma' \neq \{0\}$.

More drastically, we shall assume the *Continuum Hypothesis* CH, in order to make our job a little easier. Theorem 5.11 belongs to a class of mathematical statements with the property that any proof of a statement in that class from ZFC + CH can be converted into a, possibly much longer, proof of the same statement from ZFC alone. We suggest to the reader to take this fact on faith for the moment. After the proof of Corollary 5.15 we return to this issue. Later in this chapter we obtain by similar methods a stronger result, the Equivalence Theorem, but without using this fact about ZFC.

After replacing $\mathcal{K}$ and $\mathcal{K}'$ by elementarily equivalent lc-valued fields, we may and shall assume that $\mathcal{K}$ and $\mathcal{K}'$ are saturated of cardinality $\aleph_1$. This is exactly where CH gets used. In particular, $\boldsymbol{k}$ and $\boldsymbol{k}'$ are saturated of cardinality $\aleph_1$, so we have an isomorphism $f_r : \boldsymbol{k} \cong \boldsymbol{k}'$ of fields. Likewise, Γ and Γ' are saturated of cardinality $\aleph_1$, so we have an isomorphism $f_v : \Gamma \cong \Gamma'$ of ordered abelian groups. To obtain Theorem 5.11 it suffices to establish:

Claim. There is a field isomorphism $f : K \cong K'$ which together with f_r and f_v yields an isomorphism $(f, f_r, f_v) : \mathcal{K} \cong \mathcal{K}'$.

We shall obtain such f by a back-and-forth construction. A *good subfield of* K is by definition a subfield E of K such that $i\boldsymbol{k} \subseteq E$, $s\big(v(E^\times)\big) \subseteq E$, and $v(E^\times)$ is countable. (Note that then E is the underlying field of an lc-valued subfield of $\mathcal{K}$.) Likewise, we define the notion *good subfield of* K'. A *good map* is an isomorphism $e : E \cong E'$ between good subfields E and E' of K and K', such that $e(E \cap \mathcal{O}_v) = E' \cap \mathcal{O}_{v'}$ and

$$f_r\big(\pi(a)\big) = \pi'\big(e(a)\big) \text{ for } a \in E \cap \mathcal{O}_v, \qquad e\big(i(r)\big) = i'\big(f_r(r)\big) \text{ for } r \in \boldsymbol{k},$$

$$f_v\big(v(a)\big) = v'\big(e(a)\big) \text{ for } a \in E^\times, \qquad e\big(s(\gamma)\big) = s'\big(f_v(\gamma)\big) \text{ for } \gamma \in v(E^\times).$$

Note that $i(\boldsymbol{k})$ and $i'(\boldsymbol{k}')$ are good subfields of K and K' and that

$$i(r) \mapsto i'\big(f_{\mathrm{r}}(r)\big) \; : \; i(\boldsymbol{k}) \to i'(\boldsymbol{k}') \quad (r \in \boldsymbol{k})$$

is a good map. Thus the claim above is a consequence of:
Back-and-Forth. The set of good maps has the back-and-forth property.

To prove *Back-and-Forth*, consider a good map $e : E \cong E'$. We view E and E' as valued subfields of $(K, \mathcal{O}_v)$ and $(K', \mathcal{O}_{v'},)$. Thus the residue field of E equals that of $(K, \mathcal{O}_v)$ (and likewise with E'), and $e : E \to E'$ is a valued field isomorphism. Before addressing *Back-and-Forth* we indicate two basic ways in which e can be extended.

(1) *Adjunction of an element of* Γ *that has prime torsion modulo* $v(E^\times)$. Let $\delta \in \Gamma$ be such that $\delta \notin v(E^\times)$ but $p\delta \in v(E^\times)$ where p is a prime number. Take $x = s(\delta)$, so $x^p = s(p\delta) \in E^\times$. Then by Lemma 5.6,

$$v\big(E(x)^\times\big) \;=\; v(E^\times) + \mathbb{Z}\delta \;=\; \bigcup_{i=0}^{p-1} \big(vE^\times + i\delta\big) \qquad \text{(disjoint union).}$$

Since $s\big(v(E^\times) + \mathbb{Z}\delta\big) = s\big(v(E^\times)\big) \cdot x^{\mathbb{Z}} \subseteq E(x)$, it follows that $E(x)$ is a good subfield of K. Put $x' := s'\big(f_v(\delta)\big) \in K'$. Then $E'(x')$ is likewise a good subfield of K', and Lemma 5.6 shows we have a good map $E(x) \to E'(x')$ that extends e and maps x to x'.

(2) *Adjunction of an element of* Γ *that has no torsion modulo* $v(E^\times)$. Let $\gamma \in \Gamma$ be such that $n\gamma \notin v(E^\times)$ for all $n \geq 1$. Take $x = s(\gamma)$. Then x is transcendental over E and $v\big(E(x)^\times\big) = v(E^\times) \oplus \mathbb{Z}\gamma$, by Corollary 3.20 and Lemma 3.23. Since

$$s\big(v(E^\times) \oplus \mathbb{Z}\gamma\big) \;=\; s\big(v(E^\times)\big) \cdot x^{\mathbb{Z}} \;\subseteq\; E(x),$$

$E(x)$ is a good subfield of K. Put $x' := s'\big(f_v(\gamma)\big) \in K'$. Then $E'(x')$ is likewise a good subfield of K' with x' transcendental over E', so Lemma 3.23 yields a good map $E(x) \to E'(x')$ that extends e and maps x to x'.

Let Δ be any countable subgroup of Γ that contains $v(E^\times)$. We now show how to extend our good map e to a good map whose domain has value group Δ. Clearly Δ is the union $\bigcup_{k=0}^{\infty} \Delta_k$ of an increasing sequence

$$v(E^\times) = \Delta_0 \subseteq \Delta_1 \subseteq \Delta_2 \subseteq \cdots$$

of subgroups Δ_k of Δ such that for each $k \in \mathbb{N}$,

(i) either $\Delta_{k+1} = \Delta_k$,
(ii) or $\Delta_{k+1} = \Delta_k + \mathbb{Z}\delta_k$ with $\delta_k \in \Delta \setminus \Delta_k$ and $p\delta_k \in \Delta_k$ for some prime p,
(iii) or $\Delta_{k+1} = \Delta_k + \mathbb{Z}\delta_k$ and $n\delta_k \notin \Delta_k$ for all $n \geq 1$.

Then by repeated application of (1) and (2) and taking a union we see that $E_\Delta := E\big(s(\delta_k) : k = 0, 1, 2, \ldots\big)$ is a good subfield of K with $v(E_\Delta^\times) = \Delta$, and that e extends (uniquely) to a good map with domain E_Δ.
In addition to (1) and (2), Lemma 4.30 provides a third way to extend good maps. We now combine these three ways of extending e to establish *Back-and-Forth*. Consider an element $x \in K$. We wish to find a good map extending e that contains x in its domain. By Corollary 5.9 the group $v(E(x)^\times)$ is countable, so by the construction above we can extend e to a good map e_1 with domain E_1 such that $v(E_1^\times) = v(E(x)^\times)$. In the same way we extend e_1 to a good map e_2 with domain E_2 such that $v(E_2^\times) = v(E_1(x)^\times)$. By iterating this construction and taking a union, and renaming this union as E, we reduce to the case that $E(x)$ as a valued subfield of $(K, \Gamma; v)$ is an immediate extension of E, since the residue field of E equals that of $(K, \mathcal{O}_v)$. This valued field $E(x)$ has an immediate maximal valued field extension, which by Lemma 4.30 can be taken inside $(K, \Gamma; v)$; let $E_\bullet$ be the underlying field of such an immediate maximal extension of $E(x)$ inside $(K, \Gamma; v)$. Since $E_\bullet$ as valued field is also an immediate maximal extension of E, we can use Lemma 4.30 to extend e to a valued field isomorphism $e_\bullet$ from $E_\bullet$ onto a valued subfield $E'_\bullet$ of $(K', \Gamma'; v')$. Then $e_\bullet$ is a good map with x in its domain.

This takes care of the *Forth* part of *Back-and-Forth*. The *Back* part is done likewise by interchanging the role of E and E'. □

Note that Theorem 1.1 from the Introduction is a special case of Theorem 5.1, since $\boldsymbol{k}[[t]]$ is the valuation ring of the Laurent series field $\boldsymbol{k}((t))$ (where $\boldsymbol{k}$ is a field). Another important special case says that any henselian $\mathcal{K} = (K, \boldsymbol{k}, \Gamma; \cdots)$ of equicharacteristic zero is elementarily equivalent to the Hahn field $\boldsymbol{k}((t^\Gamma))$. More precisely, construe each Hahn field $\boldsymbol{k}((t^\Gamma))$ as the lc-valued field

$$(\boldsymbol{k}((t^\Gamma)), \boldsymbol{k}, \Gamma; \pi, v, i, s)$$

where $v : \boldsymbol{k}((t^\Gamma))^\times \to \Gamma$ is the usual valuation, $\pi : \mathcal{O}_v \to \boldsymbol{k}$ assigns to each $a = \sum a_\gamma t^\gamma \in \mathcal{O}_v$ its constant term a_0, and where $i : \boldsymbol{k} \to \mathcal{O}_v$ is the natural inclusion map and $s : \Gamma \to \boldsymbol{k}((t^\Gamma))^\times$ is given by $s(\gamma) = t^\gamma$.

Corollary 5.12. *If $\mathcal{K}$ is henselian and of equicharacteristic zero, then*

$$\mathcal{K} \equiv \boldsymbol{k}((t^\Gamma)) \qquad \text{(as lc-valued fields).}$$

Elementary classifications like Theorem 5.11 often imply a seemingly stronger result to the effect that each sentence in a certain language is equivalent in a suitable theory to a sentence of a special form. We now turn our attention to this aspect. It involves the following two elementary facts on boolean algebras.

Lemma 5.13. *Let B be a boolean algebra and A and A' boolean subalgebras of B. Then the boolean subalgebra of B generated by $A \cup A'$ equals*

$$\{(a_1 \wedge a_1') \vee \cdots \vee (a_k \wedge a_k') : k \in \mathbb{N},\ a_1, \ldots, a_k \in A,\ a_1', \ldots, a_k' \in A'\}.$$

We leave the proof as a routine exercise. The next lemma is often useful.

Lemma 5.14. *Let B be a boolean algebra and $S(B)$ its Stone space of ultrafilters. Let Ψ be a subset of B and suppose the map $F \mapsto F \cap \Psi : S(B) \to \mathcal{P}(\Psi)$ is injective. Then Ψ generates the boolean algebra B.*

Proof. Let B_Ψ be the boolean subalgebra of B generated by Ψ. The inclusion $B_\Psi \hookrightarrow B$ induces the *surjective* map $F \mapsto F \cap B_\Psi : S(B) \twoheadrightarrow S(B_\Psi)$. This map is also *injective*: if $F_1, F_2 \in S(B)$ and $F_1 \cap B_\Psi = F_2 \cap B_\Psi$, then $F_1 \cap \Psi = F_2 \cap \Psi$, hence $F_1 = F_2$ by the hypothesis of the lemma. The bijectivity of $F \mapsto F \cap B_\Psi : S(B) \to S(B_\Psi)$ yields $B = B_\Psi$ by the Stone representation theorem. □

Let L be the 3-sorted language of lc-valued fields with 1-sorted sublanguages L_r for residue fields and L_v for value groups. Let T be the L-theory of henselian lc-valued fields with residue field of characteristic 0. Then:

Corollary 5.15. *For each L-sentence σ there are sentences $\sigma_r^1, \dots, \sigma_r^k$ in L_r and sentences $\sigma_v^1, \dots, \sigma_v^k$ in the language L_v such that*

$$T \vdash \sigma \longleftrightarrow (\sigma_r^1 \wedge \sigma_v^1) \vee \cdots \vee (\sigma_r^k \wedge \sigma_v^k).$$

Proof. Let B be the boolean algebra of L-sentences modulo T-equivalence. Let A be its boolean subalgebra of L_r-sentences modulo T-equivalence and A' its boolean subalgebra of L_v-sentences modulo T-equivalence. Set $\Psi := A \cup A'$, a subset of B. Using the familiar bijective correspondence between ultrafilters of B and complete L-theories extending T, Theorem 5.11 shows that the map $F \mapsto F \cap \Psi : S(B) \to \mathcal{P}(\Psi)$ from the Stone space $S(B)$ of B into the power set of Ψ is injective. Thus Ψ generates B by Lemma 5.14. It remains to use Lemma 5.13. □

Corollary 5.15 trivially implies Theorem 5.11 from which it was derived. This is relevant in connection with the set-theoretic issue around CH in the proof of Theorem 5.11. The point is that the statement of Corollary 5.15 is of arithmetic nature: we have an explicit axiomatization for the theory T, and so by Gödel numbering, this corollary becomes an arithmetic statement, that is, expressible by a sentence in the language of first-order Peano arithmetic. Any proof of an arithmetic statement from ZFC +CH can be converted into a proof from just ZF (so without Axiom of Choice, which, however, was involved in deriving Corollary 5.15 from Theorem 5.11). In fact, much stronger forms of this fact about ZF are valid. We refer to Chap. 9 of [50] for these logical issues around ZF which have to do with the phenomenon of *absoluteness*.

The equivalence displayed in Corollary 5.15 holds in all henselian lc-valued fields of equicharacteristic 0, and thus also in all henselian lc-valued fields with residue characteristic $p > N$ where $N \in \mathbb{N}$ depends only on σ.

When we fix the value group to be $\mathbb{Z}$, Corollary 5.15 becomes:

Corollary 5.16. *For each L-sentence σ there is a sentence σ_r in L_r such that for all henselian $\mathcal{K}$ of equicharacteristic 0 with value group $\mathbb{Z}$,*

$$\mathcal{K} \models \sigma \iff \boldsymbol{k} \models \sigma_r.$$

It is convenient to state some of the above also for (one-sorted) valued fields (K, A) without a lifting or cross-section as part of the structure.

Corollary 5.17. *Let σ be a sentence in the language of rings with an extra unary relation symbol. Then there are sentences $\sigma_r^1, \ldots, \sigma_r^k$ in L_r and sentences $\sigma_v^1, \ldots, \sigma_v^k$ in L_v with the property that for every henselian valued field (K, A) of equicharacteristic* 0*:*

$$(K, A) \models \sigma \iff \boldsymbol{k}_A \models \sigma_r^i \text{ and } \Gamma_A \models \sigma_v^i, \text{ for some } i \in \{1, \ldots, k\}.$$

It should be clear how this follows from Corollary 5.15 in view of Remark 5.10. Note also that the equivalence holds for all henselian valued fields with residue characteristic $p > N$ where N depends only on σ. Thus we have established Theorem 1.2 and the Ax-Kochen Principle 2.20.

There remain some claims in the Introduction, for example,

$$\mathbb{Q}^{ac}[[t]] \preccurlyeq \mathbb{C}[[t]] \quad \text{(as rings)},$$

which belong to the circle of AKE-results, but where some extra work is needed. We also want to have equivalences like the above not just for sentences, but also for formulas (with applications to the structure of definable sets). These issues will be addressed in the rest of this chapter, with the help of the main result of these notes, the Equivalence Theorem 5.21. This theorem tells us when two henselian valued fields of equicharacteristic zero are elementarily equivalent over a common substructure.

We begin with a short section on angular component maps. The presence of such maps simplifies the proof of the Equivalence Theorem, but in the aftermath we can often discard these maps again, by Corollary 5.18 below. The use of angular component maps rather than cross-sections in matters of this nature was initiated by Denef [18] and continued by Pas [40].

5.4 Angular Component Maps

In the rest of this chapter we consider a valued field as a 3-sorted structure $\mathcal{K} = (K, \boldsymbol{k}, \Gamma; \pi, v)$ as indicated in the beginning of Sect. 5.3.

Let $\mathcal{K} = (K, \boldsymbol{k}, \Gamma; \pi, v)$ be a valued field. An *angular component map* on $\mathcal{K}$ is a multiplicative group morphism

$$\text{ac}\colon K^\times \to \boldsymbol{k}^\times$$

such that $\text{ac}(a) = \pi(a)$ whenever $v(a) = 0$; we extend it to $\text{ac}\colon K \to \boldsymbol{k}$ by setting $\text{ac}(0) = 0$ (so $\text{ac}(ab) = \text{ac}(a)\,\text{ac}(b)$ for all $a, b \in K$), and also refer to this extension as an angular component map on $\mathcal{K}$. For example, construing a Hahn field $\boldsymbol{k}((t^\Gamma))$ as a valued field

$$\mathcal{K} = \big(\boldsymbol{k}((t^\Gamma)), \boldsymbol{k}, \Gamma; \pi, v\big)$$

in the natural way, we have the angular component map $\mathrm{ac} : \boldsymbol{k}((t^\Gamma)) \to \boldsymbol{k}$ on $\mathcal{K}$ given by $\mathrm{ac}(ct^\gamma + g) = ct^\gamma$ for $c \in \boldsymbol{k}^\times$, $\gamma \in \Gamma$, and $g \in \boldsymbol{k}((t^\Gamma))$ with $v(g) > \gamma$.

A cross-section s on the valued field $\mathcal{K}$ yields an angular component map ac on $\mathcal{K}$ by setting $\mathrm{ac}(x) = \pi\big(x/s(v(x))\big)$ for $x \in K^\times$. Thus by Corollary 5.5 on cross-sections we have:

Corollary 5.18. *Every valued field $\mathcal{K}$ has an elementary extension $\mathcal{K}^*$ with an angular component map on it.*

The reader might find it instructive to note that an angular component map on a valued field $\mathcal{K}$ corresponds to a splitting of the short exact sequence

$$1 \to \boldsymbol{k}^\times \to K^\times/(1+\mathfrak{m}) \to \Gamma \to 0$$

and to compare this with the short exact sequence in the proof of Lemma 4.31.

Let $\mathcal{K} = (K, \boldsymbol{k}, \Gamma; \pi, v)$ be a valued field. We let $\mathcal{O} := \mathcal{O}_v$ be its valuation ring. Any subfield E of K is viewed as a valued subfield of $\mathcal{K}$ with valuation ring $\mathcal{O}_E := \mathcal{O} \cap E$. In the proof of the Equivalence Theorem below we use the following easy cardinality bounds, which are immediate from Lemma 3.24.

Corollary 5.19. *Let E and F be subfields of K with $E \subseteq F$ such that F has countable transcendence degree over E. If $\pi(\mathcal{O}_E)$ is infinite, then $|\pi(\mathcal{O}_F)| = |\pi(\mathcal{O}_E)|$. If $v(E^\times) \neq \{0\}$, then $|v(F^\times)| = |v(E^\times)|$.*

5.5 The Equivalence Theorem

In this section we consider 3-sorted structures

$$\mathcal{K} = \big(K, \boldsymbol{k}, \Gamma; \pi, v, \mathrm{ac}\big)$$

where $\big(K, \boldsymbol{k}, \Gamma; \pi, v\big)$ is a valued field and $\mathrm{ac} : K \to \boldsymbol{k}$ is an angular component map on $\big(K, \boldsymbol{k}, \Gamma; \pi, v\big)$. Such a structure will be called an *ac-valued field*. We set $\mathcal{O} := \mathcal{O}_v$, and for any subfield E of K we put $\mathcal{O}_E := \mathcal{O} \cap E$.

A *good substructure* of $\mathcal{K} = (K, \boldsymbol{k}, \Gamma; \pi, v, \mathrm{ac})$ is a triple $\mathcal{E} = (E, \boldsymbol{k}_\mathcal{E}, \Gamma_\mathcal{E})$ such that:

(1) E is a subfield of K,
(2) $\boldsymbol{k}_\mathcal{E}$ is a subfield of $\boldsymbol{k}$ with $\mathrm{ac}(E) \subseteq \boldsymbol{k}_\mathcal{E}$ (hence $\pi(\mathcal{O}_E) \subseteq \boldsymbol{k}_\mathcal{E}$),
(3) $\Gamma_\mathcal{E}$ is an ordered abelian subgroup of Γ with $v(E^\times) \subseteq \Gamma_\mathcal{E}$.

For good substructures $\mathcal{E}_1 = (E_1, \boldsymbol{k}_1, \Gamma_1)$ and $\mathcal{E}_2 = (E_2, \boldsymbol{k}_2, \Gamma_2)$ of $\mathcal{K}$, we define $\mathcal{E}_1 \subseteq \mathcal{E}_2$ to mean that $E_1 \subseteq E_2$, $\boldsymbol{k}_1 \subseteq \boldsymbol{k}_2$, $\Gamma_1 \subseteq \Gamma_2$. If E is a subfield of K with $\mathrm{ac}(E) = \pi(\mathcal{O}_E)$, then $\big(E, \pi(\mathcal{O}_E), v(E^\times)\big)$ is a good substructure of $\mathcal{K}$, and if in addition $F \supseteq E$ is a subfield of K such that $v(F^\times) = v(E^\times)$, then $\mathrm{ac}(F) = \pi(\mathcal{O}_F)$. Throughout this subsection

$$\mathcal{K} = (K, \boldsymbol{k}, \Gamma; \pi, v, \mathrm{ac}), \qquad \mathcal{K}' = (K', \boldsymbol{k}', \Gamma'; \pi', v', \mathrm{ac}')$$

are ac-valued fields, with valuation rings $\mathcal{O}$ and $\mathcal{O}'$, and

$$\mathcal{E} = (E, \boldsymbol{k}_{\mathcal{E}}, \Gamma_{\mathcal{E}}), \qquad \mathcal{E}' = (E', \boldsymbol{k}_{\mathcal{E}'}, \Gamma_{\mathcal{E}'})$$

are good substructures of $\mathcal{K}$, $\mathcal{K}'$ respectively.

A *good map* $\mathbf{f} : \mathcal{E} \to \mathcal{E}'$ is a triple $\mathbf{f} = (f, f_{\mathrm{r}}, f_{\mathrm{v}})$ consisting of field isomorphisms $f : E \to E'$ and $f_{\mathrm{r}} : \boldsymbol{k}_{\mathcal{E}} \to \boldsymbol{k}_{\mathcal{E}'}$, and an ordered group isomorphism $f_{\mathrm{v}} : \Gamma_{\mathcal{E}} \to \Gamma_{\mathcal{E}'}$, such that

(i) $f_{\mathrm{r}}(\mathrm{ac}(a)) = \mathrm{ac}'(f(a))$ for all $a \in E$, and f_{r} is elementary as a partial map between the fields $\boldsymbol{k}$ and $\boldsymbol{k}'$.
(ii) $f_{\mathrm{v}}(v(a)) = v'(f(a))$ for all $a \in E^\times$, and f_{v} is elementary as a partial map between the ordered abelian groups Γ and Γ'.

Let $\mathbf{f} : \mathcal{E} \to \mathcal{E}'$ be a good map as above. Then the field part $f : E \to E'$ of $\mathbf{f}$ is a valued field isomorphism, and f_{r} and f_{v} agree on $\pi(\mathcal{O}_E)$ and $v(E^\times)$ with the maps $\pi(\mathcal{O}_E) \to \pi'(\mathcal{O}'_{E'})$ and $v(E^\times) \to v'(E'^\times)$ induced by f. We say that a good map $\mathbf{g} = (g, g_{\mathrm{r}}, g_{\mathrm{v}}) : \mathcal{F} \to \mathcal{F}'$ *extends* $\mathbf{f}$ if $\mathcal{E} \subseteq \mathcal{F}$, $\mathcal{E}' \subseteq \mathcal{F}'$, and g, g_{r}, g_{v} extend f, f_{r}, f_{v}, respectively. The *domain* of $\mathbf{f}$ is $\mathcal{E}$.

The next two lemmas show that part of condition (i) above is automatically satisfied by certain extensions of good maps.

Lemma 5.20. *Let $\mathbf{f} : \mathcal{E} \to \mathcal{E}'$ be a good map, and let $F \supseteq E$ and $F' \supseteq E'$ be subfields of K and K', respectively, such that $\pi(\mathcal{O}_F) \subseteq \boldsymbol{k}_{\mathcal{E}}$ and $v(F^\times) = v(E^\times)$. Let $g : F \to F'$ be a valued field isomorphism such that g extends f and $f_{\mathrm{r}}(\pi(u)) = \pi'(g(u))$ for all $u \in \mathcal{O}_F$. Then $\mathrm{ac}(F) \subseteq \boldsymbol{k}_{\mathcal{E}}$ and $f_{\mathrm{r}}(\mathrm{ac}(a)) = \mathrm{ac}'(g(a))$ for all $a \in F$.*

Proof. Let $a \in F$. Then $a = a_1 u$ where $a_1 \in E$ and $u \in \mathcal{O}_F$, $v(u) = 0$, so $\mathrm{ac}(a) = \mathrm{ac}(a_1)\pi(u) \in \boldsymbol{k}_{\mathcal{E}}$. It follows easily that $f_{\mathrm{r}}(\mathrm{ac}(a)) = \mathrm{ac}'(g(a))$. □

We are now ready to state and prove the Equivalence Theorem:

Theorem 5.21. *Suppose* $\mathrm{char}(\boldsymbol{k}) = 0$, *and $\mathcal{K}$, $\mathcal{K}'$ are henselian. Then any good map $\mathcal{E} \to \mathcal{E}'$ is a partial elementary map between $\mathcal{K}$ and $\mathcal{K}'$.*

Proof. The theorem holds trivially for $\Gamma = \{0\}$, so assume that $\Gamma \neq \{0\}$. Let $\mathbf{f} = (f, f_{\mathrm{r}}, f_{\mathrm{v}}) : \mathcal{E} \to \mathcal{E}'$ be a good map. By passing to suitable elementary extensions of $\mathcal{K}$ and $\mathcal{K}'$ we arrange that $\mathcal{K}$ and $\mathcal{K}'$ are κ-saturated, where κ is an uncountable cardinal such that $|\boldsymbol{k}_{\mathcal{E}}|$, $|\Gamma_{\mathcal{E}}| < \kappa$. Call a good substructure $\mathcal{E}_1 = (E_1, \boldsymbol{k}_1, \Gamma_1)$ of

$\mathcal{K}$ *small* if $|\boldsymbol{k}_1|, |\Gamma_1| < \kappa$. We shall prove that the good maps with small domain form a back-and-forth system between $\mathcal{K}$ and $\mathcal{K}'$. (This clearly suffices to obtain the theorem.) In other words, we shall prove that under the present assumptions on $\mathcal{E}$, $\mathcal{E}'$ and $\mathbf{f}$, there is for each $a \in K$ a good map $\mathbf{g}$ extending $\mathbf{f}$ such that $\mathbf{g}$ has small domain $\mathcal{F} = (F, \dots)$ with $a \in F$. In addition to Lemma 4.30, we need five basic extension procedures:

(1) *Given $\alpha \in \boldsymbol{k}$, arranging that $\alpha \in \boldsymbol{k}_{\mathcal{E}}$.* By saturation and the definition of "good map" this can be achieved without changing $f, f_v, E, \Gamma_{\mathcal{E}}$ by extending f_r to a partial elementary map between $\boldsymbol{k}$ and $\boldsymbol{k}'$ with α in its domain.

(2) *Given $\gamma \in \Gamma$, arranging that $\gamma \in \Gamma_{\mathcal{E}}$.* This follows in the same way.

(3) *Arranging $\boldsymbol{k}_{\mathcal{E}} = \pi(\mathcal{O}_E)$.* Suppose $\alpha \in \boldsymbol{k}_{\mathcal{E}}$, $\alpha \notin \pi(\mathcal{O}_E)$; set $\alpha' := f_r(\alpha)$.

Consider first the case that α is transcendental over $\pi(\mathcal{O}_E)$. Pick $a \in \mathcal{O}$ and $a' \in \mathcal{O}'$ such that $\bar{a} = \alpha$ and $\bar{a'} = \alpha'$, and then Lemmas 3.22 and 5.20 yield a good map $\mathbf{g} = (g, f_r, f_v)$ with small domain $(E(a), \boldsymbol{k}_{\mathcal{E}}, \Gamma_{\mathcal{E}})$ such that $\mathbf{g}$ extends $\mathbf{f}$ and $g(a) = a'$.

Next, assume that α is algebraic over $\pi(\mathcal{O}_E)$. Let $P(x) \in \mathcal{O}_E[x]$ be monic such that $\bar{P}(x)$ is the minimum polynomial of α over $\pi(\mathcal{O}_E)$. Pick $a \in \mathcal{O}$ such that $\bar{a} = \alpha$. Then $v\big(P(a)\big) > 0$ and $v\big(P'(a)\big) = 0$, so we have $b \in \mathcal{O}$ such that $P(b) = 0$ and $\bar{b} = \bar{a} = \alpha$. Likewise, we obtain $b' \in \mathcal{O}'$ such that $f(P)(b') = 0$ and $\bar{b'} = \alpha'$, where $f(P)$ is the polynomial over $\mathcal{O}'_{E'}$ that corresponds to P under f. By Lemmas 3.21 and 5.20 we obtain a good map extending $\mathbf{f}$ with small domain $(E(b), \boldsymbol{k}_{\mathcal{E}}, \Gamma_{\mathcal{E}})$ and sending b to b'.

By iterating these steps we can arrange $\boldsymbol{k}_{\mathcal{E}} = \pi(\mathcal{O}_E)$; it is easy to check that this condition is preserved in the extension procedures (4) and (5) below. We do assume in the rest of the proof that $\boldsymbol{k}_{\mathcal{E}} = \pi(\mathcal{O}_E)$.
In the extension procedures of (3) the value group $v(E^{\times})$ does not change. Also $\Gamma_{\mathcal{E}}$ does not change in (3), but at this stage we can have $\Gamma_{\mathcal{E}} \neq v(E^{\times})$.

(4) *Towards arranging $\Gamma_{\mathcal{E}} = v(E^{\times})$; the case of no torsion modulo $v(E^{\times})$.*
Suppose $\gamma \in \Gamma_{\mathcal{E}}$ has no torsion modulo $v(E^{\times})$, that is, $n\gamma \notin v(E^{\times})$ for all $n > 0$. Take $a \in K$ such that $v(a) = \gamma$. Replacing a by a/u with $u \in U(\mathcal{O})$ such that $\mathrm{ac}(u) = \mathrm{ac}(a)$, we arrange that $\mathrm{ac}(a) = 1$. In the same way we obtain $a' \in K'$ such that $v'(a') = \gamma' := f_v(\gamma)$ and $\mathrm{ac}'(a') = 1$. Then by Lemma 3.23 we have an isomorphism of valued fields $g : E(a) \to E'(a')$ extending f with $g(a) = a'$. Then (g, f_r, f_v) is a good map with small domain $(E(a), \boldsymbol{k}_{\mathcal{E}}, \Gamma_{\mathcal{E}})$.

(5) *Towards arranging $\Gamma_{\mathcal{E}} = v(E^{\times})$; the case of prime torsion modulo $v(E^{\times})$.*
Let $\gamma \in \Gamma_{\mathcal{E}} \setminus v(E^{\times})$ with $\ell\gamma \in v(E^{\times})$, where ℓ is a prime number. Pick $b \in E$ such that $v(b) = \ell\gamma$. Since $\mathrm{ac}(b) \in \pi(\mathcal{O}_E)$ we can replace b by b/u with $u \in U(\mathcal{O}_E)$ such that $\mathrm{ac}(u) = \mathrm{ac}(b)$ to arrange that $\mathrm{ac}(b) = 1$. We shall find $c \in K$ such that $c^{\ell} = b$ and $\mathrm{ac}(c) = 1$. As in (4) we have $a \in K$ such that $v(a) = \gamma$ and $\mathrm{ac}(a) = 1$. Then the polynomial $P(x) := x^{\ell} - b/a^{\ell}$ over $\mathcal{O}$ satisfies $v\big(P(1)\big) > 0$ and $v\big(P'(1)\big) = 0$. This gives $u \in K$ such that $P(u) = 0$ and $\bar{u} = 1$. Now let

$c = au$. Clearly $c^\ell = b$ and $\mathrm{ac}(c) = 1$. Likewise we find $c' \in K'$ such that $c'^\ell = f(b)$ and $\mathrm{ac}'(c') = 1$. Then by Lemma 5.6 the map $\mathbf{f}$ extends to a good map with domain $(E(c), \boldsymbol{k}_\mathcal{E}, \Gamma_\mathcal{E})$ sending c to c'.

By iterating (4) and (5) we can arrange $\Gamma_\mathcal{E} = v(E^\times)$. It is easy to check that this condition is preserved in the extension procedures of (3).

Let $a \in K$ be given. We need to extend $\mathbf{f}$ to a good map whose domain is small and contains a. We may and do assume $\boldsymbol{k}_\mathcal{E} = \pi(\mathcal{O}_E)$, $\Gamma_\mathcal{E} = v(E^\times)$. Let $\boldsymbol{k}_1 = \pi(\mathcal{O}_{E(a)}) \subseteq \boldsymbol{k}$ be the residue field of $E(a)$ and $\Gamma_1 := v\big(E(a)^\times\big)$ the value group of $E(a)$, so $|\boldsymbol{k}_1| < \kappa$ and $|\Gamma_1| < \kappa$ by Lemma 5.19. Using (1) and (2) we extend $\mathbf{f}$ to a good map $(f, f_{1,\mathrm{r}}, f_{1,\mathrm{v}})$ with domain $(E, \boldsymbol{k}_1, \Gamma_1)$. By iterating (3)–(5) we next extend $(f, f_{1,\mathrm{r}}, f_{1,\mathrm{v}})$ to a good map $\mathbf{f}_1 = (f_1, f_{1,\mathrm{r}}, f_{1,\mathrm{v}})$ with small domain $\mathcal{E}_1$ such that

$$\mathcal{E} \subseteq \mathcal{E}_1 = (E_1, \boldsymbol{k}_1, \Gamma_1), \qquad \boldsymbol{k}_1 = \pi(\mathcal{O}_{E_1}), \qquad \Gamma_1 = v(E_1^\times).$$

In the same way we extended $\mathbf{f}$ to $\mathbf{f}_1$, we extend $\mathbf{f}_1$ to $\mathbf{f}_2$ with small domain $\mathcal{E}_2 = (E_2, \boldsymbol{k}_2, \Gamma_2)$ such that

$$\boldsymbol{k}_2 = \pi(\mathcal{O}_{E_1(a)}) = \pi(\mathcal{O}_{E_2}), \qquad \Gamma_2 = v\big(E_1(a)^\times\big) = v(E_2^\times).$$

Continuing this way and taking a union of the resulting domains and good maps gives a good substructure $\mathcal{E}_\infty = (E_\infty, \boldsymbol{k}_\infty, \Gamma_\infty)$ such that

$$\mathcal{E} \subseteq \mathcal{E}_\infty, \quad \boldsymbol{k}_\infty = \pi\big(\mathcal{O}_{E_\infty(a)}\big) = \pi(\mathcal{O}_{E_\infty}), \quad \Gamma_\infty = v\big(E_\infty(a)^\times\big) = v(E_\infty^\times),$$

together with an extension of $\mathbf{f}$ to a good map $\mathbf{f}_\infty = (f_\infty, \dots)$ with domain $\mathcal{E}_\infty$. Note that then $E_\infty(a)$ is an immediate extension of E_∞. By Lemma 4.30 the valued subfield $E_\infty(a)$ of $\mathcal{K}$ has an immediate maximal valued field extension F inside $\mathcal{K}$. Then F is an immediate maximal extension of E_∞ as well. This gives a good substructure $\mathcal{F} = (F, \boldsymbol{k}_\infty, \Gamma_\infty)$ of $\mathcal{K}$. Likewise, the valued subfield $f_\infty(E_\infty)$ of $\mathcal{K}'$ has an immediate maximal valued field extension F' in $\mathcal{K}'$. By Corollary 4.29 we can extend $\mathbf{f}_\infty$ to a good map $\mathcal{F} \to \mathcal{F}' = (F', \dots)$. It remains to note that $a \in F$. □

5.6 Definable Sets

Here we derive various consequences of the Equivalence Theorem. We use the symbols $\equiv$ and $\preceq$ for the relations of elementary equivalence and being an elementary submodel, in the setting of many-sorted structures, and "definable" means "definable with parameters from the ambient structure". Let $\mathcal{L}$ be the 3-sorted language of valued fields, with sorts f (the field sort), r (the residue sort), and v (the value group sort). We view a valued field $(K, \boldsymbol{k}, \Gamma; \dots)$ as an $\mathcal{L}$-structure, with f-variables ranging over K, r-variables over $\boldsymbol{k}$, and v-variables over Γ. Augmenting $\mathcal{L}$ with a function symbol ac of sort (f, r) gives the language $\mathcal{L}(\mathrm{ac})$ of ac-valued fields. In this section

$$\mathcal{K} = (K, \boldsymbol{k}, \Gamma; \ldots), \qquad \mathcal{K}' = (K', \boldsymbol{k}', \Gamma'; \ldots)$$

are henselian ac-valued fields of equicharacteristic 0; they are considered as $\mathcal{L}(\mathrm{ac})$-structures in the obvious way.

Corollary 5.22. *$\mathcal{K} \equiv \mathcal{K}'$ if and only if $\boldsymbol{k} \equiv \boldsymbol{k}'$ as fields and $\Gamma \equiv \Gamma'$ as ordered abelian groups.*

Proof. The "only if" direction is obvious. Suppose $\boldsymbol{k} \equiv \boldsymbol{k}'$ as fields, and $\Gamma \equiv \Gamma'$ as ordered groups. This gives good substructures $\mathcal{E} := (\mathbb{Q}, \mathbb{Q}, \{0\})$ of $\mathcal{K}$, and $\mathcal{E}' := (\mathbb{Q}, \mathbb{Q}, \{0\})$ of $\mathcal{K}'$, and an obvious good map $\mathcal{E} \to \mathcal{E}'$. Now apply Theorem 5.21. □

Thus $\mathcal{K}$ is elementarily equivalent to the Hahn field $\boldsymbol{k}((t^{\Gamma}))$ equipped with the angular component map defined in Sect. 5.4.

Corollary 5.23. *Let $\mathcal{E} = (E, \boldsymbol{k}_E, \Gamma_E; \ldots)$ be a henselian ac-valued subfield of $\mathcal{K}$ such that $\boldsymbol{k}_E \preccurlyeq \boldsymbol{k}$ as fields, and $\Gamma_E \preccurlyeq \Gamma$ as ordered abelian groups. Then $\mathcal{E} \preccurlyeq \mathcal{K}$.*

Proof. Take an elementary extension $\mathcal{K}'$ of $\mathcal{E}$. Then $(E, \boldsymbol{k}_E, \Gamma_E)$ is a good substructure of both $\mathcal{K}$ and $\mathcal{K}'$, and the identity on $(E, \boldsymbol{k}_E, \Gamma)$ is a good map. Thus $\mathcal{K} \equiv_{\mathcal{E}} \mathcal{K}'$ by Theorem 5.21. Since $\mathcal{E} \preccurlyeq \mathcal{K}'$, this gives $\mathcal{E} \preccurlyeq \mathcal{K}$. □

The proofs of these corollaries use only little of the power of the Equivalence Theorem, but now we turn to something that uses much more: a relative elimination of quantifiers for the $\mathcal{L}(\mathrm{ac})$-theory T of henselian ac-valued fields of equicharacteristic 0. We specify that the function symbols π and v of $\mathcal{L}(\mathrm{ac})$ are to be interpreted as *total* functions in any $\mathcal{K}$ as follows: extend $\pi : \mathcal{O} \to \boldsymbol{k}$ to $\pi : K \to \boldsymbol{k}$ by $\pi(a) = 0$ for $a \notin \mathcal{O}$, and extend $v : K^{\times} \to \Gamma$ to $v : K \to \Gamma$ by $v(0) = 0$. (This is admittedly artificial, but harmless.)

Let $\mathcal{L}_{\mathrm{r}}$ be the sublanguage of $\mathcal{L}(\mathrm{ac})$ involving only the sort r, that is, $\mathcal{L}_{\mathrm{r}}$ is a copy of the language $\{0, 1, +, -, \cdot\}$ of rings. Let $\mathcal{L}_{\mathrm{v}}$ be the sublanguage of $\mathcal{L}(\mathrm{ac})$ involving only the sort v, that is, $\mathcal{L}_{\mathrm{v}}$ is the language $\{0, +, -, <\}$ of ordered abelian groups.[1] Let $x = (x_1, \ldots, x_l)$ be a tuple of distinct f-variables, $y = (y_1, \ldots, y_m)$ a tuple of distinct r-variables, and $z = (z_1, \ldots, z_n)$ a tuple of distinct v-variables. Define a *special* r-*formula in* (x, y) to be an $\mathcal{L}(\mathrm{ac})$-formula

$$\psi(x, y) := \psi'\big(\mathrm{ac}(q_1(x)), \ldots, \mathrm{ac}(q_k(x)), y\big)$$

where $k \in \mathbb{N}$, $\psi'(u_1, \ldots, u_k, y)$ is an $\mathcal{L}_{\mathrm{r}}$-formula, and $q_1(x), \ldots, q_k(x) \in \mathbb{Z}[x]$. Also, a *special* v-*formula in* (x, z) is an $\mathcal{L}(\mathrm{ac})$-formula

[1] One should of course regard $\mathcal{L}_{\mathrm{r}}$ and $\mathcal{L}_{\mathrm{v}}$ as disjoint.

$$\theta(x,z) \ := \ \theta'\big(v(q_1(x)),\dots,v(q_k(x)),z\big)$$

where $k \in \mathbb{N}$, $\theta'(v_1,\dots,v_k,y)$ is an $\mathcal{L}_v$-formula, and $q_1(x),\dots,q_k(x) \in \mathbb{Z}[x]$. Note that these special formulas do not have quantified f-variables. We can now state our relative quantifier elimination:

Corollary 5.24. *Every* $\mathcal{L}(\mathrm{ac})$*-formula* $\phi(x,y,z)$ *is* T*-equivalent to*

$$\big(\psi_1(x,y)\wedge\theta_1(x,z)\big)\vee\cdots\vee\big(\psi_N(x,y)\wedge\theta_N(x,z)\big)$$

for some $N \in \mathbb{N}$ *and some special* r*-formulas* $\psi_1(x,y),\dots,\psi_N(x,y)$ *in* (x,y)*, and some special* v*-formulas* $\theta_1(x,z),\dots,\theta_N(x,z)$ *in* (x,z)*.*

If a reader wonders about the absence of formulas $q(x)=0$ (with $q(x)\in\mathbb{Z}[x]$) in the display, the explanation is that these are absorbed into special r-formulas using our convention that for $a \in K$ we have $\mathrm{ac}(a)=0 \Leftrightarrow a=0$.

Proof. We apply Lemmas 5.13 and 5.14 to the boolean algebra B of $\mathcal{L}(\mathrm{ac})$-formulas $\phi(x,y,z)$ modulo T-equivalence, its subalgebra A of special formulas $\psi(x,y)$ in (x,y) modulo T-equivalence, its subalgebra A' of special formulas $\theta(x,z)$ in (x,z) modulo T-equivalence, and $\Psi = A \cup A'$.

Let $\psi(x,y)$ and $\theta(x,z)$ range over special formulas as described above. For a model $\mathcal{K}=(K,\boldsymbol{k},\Gamma;\dots)$ of T and $a\in K^l$, $r\in\boldsymbol{k}^m$, $\gamma\in\Gamma^n$, let

$$\begin{aligned}\mathrm{tp}_{\mathrm{r}}^{\mathcal{K}}(a,r) \ &:= \ \{\psi(x,y) : \mathcal{K}\models\psi(a,r)\}\\ \mathrm{tp}_{\mathrm{v}}^{\mathcal{K}}(a,\gamma) \ &:= \ \{\theta(x,z) : \mathcal{K}\models\theta(a,\gamma)\}.\end{aligned}$$

Let $\mathcal{K}$ and $\mathcal{K}'$ be any models of T, and let

$$(a,r,\gamma)\in K^l\times\boldsymbol{k}^m\times\Gamma^n, \qquad (a',r',\gamma')\in K'^l\times\boldsymbol{k}'^m\times\Gamma'^n$$

be such that $\mathrm{tp}_{\mathrm{r}}^{\mathcal{K}}(a,r)=\mathrm{tp}_{\mathrm{r}}^{\mathcal{K}'}(a',r')$ and $\mathrm{tp}_{\mathrm{v}}^{\mathcal{K}}(a,\gamma)=\mathrm{tp}_{\mathrm{v}}^{\mathcal{K}'}(a',\gamma')$. By the lemmas mentioned it suffices to show that under these assumptions we have

$$\mathrm{tp}^{\mathcal{K}}(a,r,\gamma) = \mathrm{tp}^{\mathcal{K}'}(a',r',\gamma').$$

Let $\mathcal{E} := (E,\boldsymbol{k}_{\mathcal{E}},\Gamma_{\mathcal{E}})$ where $E := \mathbb{Q}(a)$, $\boldsymbol{k}_{\mathcal{E}}$ is the subfield of $\boldsymbol{k}$ generated by $\mathrm{ac}(E)$ and r, and $\Gamma_{\mathcal{E}}$ is the ordered subgroup of Γ generated by γ over $v(E^\times)$, so $\mathcal{E}$ is a good substructure of $\mathcal{K}$. Likewise we define the good substructure $\mathcal{E}'$ of $\mathcal{K}'$. For each $q(x)\in\mathbb{Z}[x]$ we have $q(a)=0$ iff $\mathrm{ac}(q(a))=0$, and also $q(a')=0$ iff $\mathrm{ac}'(q(a'))=0$, and thus $q(a)=0$ iff $q(a')=0$. In view of this fact, the assumptions give us a good map $\mathcal{E}\to\mathcal{E}'$ sending a to a', γ to γ' and r to r'. It remains to apply Theorem 5.21. □

In the proof above it is important that our notion of a good substructure $\mathcal{E} = (E,\boldsymbol{k}_{\mathcal{E}},\Gamma_{\mathcal{E}})$ did not require $\boldsymbol{k}_{\mathcal{E}}=\pi(\mathcal{O}_E)$ or $\Gamma_{\mathcal{E}}=v(E^\times)$. Related to it is that in Corollary 5.24 we have a separation of r- and v-variables; this makes the next result almost obvious.

Corollary 5.25. *Each subset of $\boldsymbol{k}^m \times \Gamma^n$ definable in $\mathcal{K}$ is a finite union of rectangles $Y \times Z$ with $Y \subseteq \boldsymbol{k}^m$ definable in the field $\boldsymbol{k}$ and $Z \subseteq \Gamma^n$ definable in the ordered abelian group Γ.*

Proof. By Corollary 5.24 and using its notations it is enough to observe that for $a \in K^l$, a special r-formula $\psi(x, y)$ in (x, y), and a special v-formula $\theta(x, z)$ in (x, z), the set $\{r \in \boldsymbol{k}^m : \mathcal{K} \models \psi(a, r)\}$ is definable in the field $\boldsymbol{k}$, and the set $\{\gamma \in \Gamma^n : \mathcal{K} \models \theta(a, \gamma)\}$ is definable in the ordered abelian group Γ. □

Corollary 5.25 says in particular that the relations on $\boldsymbol{k}$ definable in $\mathcal{K}$ are definable in the field $\boldsymbol{k}$, and likewise, the relations on Γ definable in $\mathcal{K}$ are definable in the ordered abelian group Γ. In terms of the model-theoretic notions of *stably embedded* and *orthogonal*, this gives:

(i) $\boldsymbol{k}$ is stably embedded in $\mathcal{K}$,
(ii) Γ is stably embedded in $\mathcal{K}$,
(iii) $\boldsymbol{k}$ and Γ are orthogonal in $\mathcal{K}$.

By Corollary 5.18 we can get rid of angular component maps in Corollaries 5.22 and 5.25: these go through if we replace "ac-valued" by "valued". In particular, any henselian valued field, with residue field $\boldsymbol{k}$ of characteristic 0 and value group Γ, is elementarily equivalent to the Hahn field $\boldsymbol{k}((t^\Gamma))$. (We already knew some of this from Sect. 5.3, but there we used a nontrivial fact about ZFC, in contrast to the treatment just given.)

6 Rings of Witt Vectors

We now aim for results in the mixed characteristic case that are analogous to the "equicharacteristic 0" theorems in the previous chapter. One difference is that in the mixed characteristic case we don't have a lifting of the residue field. For complete discrete valuation rings with perfect residue field we do have a nice substitute, namely the Teichmüller map.

After introducing the Teichmüller map we are naturally led to Witt vectors which allow us to construct, for each perfect field $\boldsymbol{k}$ of characteristic $p > 0$, a complete discrete valuation ring $\mathrm{W}[\boldsymbol{k}]$ with residue field $\boldsymbol{k}$. This construction is functorial and has various other good properties. If $\boldsymbol{k}$ is the prime field of p elements, then $\mathrm{W}[\boldsymbol{k}] \cong \mathbb{Z}_p$. In the last chapter of these notes we shall determine the elementary theory of the ring $\mathrm{W}[\boldsymbol{k}]$ in terms of the elementary theory of the perfect field $\boldsymbol{k}$; this covers the elementary theory of the ring $\mathbb{Z}_p$ of p-adic integers as a special case. It is possible to determine the latter without introducing Witt vectors, but $\mathrm{W}[\boldsymbol{k}]$ is of considerable interest also for other $\boldsymbol{k}$. The present chapter is purely algebraic in nature and largely borrowed from [48].

Throughout we fix a prime number p. Recall that $\mathbb{F}_p = \mathbb{Z}/p\mathbb{Z}$ is the field of p elements. For integers a and $N \geq 1$ we let $a \bmod N$ be the image of a in the residue ring $\mathbb{Z}/N\mathbb{Z}$. Also "ring" means "commutative ring with 1" and ring morphisms preserve 1 by definition.

6.1 The Teichmüller Map

As we have seen, we cannot lift the residue fields of (henselian) local rings like $\mathbb{Z}/p^2\mathbb{Z}$ and $\mathbb{Z}_p$ for the simple reason that these rings do not contain any subfield. Fortunately, it turns out that there is a canonical system of representatives for the residue field in these rings, but it only preserves the multiplicative structure of the residue field, not its additive structure. For $\mathbb{Z}/p^2\mathbb{Z}$ it is the map

$$\tau : \mathbb{F}_p = \mathbb{Z}/p\mathbb{Z} \to \mathbb{Z}/p^2\mathbb{Z}, \quad \tau(a \bmod p) = a^p \bmod p^2, \quad a \in \mathbb{Z}.$$

To see that such a map τ exists, use the fact that for all $a, b \in \mathbb{Z}$,

$$a \equiv b \mod p \implies a^p \equiv b^p \mod p^2.$$

Because also $a \equiv a^p \mod p$ for all $a \in \mathbb{Z}$, this map τ is a system of representatives for the residue field $\mathbb{F}_p$, in the sense that for each $x \in \mathbb{F}_p$ the image of $\tau(x)$ in $\mathbb{F}_p$ is x. Note also that $\tau(xy) = \tau(x)\tau(y)$ for all $x, y \in \mathbb{F}_p$. For the ring $\mathbb{Z}/p^3\mathbb{Z}$, the canonical system of representatives for its residue field $\mathbb{F}_p$ is given by $a \bmod p \mapsto a^{p^2} \bmod p^3$, $(a \in \mathbb{Z})$. Below we extend these observations to complete local rings with perfect residue field of characteristic p.

A field $\boldsymbol{k}$ of characteristic p is said to be *perfect* if its *Frobenius endomorphism* $x \mapsto x^p$ is an automorphism. Given any local ring A with residue field $\boldsymbol{k}$ of characteristic p, one shows easily by induction on n that for all $x, y \in A$,

$$(*) \qquad x \equiv y \mod \mathfrak{m} \implies x^{p^n} \equiv y^{p^n} \mod \mathfrak{m}^{n+1}.$$

Theorem 6.1. *Let A be a complete local ring with maximal ideal $\mathfrak{m}$ and perfect residue field $\boldsymbol{k}$ of characteristic p. Then there is a unique map $\tau : \boldsymbol{k} \to A$ such that for all $x \in \boldsymbol{k}$, $\overline{\tau(x)} = x$ and $\tau(x^p) = \tau(x)^p$. This (Teichmüller) map τ has the following properties:*

(i) $\tau(\boldsymbol{k}) = \{a \in A :$ *a is a p^nth power in A, for each n*$\}$;
(ii) $\tau(xy) = \tau(x)\tau(y)$ *for all $x, y \in \boldsymbol{k}$*, $\tau(0) = 0$, $\tau(1) = 1$, *and* $\tau(\boldsymbol{k}^\times) \subseteq U(A)$;
(iii) *if $p \cdot 1 = 0$ in A, then τ is a ring embedding.*

Proof. Let $x \in \boldsymbol{k}$, and choose for each n an $a_n \in A$ such that

$$x = (\overline{a_n})^{p^n}, \qquad \text{(possible since } \boldsymbol{k} \text{ is perfect).}$$

Then $a_{n+1}^p \equiv a_n \mod \mathfrak{m}$, so $a_{n+1}^{p^{n+1}} \equiv a_n^{p^n} \mod \mathfrak{m}^{n+1}$ by (*) above. Hence the sequence $\{a_n^{p^n}\}$ converges in the $\mathfrak{m}$-adic norm to an element in A. This limit does not depend on the choice of $\{a_n\}$: if $\{b_n\}$ is another choice, then $a_n \equiv b_n \mod \mathfrak{m}$, so $a_n^{p^n} \equiv b_n^{p^n} \mod \mathfrak{m}^{n+1}$ by (*) above, and thus $\{a_n^{p^n}\}$ and $\{b_n^{p^n}\}$ have the same limit. We define $\tau(x)$ to be this limit. It is easy to check that for all $x \in \boldsymbol{k}$, $\overline{\tau(x)} = x$ and $\tau(x^p) = \tau(x)^p$.

To show that these two identities uniquely determine τ, consider a map $\iota : \boldsymbol{k} \to A$ such that $\overline{\iota(x)} = x$ and $\iota(x^p) = \iota(x)^p$ for all $x \in \boldsymbol{k}$. For $x \in \boldsymbol{k}$ and $\{a_n\}$ as above, put $b_n := \iota(\overline{a_n})$, so $\iota(x) = b_n^{p^n}$, so $x = (\overline{b_n})^{p^n}$ (for all n), and thus

$$\tau(x) = \lim_n b_n^{p^n} = \iota(x).$$

To prove (i), each element of $\tau(\boldsymbol{k})$ is a p^nth power in A for each n, since each element of $\boldsymbol{k}$ is a p^nth power in $\boldsymbol{k}$ for each n. Conversely, let $a \in A$ be a p^nth power in A for each n. Choose for each n an $a_n \in A$ such that $a = a_n^{p^n}$. Then the construction of τ above shows that for $x := \bar{a}$ we have $\tau(x) = a$.

This construction also yields (ii), and (iii) follows from (i) by noting that if $p \cdot 1 = 0$ in A, then $(a+b)^{p^n} = a^{p^n} + b^{p^n}$ for all $a, b \in A$ and all n. □

The reader should verify that for $A = \mathbb{Z}/p^2\mathbb{Z}$ and $A = \mathbb{Z}/p^3\mathbb{Z}$ the Teichmüller map is the one given at the beginning of this section.

Exercise. With the same assumptions as in the Theorem, show that $\tau(\boldsymbol{k})$ is the unique set $S \subseteq A$ such that $s^p \in S$ for each $s \in S$, and S is mapped bijectively onto $\boldsymbol{k}$ by the residue class map. Show that for $A = \mathbb{Z}_p$ one has $\tau(\mathbb{F}_p^\times) = \{x \in \mathbb{Z}_p : x^{p-1} = 1\}$.

A *local p-ring* is a complete local ring A with maximal ideal pA and perfect residue field A/pA. If in addition $p^n \neq 0$ in A for all n, then we say that A is *strict*. (Here $p := p \cdot 1 \in A$, and below we often commit the same abuse of notation by letting p denote the element $p \cdot 1$ in a ring.)

Corollary 6.2. *Let A be a local p-ring, and $\boldsymbol{k} := A/pA$. If $p^n \neq 0$ and $p^{n+1} = 0$ in A, then for every $a \in A$ we have*

$$a = \sum_{i=0}^{n} \tau(x_i) p^n$$

for a unique tuple $(x_0, \ldots, x_n) \in \boldsymbol{k}^{n+1}$. If A is strict, then there is for every $a \in A$ a unique sequence $\{x_n\} \in \boldsymbol{k}^{\mathbb{N}}$ such that

$$a = \sum_{n=0}^{\infty} \tau(x_n) p^n.$$

Let A be a strict local p-ring and set $\boldsymbol{k} := A/pA$. For $a \in A$ we call $\{x_n\}$ as above the *Teichmüller vector of* a. Note that the map

$$a \mapsto \text{Teichmüller vector of } a \;:\; A \to \boldsymbol{k}^{\mathbb{N}}$$

is a bijection. Given the Teichmüller vectors of $a, b \in A$, how do we obtain the Teichmüller vectors of $a + b$ and ab? It turns out that there is a good answer to this question, but things become more transparent if we change from Teichmüller vectors to Witt vectors: Let $a \in A$; then the unique sequence $\{x_n\} \in \boldsymbol{k}^{\mathbb{N}}$ such that $a = \sum_{n=0}^{\infty} \tau(x_n^{p^{-n}})p^n$ is called the *Witt vector of a* (and then $\{x_n^{p^{-n}}\}$ is its Teichmüller vector). Note that if $\boldsymbol{k} = \mathbb{F}_p$, then there is no difference between the Teichmüller vector and the Witt vector of a.

We also want to show that, given any perfect field $\boldsymbol{k}$, there is exactly one strict local p-ring with residue field isomorphic to $\boldsymbol{k}$, up to isomorphism. To construct such a ring from $\boldsymbol{k}$ we shall take the set $\boldsymbol{k}^{\mathbb{N}}$ of Witt vectors over $\boldsymbol{k}$ and define a suitable addition and multiplication on it making it a strict local p-ring $\mathrm{W}[\boldsymbol{k}]$. This is done in the next section.

Exercise. For any ring R, let $\mu(R) := \{x \in R : x^n = 1 \text{ for some } n \geq 1\}$ be the group of roots of unity in R, so $\mu(R)$ is a subgroup of $U(R)$. Let A be a strict local p-ring. Show that if $p \neq 2$, then τ maps $\mu(\boldsymbol{k})$ bijectively onto $\mu(A)$.

6.2 *Witt Vectors*

Under the familiar bijection $\mathbb{F}_p^2 \longrightarrow \mathbb{Z}/p^2\mathbb{Z}$ given by

$$(i \bmod p,\ j \bmod p) \mapsto (i + jp) \bmod p^2, \qquad i, j \in \{0, \dots, p-1\}$$

the addition and multiplication of $\mathbb{Z}/p^2\mathbb{Z}$ do not correspond to algebraically natural operations on $\mathbb{F}_p^2$. One motivation for Witt vectors is to find a bijection $\mathbb{F}_p^2 \longrightarrow \mathbb{Z}/p^2\mathbb{Z}$ that makes the addition map of the ring $\mathbb{Z}/p^2\mathbb{Z}$,

$$+ : \mathbb{Z}/p^2\mathbb{Z} \times \mathbb{Z}/p^2\mathbb{Z} \longrightarrow \mathbb{Z}/p^2\mathbb{Z}$$

correspond to an explicit polynomial map

$$\mathbb{F}_p^4 = \mathbb{F}_p^2 \times \mathbb{F}_p^2 \longrightarrow \mathbb{F}_p^2,$$

and likewise with multiplication on $\mathbb{Z}/p^2\mathbb{Z}$. Corollary 6.2 applied to $A = \mathbb{Z}/p^2\mathbb{Z}$ turns out to provide just the right (Witt) bijection $\mathbb{F}_p^2 \longrightarrow \mathbb{Z}/p^2\mathbb{Z}$:

$$(a \bmod p,\ a' \bmod p) \mapsto (a^p + pa') \bmod p^2, \qquad a, a' \in \mathbb{Z}.$$

To see which binary operations on $\mathbb{F}_p^2$ correspond under this bijection to addition and multiplication on $\mathbb{Z}/p^2\mathbb{Z}$ we note that for integers a, a', b, b',

$$(a^p + pa') + (b^p + pb') = (a+b)^p + p\big(a' + b' - \sum_{i=1}^{p-1} c(p,i)a^i b^{p-i}\big),$$

$$(a^p + pa') \times (b^p + pb') = (ab)^p + p\big(a^p b' + a'b^p + pa'b'\big),$$

where $c(p,i)$ is the integer $\binom{p}{i}/p$ for $i = 1, \ldots, p-1$. Thus addition on $\mathbb{Z}/p^2\mathbb{Z}$ corresponds under the Witt bijection to the binary operation on $\mathbb{F}_p^2$ given by

$$\big((x_0, x_1), (y_0, y_1)\big) \mapsto \Big(x_0 + y_0,\ x_1 + y_1 - \sum_{i=1}^{p-1} c(p,i) x_0^i y_0^{p-i}\Big) \ :\ \mathbb{F}_p^2 \times \mathbb{F}_p^2 \to \mathbb{F}_p^2.$$

Likewise, multiplication on $\mathbb{Z}/p^2\mathbb{Z}$ corresponds under this bijection to

$$\big((x_0, x_1), (y_0, y_1)\big) \mapsto \big(x_0 y_0,\ x_0^p y_1 + x_1 y_0^p + p x_1 y_1\big) \ :\ \mathbb{F}_p^2 \times \mathbb{F}_p^2 \to \mathbb{F}_p^2.$$

(Of course, in $\mathbb{F}_p$ we have the identity $x_0^p y_1 + x_1 y_0^p + p x_1 y_1 = x_0 y_1 + x_1 y_0$, but we prefer to state it the way we did because later we shall replace $\mathbb{F}_p$ by an arbitrary ring.) So our bijection turns the ring $\mathbb{Z}/p^2\mathbb{Z}$ into an algebraic-geometric object living in $\mathbb{F}_p$ so to say.

Next we try to extend all this to $\mathbb{Z}/p^3\mathbb{Z}$, $\mathbb{Z}/p^4\mathbb{Z}$, and so on, all the way up to $\mathbb{Z}_p = \varprojlim \mathbb{Z}/p^n\mathbb{Z}$. Applying Corollary 6.2 to $A = \mathbb{Z}/p^3\mathbb{Z}$ yields the bijection

$$(a \bmod p,\ a' \bmod p,\ a'' \bmod p) \mapsto (a^{p^2} + p a'^p + p^2 a'') \bmod p^3, \quad a, a', a'' \in \mathbb{Z}.$$

from $\mathbb{F}_p^3$ onto $\mathbb{Z}/p^3\mathbb{Z}$. Again, under this bijection the addition and multiplication of $\mathbb{Z}/p^3\mathbb{Z}$ correspond to binary operations on $\mathbb{F}_p^3$ given by easily specifiable polynomials over $\mathbb{F}_p$.

Just as crucial is that this Witt bijection between $\mathbb{F}_p^3$ and $\mathbb{Z}/p^3\mathbb{Z}$ is compatible with the earlier one between $\mathbb{F}_p^2$ and $\mathbb{Z}/p^2\mathbb{Z}$ in the sense that we have a commuting diagram

$$\begin{array}{ccc} \mathbb{F}_p^3 & \longrightarrow & \mathbb{F}_p^2 \\ \downarrow & & \downarrow \\ \mathbb{Z}/p^3\mathbb{Z} & \longrightarrow & \mathbb{Z}/p^2\mathbb{Z} \end{array}$$

where the vertical arrows are the Witt bijections, and horizontal arrows are given by

$$\begin{aligned} (x_0, x_1, x_2) &\mapsto (x_0, x_1) && \text{(top)}, \\ a \bmod p^3 &\mapsto a \bmod p^2,\ a \in \mathbb{Z} && \text{(bottom)}. \end{aligned}$$

It is also important that these constructions are *functorial*. Indeed, for any ring R we can define the ring $\mathrm{W}_2[R]$ as follows: its underlying set is R^2, its addition is given by

$$\big((x_0, x_1), (y_0, y_1)\big) \mapsto \Big(x_0 + y_0,\ x_1 + y_1 - \sum_{i=1}^{p-1} c(p,i) x_0^i y_0^{p-i}\Big) \ :\ R^2 \times R^2 \to R^2$$

and its multiplication by

$$\big((x_0, x_1), (y_0, y_1)\big) \mapsto (x_0 y_0,\ x_0^p y_1 + x_1 y_0^p + p x_1 y_1) \ :\ R^2 \times R^2 \to R^2.$$

The zero element of $W_2[R]$ is $(0,0)$, and its multiplicative identity is $(1,0)$. One can view some of the above as providing an explicit isomorphism from the ring $W_2[\mathbb{F}_p]$ onto the ring $\mathbb{Z}/p^2\mathbb{Z}$. Likewise, we can define a ring $W_3[R]$, and so on, but at this point we may as well go all the way, and define $W[R]$.

The Witt Polynomials. Fix distinct indeterminates

$$X_0, X_1, X_2, \dots, Y_0, Y_1, Y_2, \dots.$$

The polynomials $W_0(X_0),\ W_1(X_0, X_1),\ \dots,\ W_n(X_0, \dots, X_n), \dots$ are elements of the polynomial ring $\mathbb{Z}[X_0, X_1, X_2, \dots]$ defined as follows:

$$\begin{aligned}
W_0 &= X_0,\\
W_1 &= X_0^p + pX_1,\\
W_2 &= X_0^{p^2} + pX_1^p + p^2 X_2,\\
&\dots\dots\\
W_n &= X_0^{p^n} + pX_1^{p^{n-1}} + \dots + p^n X_n = \sum_{i=0}^{n} p^i X_i^{p^{n-i}},\\
&\dots\dots
\end{aligned}$$

Note that $W_{n+1}(X_0, \dots, X_{n+1}) = W_n(X_0^p, \dots, X_n^p) + p^{n+1} X_{n+1}$. Let R be a ring and consider the map $\theta : R^{\mathbb{N}} \longrightarrow R^{\mathbb{N}}$ defined by

$$\theta(x_0, x_1, x_2, \dots) = \big(W_0(x_0), W_1(x_0, x_1), \dots, W_n(x_0, \dots, x_n), \dots\big).$$

Exercise. Show that $\theta(1, 0, 0, 0, \dots) = (1, 1, 1, 1, \dots)$. Show also:

$$\begin{aligned}
\theta(-1, 0, 0, 0, \dots) &= (-1, -1, -1, -1, \dots) \text{ if } p \text{ is odd},\\
\theta(-1, -1, -1, -1, \dots) &= (-1, -1, -1, -1, \dots) \text{ if } p = 2.
\end{aligned}$$

Below, $x = (x_0, x_1, x_2, \dots)$ and $y = (y_0, y_1, y_2, \dots)$ range over $R^{\mathbb{N}}$.

Lemma 6.3. *Suppose $p \cdot 1 \in U(R)$. Then θ is a bijection.*

Proof. We have $\theta(x) = y$ if and only if

$$x_0 = y_0,\ px_1 = y_1 - x_0^p,\ \dots,\ p^n x_n = y_n - x_0^{p^n} - \dots - p^{n-1} x_{n-1}^p, \dots,$$

so there is exactly one such x for each y. □

Remark. By the same argument, θ is injective if $pa \neq 0$ for all $a \in R \setminus \{0\}$.

Lemma 6.4. *Suppose $pa \neq 0$ for all nonzero $a \in R$, let m be given, and let $x_0, x_1, \ldots, x_n, y_0, y_1, \ldots, y_n \in R$. Then the following are equivalent:*

(1) $x_i \equiv y_i \mod p^m R$ *for* $i = 0, \ldots, n$;
(2) $W_i(x_0, \ldots, x_i) \equiv W_i(y_0, \ldots, y_i) \mod p^{m+i} R$ *for* $i = 0, \ldots, n$.

Proof. That (1) implies (2) is because for all $a, b \in R$ and $i \in \mathbb{N}$,

$$a \equiv b \mod p^m R \Longrightarrow a^{p^i} \equiv a^{p^i} \mod p^{m+i} R.$$

The converse follows by induction on i, using the assumption on p. □

Lemma 6.5. *Let $A := \mathbb{Z}[X_0, X_1, \ldots, Y_0, Y_1, \ldots]$. There are polynomials $S_0, S_1, S_2 \ldots, P_0, P_1, P_2, \cdots \in A$ such that for all n,*

$$\begin{aligned} W_n(X_0, \ldots, X_n) + W_n(Y_0, \ldots, Y_n) &= W_n(S_0, \ldots, S_n), \\ W_n(X_0, \ldots, X_n) \times W_n(Y_0, \ldots, Y_n) &= W_n(P_0, \ldots, P_n). \end{aligned}$$

These conditions determine the sequences $\{S_n\}$ and $\{P_n\}$ uniquely, and for each n we have $S_n, P_n \in \mathbb{Z}[X_0, \ldots, X_n, Y_0, \ldots, Y_n]$.

Proof. We make $p \in A$ a unit by working in the larger domain

$$R := \{a/p^m : a \in A, m = 0, 1, 2 \ldots\} \subseteq \mathbb{Q}[X_0, X_1, \ldots, Y_0, Y_1, \ldots].$$

Then the bijectivity of the map θ of Lemma 6.3 shows that there are unique sequences $\{S_n\}$ and $\{P_n\}$ in R such that the identities above hold. Note that $S_0 = X_0 + Y_0$. Assume for a certain $n \geq 1$ that $S_i \in \mathbb{Z}[X_0, \ldots, X_i, Y_0, \ldots, Y_i]$ for $i = 0, \ldots, n-1$. We have

$$\begin{aligned} W_n(S_0, \ldots, S_n) &= W_{n-1}(S_0^p, \ldots, S_{n-1}^p) + p^n S_n \\ &= W_n(X_0, \ldots, X_n) + W_n(Y_0, \ldots, Y_n) \\ &= W_{n-1}(X_0^p, \ldots, X_{n-1}^p) + W_{n-1}(Y_0^p, \ldots, Y_{n-1}^p) + p^n(X_n + Y_n) \\ &= W_{n-1}\big(S_0(X_0^p, Y_0^p), \ldots, S_{n-1}(X_0^p, \ldots, X_{n-1}^p, Y_0^p, \ldots, Y_{n-1}^p)\big) \\ &\quad + p^n(X_n + Y_n). \end{aligned}$$

Also, $S_i^p \equiv S_i(X_0^p, \ldots, X_i^p, Y_0^p, \ldots, Y_i^p) \mod pA$ for $i = 0, \ldots, n-1$, hence by Lemma 6.4 the polynomial $W_{n-1}(S_0^p, \ldots, S_{n-1}^p)$ is congruent modulo $p^n A$ to

$$W_{n-1}\big(S_0(X_0^p, Y_0^p), \ldots, S_{n-1}(X_0^p, \ldots, X_{n-1}^p, Y_0^p, \ldots, Y_{n-1}^p)\big).$$

It follows that all coefficients of S_n are in $\mathbb{Z}$. The proof for the P_n is similar, starting with $P_0 = X_0 Y_0$. □

Of course, the letters S and P were chosen to remind the reader of *sum* and *product*. The polynomials S_0, S_1, P_0, P_1 are easy to write down:

$$S_0 = X_0 + Y_0, \qquad S_1 = X_1 + Y_1 - \sum_{i=1}^{p-1} c(p,i) X_0^i Y_0^{p-i},$$

$$P_0 = X_0 Y_0, \qquad P_1 = X_0^p Y_1 + X_1 Y_0^p + pX_1 Y_1.$$

It is also easy to check that the constant terms of S_n and P_n are zero.

Let $\mathrm{W}[R]$ be the set $R^{\mathbb{N}}$ equipped with the binary operations

$$+ : \mathrm{W}[R] \times \mathrm{W}[R] \to \mathrm{W}[R], \qquad \cdot : \mathrm{W}[R] \times \mathrm{W}[R] \to \mathrm{W}[R],$$

defined as follows:

$$x + y := \big(S_0(x_0, y_0), S_1(x_0, x_1, y_0, y_1), \dots, S_n(x_0, \dots, x_n, y_0, \dots, y_n), \dots\big),$$

$$x \cdot y := \big(P_0(x_0, y_0), P_1(x_0, x_1, y_0, y_1), \dots, P_n(x_0, \dots, x_n, y_0, \dots, y_n), \dots\big).$$

We also define the elements $\mathbf{0}$ and $\mathbf{1}$ of $W[R]$ by $\mathbf{0} = (0, 0, 0, \dots)$ and $\mathbf{1} = (1, 0, 0, \dots)$. For any ring morphism $\phi : R \to R'$ we define

$$\mathrm{W}[\phi] : \mathrm{W}[R] \to \mathrm{W}[R'], \qquad \mathrm{W}[\phi](x) = \big(\phi(x_0), \phi(x_1), \phi(x_2), \dots\big),$$

so $\mathrm{W}[\phi]$ is a homomorphism for the addition and multiplication operations on $\mathrm{W}[R]$ and $\mathrm{W}[R']$, and $\mathrm{W}[\phi](\mathbf{0}) = \mathbf{0}$, $\mathrm{W}[\phi](\mathbf{1}) = \mathbf{1}$.

Corollary 6.6. *With these operations,* $\mathrm{W}[R]$ *is a ring with zero element* $\mathbf{0}$ *and multiplicative identity* $\mathbf{1}$.

Proof. By the previous lemma, the map $\theta : \mathrm{W}[R] \to R^{\mathbb{N}}$ is a homomorphism for addition and multiplication, where these operations are defined componentwise on $R^{\mathbb{N}}$. Note also that $\theta(\mathbf{0}) = \mathbf{0}$ and $\theta(\mathbf{1}) = (1, 1, 1, \dots)$ are the zero and multiplicative identity of the ring $R^{\mathbb{N}}$. Thus by the exercise preceding Lemma 6.3 we have a subring $\theta(\mathrm{W}[R])$ of $R^{\mathbb{N}}$. In view of the remark following Lemma 6.3, we conclude that if $pa \neq 0$ for all nonzero $a \in R$, then $\mathrm{W}[R]$ is a ring with zero element $\mathbf{0}$ and multiplicative identity $\mathbf{1}$. In particular, $\mathrm{W}[A]$ is a ring with zero element $\mathbf{0}$ and multiplicative identity $\mathbf{1}$ for any polynomial ring $A = \mathbb{Z}[(X_i)_{i\in I}]$ in any family of distinct indeterminates $(X_i)_{i\in I}$. The general case now follows by taking a surjective ring morphism $\phi : A \to R$ where A is such a polynomial ring, and using the surjective map $\mathrm{W}[\phi] : \mathrm{W}[A] \to \mathrm{W}[R]$. □

We call $\mathrm{W}[R]$ the *ring of Witt vectors over R*. The map θ is a ring morphism of $\mathrm{W}[R]$ into the product ring $R^{\mathbb{N}}$. Note also that if $\phi : R \to R'$ is a ring morphism, so is $\mathrm{W}[\phi] : \mathrm{W}[R] \to \mathrm{W}[R']$. In this way we have defined an endofunctor W on the category of rings (whose morphisms are the ring morphisms). Below we consider

$x = (x_0, x_1, \dots)$ and $y = (y_0, y_1, \dots)$ as elements of the ring $\mathrm{W}[R]$ rather than of the product ring $R^{\mathbb{N}}$. We set $x^{(i)} := \theta(x)_i = W_i(x_0, \dots, x_i) \in R$ for $i \in \mathbb{N}$, and $x^{(i)} := 0 \in R$ for negative $i \in \mathbb{Z}$.

Exercise. Show that if p is odd, then the additive inverse of x in $\mathrm{W}[R]$ is given by $-x = (-x_0, -x_1, -x_2, \dots)$.

We now define two further (unary) operations on $\mathrm{W}[R]$:

$$\mathbf{V} : \mathrm{W}[R] \to \mathrm{W}[R], \quad \mathbf{V}(x) := (0, x_0, x_1, x_2, \dots),$$
$$\mathbf{F} : \mathrm{W}[R] \to \mathrm{W}[R], \quad \mathbf{F}(x) := (x_0^p, x_1^p, x_2^p, \dots).$$

The map $\mathbf{V}$ is a *shift* operation ("Verschiebung" in German). As an operation on the ring $R^{\mathbb{N}}$, the shift map $\mathbf{V}$ is clearly an endomorphism of its additive group. As an operation on $\mathrm{W}[R]$ the shift map $\mathbf{V}$ is additive as well, but this requires an argument:

Lemma 6.7. $\big(\mathbf{V}(x)\big)^{(n)} = p \cdot x^{(n-1)}$ *in* R, $\mathbf{V}(x+y) = \mathbf{V}(x) + \mathbf{V}(y)$ *in* $\mathrm{W}[R]$, *and* $\big(\mathbf{V}^m(x)\big)^{(n)} = p^m \cdot x^{(n-m)}$ *in* R.

Proof. For $n \geq 1$ we have

$$W_n(0, x_0 \dots, x_{n-1}) = p x_0^{p^{n-1}} + \dots + p^n x_{n-1} = p W_{n-1}(x_0, \dots, x_{n-1}),$$

which gives the first identity. Thus $\theta\big(\mathbf{V}(x)\big) = p\mathbf{V}\big(\theta(x)\big)$ with $\mathbf{V}\big(\theta(x)\big)$ in the ring $R^{\mathbb{N}}$. Hence, with $x + y$ evaluated in $\mathrm{W}[R]$, but $\theta(x) + \theta(y)$ in $R^{\mathbb{N}}$,

$$\begin{aligned}\theta\big(\mathbf{V}(x+y)\big) &= p\mathbf{V}\big(\theta(x+y)\big) = p\mathbf{V}\big(\theta(x)+\theta(y)\big)\\ &= p\mathbf{V}\big(\theta(x)\big) + p\mathbf{V}\big(\theta(y)\big) = \theta\big(\mathbf{V}(x)\big) + \theta\big(\mathbf{V}(y)\big)\\ &= \theta\big(\mathbf{V}(x)+\mathbf{V}(y)\big) \text{ with } \mathbf{V}(x)+\mathbf{V}(y) \text{ evaluated in } \mathrm{W}[R].\end{aligned}$$

Thus $\mathbf{V}(x+y) = \mathbf{V}(x) + \mathbf{V}(y)$ if R is a polynomial ring over $\mathbb{Z}$. The validity of this identity in this special case implies its validity in general since each ring is isomorphic to a quotient of a polynomial ring over $\mathbb{Z}$. The third identity follows by induction on m from the first identity. □

Next we put $[a] := (a, 0, 0, \dots) \in \mathrm{W}[R]$ for $a \in R$.

Lemma 6.8. *The following identities hold, where* $a \in R$*:*

(1) $[a]^{(n)} = a^{p^n}$, *in* R;
(2) $x = \sum_{i=0}^{n-1} \mathbf{V}^i[x_i] + \mathbf{V}^n(x_n, x_{n+1}, \dots)$, *in* $\mathrm{W}[R]$;
(3) $[a]x = (ax_0, a^p x_1, a^{p^2} x_2, \dots)$, *in* $\mathrm{W}[R]$.

Proof. Identity (1) is immediate. It easily yields identity (2) for $n = 1$, initially in the case that R is a polynomial ring over $\mathbb{Z}$, and then for any R by the usual argument. Induction gives identity (2) for all n. Identity (3) follows likewise by checking that $([a]x)^{(n)} = W_n(ax_0, \dots, a^{p^n} x_n)$. □

Remark. The decomposition in (2) is clearly unique: if $x = \sum_{i=0}^{n-1} \mathbf{V}^i[y_i] + \mathbf{V}^n z$, then $x_i = y_i$ for all $i < n$.

Lemma 6.9. *Let $y = px$ in $\mathrm{W}[R]$ and $z = \mathbf{V}(\mathbf{F}(x))$. Then $y_n \equiv z_n \mod pR$.*

Proof. We have $y^{(n)} = \theta(y)_n = p\theta(x)_n = p\big(\mathbf{F}(x)^{(n-1)} + p^n x_n\big)$ by the recursion for the Witt polynomials. Also $p \cdot \big(\mathbf{F}(x)\big)^{(n-1)} = \mathbf{V}\big(\mathbf{F}(x)\big)^{(n)} = z^{(n)}$ by the first identity of Lemma 6.7, so $y^{(n)} \equiv z^{(n)} \mod p^{n+1}R$. This holds for all n, so by Lemma 6.4 we have $y_n \equiv z_n \mod pR$ for all n in case R is a polynomial ring over $\mathbb{Z}$, and thus for all R by the usual argument. □

By Lemma 6.3 the ring $\mathrm{W}[R]$ is just the product ring $R^{\mathbb{N}}$ in disguise if R is a field of characteristic 0. The situation is much more interesting if R is a perfect field of characteristic p, and from this point on we are just going to consider this case. So in the remainder of this section $\boldsymbol{k}$ denotes a perfect field of characteristic p, and $x = (x_0, x_1, x_2, \dots)$, $y = (y_0, y_1, y_2, \dots) \in \mathrm{W}[\boldsymbol{k}]$.

Note that then $\mathbf{F} = \mathrm{W}[\Phi]$ where $\Phi : \boldsymbol{k} \to \boldsymbol{k}$ is the Frobenius automorphism. Thus $\mathbf{F}$ is an automorphism of the ring $\mathrm{W}[\boldsymbol{k}]$. The map $\mathbf{V}$ is related to $\mathbf{F}$ by

$$\mathbf{F} \circ \mathbf{V} = \mathbf{V} \circ \mathbf{F} = p \cdot \mathrm{Id},$$

as a consequence of Lemma 6.9. In other words, in $\mathrm{W}[\boldsymbol{k}]$ we have

$$px = p(x_0, x_1, x_2, \dots) = (0, x_0^p, x_1^p, x_2^p, \dots),$$

and thus $p^2 x = (0, 0, x_0^{p^2}, x_1^{p^2}, \dots)$, and so on. Also

$$(**) \qquad \mathbf{V}^m(x) \cdot \mathbf{V}^n(y) = \mathbf{V}^{m+n}\big(\mathbf{F}^n(x)\mathbf{F}^m(y)\big).$$

To see this, note that upon setting $x = \mathbf{F}^m(u)$ and $y = \mathbf{F}^n(z)$ with $u, z \in \mathrm{W}[\boldsymbol{k}]$, the identity becomes $p^m u \cdot p^n z = p^{m+n} uz$.

6.3 Rings of Witt Vectors as Discrete Valuation Rings

In this section $\boldsymbol{k}$ is a perfect field of characteristic p.

Theorem 6.10. $\mathrm{W}[\boldsymbol{k}]$ *is a complete* DVR *with fraction field of characteristic* 0, *and maximal ideal $p\mathrm{W}[\boldsymbol{k}]$. The map $x \mapsto x_0 : \mathrm{W}[\boldsymbol{k}] \to \boldsymbol{k}$ is a ring morphism with kernel $p\mathrm{W}[\boldsymbol{k}]$, and thus induces an isomorphism of the residue field of $\mathrm{W}[\boldsymbol{k}]$ with the field $\boldsymbol{k}$. Upon identifying $\boldsymbol{k}$ with the residue field via this isomorphism, the Teichmüller map $\boldsymbol{k} \to \mathrm{W}[\boldsymbol{k}]$ is given by $a \mapsto [a]$, and for $x \in \mathrm{W}[\boldsymbol{k}]$ we have*

$$x = \sum_{n=0}^{\infty} [x_n^{p^{-n}}] p^n.$$

Proof. Let x, y range over $\mathrm{W}[\boldsymbol{k}]$, and define

$$v : \mathrm{W}[\boldsymbol{k}] \setminus \{\mathbf{0}\} \to \mathbb{Z}$$

by $v(x) = n$ if $x_n \neq 0$ and $x_i = 0$ for all $i < n$. So v takes only values in $\mathbb{N}$, and we extend v to all of $\mathrm{W}[\boldsymbol{k}]$ by setting $v(\mathbf{0}) = \infty \in \mathbb{Z}_\infty$. Thus $v(x) \geq n$ iff $x = \mathbf{V}^n(u)$ for some $u \in \mathrm{W}[\boldsymbol{k}]$ (and in that case, $v(x) = n$ iff $u_0 \neq 0$). Hence $v(x+y) \geq \min(vx, vy)$ by the additivity of $\mathbf{V}$. We also get $v(xy) = v(x) + v(y)$ by the identity $(**)$ at the end of the previous section, and the fact that this identity clearly holds when $v(x) = v(y) = 0$. Thus $\mathrm{W}[\boldsymbol{k}]$ is a domain, and v is a valuation on this domain. Next, we claim that $v(x-y) \geq n$ iff $x_i = y_i$ for all $i < n$. To see this, write $x = y + z$, so the claim becomes: $v(z) \geq n$ iff $x_i = y_i$ for all $i < n$. Now decompose each of x, y, z according to identity (2) of Lemma 6.8, and use the remark following this lemma to establish the claim. From this claim we get that $\mathrm{W}[\boldsymbol{k}]$ is complete with respect to the ultranorm $|x| := p^{-v(x)}$. Identity (2) of Lemma 6.8, together with $[a] = \mathbf{F}^n[a^{p^{-n}}]$ for $a \in \boldsymbol{k}$, now yields

$$x = \sum_{n=0}^{\infty} \mathbf{V}^n[x_n] = \sum_{n=0}^{\infty} [x_n^{p^{-n}}]p^n.$$

If $|x| = 1$, then $x_0 \neq 0$, and thus by (2) and (3) of Lemma 6.8, $[x_0^{-1}]x = \mathbf{1} + y$ with $|y| < 1$, so $x \in U(\mathrm{W}[\boldsymbol{k}])$. We have now shown that $\mathrm{W}[\boldsymbol{k}]$ is a local domain with maximal ideal

$$\{x : v(x) > 0\} = \mathbf{V}\big(\mathrm{W}[\boldsymbol{k}]\big) = \mathbf{V}\big(\mathbf{F}(\mathrm{W}[\boldsymbol{k}])\big) = p\,\mathrm{W}[\boldsymbol{k}].$$

It also follows that $v(x) = n$ iff $x = p^n u$ with $u \in U(\mathrm{W}[\boldsymbol{k}])$. Thus $\mathrm{W}[\boldsymbol{k}]$ is a complete DVR with residue field isomorphic to $\boldsymbol{k}$. Since $p\cdot\mathbf{1} = (0, 1, 0, 0, \ldots) \neq \mathbf{0}$, the fraction field of $\mathrm{W}[\boldsymbol{k}]$ has characteristic 0. The rest of the theorem now follows easily. □

In particular, $\mathrm{W}[\boldsymbol{k}]$ is a strict local p-ring with residue field isomorphic to $\boldsymbol{k}$. This property determines the ring $\mathrm{W}[\boldsymbol{k}]$ up to isomorphism, as a consequence of the next theorem. (It is this consequence that we need in the proof of the AKE-results in the unramified mixed characteristic case.)

Theorem 6.11. *Let A be a strict local p-ring and let $\phi : \boldsymbol{k} \to \boldsymbol{k}_A$ be a field embedding of $\boldsymbol{k}$ into the residue field $\boldsymbol{k}_A$ of A. Then there is a unique ring morphism $f : \mathrm{W}[\boldsymbol{k}] \to A$ such that the diagram*

$$\begin{array}{ccc} \mathrm{W}[\boldsymbol{k}] & \xrightarrow{\ f\ } & A \\ \downarrow & & \downarrow \\ \boldsymbol{k} & \xrightarrow[\ \phi\ ]{} & \boldsymbol{k}_A \end{array}$$

commutes. The map f is injective. If ϕ is an isomorphism, so is f.

Proof. Let τ be the Teichmüller map of A. To simplify notation we identify $\boldsymbol{k}$ with a subfield of $\boldsymbol{k}_A$ via ϕ. We define $f : \mathrm{W}[\boldsymbol{k}] \to A$ by

$$f(x) = \sum_{i=0}^{\infty} \tau(x_i^{p^{-i}})p^i.$$

It is clear that $f(\mathbf{0}) = 0$ and $f(\mathbf{1}) = 1$. Let $x + y = z$ in $\mathrm{W}[\boldsymbol{k}]$. To prove $f(x) + f(y) = f(z)$, put

$$\begin{aligned} \mathbf{F}^{-n}x &:= \left(x_0^{p^{-n}}, x_1^{p^{-n}}, x_2^{p^{-n}}, \dots\right) \in \mathrm{W}[\boldsymbol{k}], \\ \tau(\mathbf{F}^{-n}x) &:= \left(\tau(x_0^{p^{-n}}), \tau(x_1^{p^{-n}}), \tau(x_2^{p^{-n}}), \dots\right) \in \mathrm{W}[A]. \end{aligned}$$

Then

$$\left(\tau(\mathbf{F}^{-n}x)\right)^{(n)} = \sum_{i=0}^{n} \tau(x_i^{p^{-i}})p^i \in A,$$

so

$$\lim_{n\to\infty} \left(\tau(\mathbf{F}^{-n}x)\right)^{(n)} = f(x).$$

Now $\mathbf{F}^{-n}x + \mathbf{F}^{-n}y = \mathbf{F}^{-n}z$ in $\mathrm{W}[\boldsymbol{k}]$, so for all $\nu \in \mathbb{N}$:

$$\left(\tau(\mathbf{F}^{-n}x) + \tau(\mathbf{F}^{-n}y)\right)_\nu \equiv \left(\tau(\mathbf{F}^{-n}z)\right)_\nu \mod pA,$$

and thus by Lemma 6.4:

$$\left(\tau(\mathbf{F}^{-n}x)\right)^{(n)} + \left(\tau(\mathbf{F}^{-n}y)\right)^{(n)} \equiv \left(\tau(\mathbf{F}^{-n}z)\right)^{(n)} \mod p^{n+1}A.$$

Taking the limit as n goes to infinity yields $f(x) + f(y) = f(z)$. Likewise one shows that $f(x)f(y) = f(xy)$, so f is a ring morphism lifting ϕ. The other claims of the theorem are easy consequences of Corollary 6.2. □

For $A = \mathbb{Z}_p$ and residue field $\mathbb{F}_p$ this theorem yields a ring isomorphism

$$\mathrm{W}[\mathbb{F}_p] \cong \mathbb{Z}_p, \quad x = (x_0, x_1, \dots) \mapsto \sum_{i=0}^{\infty} \tau(x_n)p^n,$$

where τ is the Teichmüller map of $\mathbb{Z}_p$ (which assigns to each $a \in \mathbb{F}_p^\times$ the unique $\zeta \in \mathbb{Z}_p$ with $\zeta^{p-1} = 1$ that has residue class a.) It is usual to identify the rings $\mathrm{W}[\mathbb{F}_p]$ and $\mathbb{Z}_p$ via this isomorphism.

We let $\mathrm{W}(\boldsymbol{k})$ be the fraction field of $\mathrm{W}[\boldsymbol{k}]$, and consider it as the *valued field* that has $\mathrm{W}[\boldsymbol{k}]$ as its valuation ring. The normalized valuation of $\mathrm{W}(\boldsymbol{k})$ with this valuation ring restricts to the valuation v on $\mathrm{W}[\boldsymbol{k}]$ defined in the proof of Theorem 6.10. In particular, for $\boldsymbol{k} = \mathbb{F}_p$ the above identification of $\mathrm{W}[\mathbb{F}_p]$ with $\mathbb{Z}_p$ extends to a valued field isomorphism $\mathrm{W}(\mathbb{F}_p) \cong \mathbb{Q}_p$ via which we identify $\mathrm{W}(\mathbb{F}_p)$ with $\mathbb{Q}_p$.

We followed here the notation in [8], but we alert the reader that this is not standard in the literature: our $\mathrm{W}[\boldsymbol{k}]$ is usually written as $\mathrm{W}(\boldsymbol{k})$, see for example [48]. For us, $\mathrm{W}(\boldsymbol{k})$ is the fraction field of $\mathrm{W}[\boldsymbol{k}]$.

Exercise. Show:

$$\{x \in \mathrm{W}[\boldsymbol{k}] : \mathbf{F}(x) = x^p\} = \{[a] : a \in \boldsymbol{k}\}.$$

With $\mathrm{Aut}(R)$ denoting the group of automorphisms of a ring R, show that

$$\phi \mapsto \mathrm{W}[\phi] : \mathrm{Aut}(\boldsymbol{k}) \to \mathrm{Aut}(\mathrm{W}[\boldsymbol{k}])$$

is a group isomorphism.

If R is a subring of the ring R', then we consider $\mathrm{W}[R]$ as a subring of $\mathrm{W}[R']$. At one point in the next chapter we shall need:

Lemma 6.12. *Suppose the valuation ring A is a subring of $\mathrm{W}[\boldsymbol{k}]$ such that $\mathfrak{m}_A \subseteq p\,\mathrm{W}[\boldsymbol{k}]$ and $\{x_0 : x \in A\}$ is a perfect subfield $\boldsymbol{k}_0$ of $\boldsymbol{k}$. Then $A \subseteq \mathrm{W}[\boldsymbol{k}_0]$.*

To avoid confusion, we mention that our use of 0 as a subscript in denoting the subfield $\boldsymbol{k}_0$ of $\boldsymbol{k}$ has nothing to do with its use as a subscript in denoting the component x_0 of a Witt vector $x = (x_0, x_1, x_2, \dots) \in \mathrm{W}[\boldsymbol{k}]$.

Proof. By the assumption on $\mathfrak{m}_A$, the valued field $(\mathrm{W}(\boldsymbol{k}), \mathrm{W}[\boldsymbol{k}])$ is a valued field extension of (K, A) where K is the fraction field of A in $\mathrm{W}(\boldsymbol{k})$. With the usual identifications, these valued fields have the same value group $\mathbb{Z}$, and the residue field of A as a subfield of the residue field $\boldsymbol{k}$ of $\mathrm{W}[\boldsymbol{k}]$ equals

$$\{x_0 : x \in A\} = \boldsymbol{k}_0 \subseteq \boldsymbol{k}.$$

By taking the closure of A in $\mathrm{W}[\boldsymbol{k}]$ we arrange in addition that A is a complete DVR. We claim that then $A = \mathrm{W}[\boldsymbol{k}_0]$. To prove this claim, first note that Theorem 6.11 gives an isomorphism $f : \mathrm{W}[\boldsymbol{k}_0] \to A$ such that $x_0 = f(x)_0$ for all $x \in \mathrm{W}[\boldsymbol{k}_0]$. Composing this with the inclusion $i : A \to \mathrm{W}[\boldsymbol{k}]$ yields a ring morphism $g = i \circ f : \mathrm{W}[\boldsymbol{k}_0] \to \mathrm{W}[\boldsymbol{k}]$ such that $x_0 = g(x)_0$ for all $x \in \mathrm{W}[\boldsymbol{k}_0]$. By the uniqueness part of Theorem 6.11, g must be the inclusion of $\mathrm{W}[\boldsymbol{k}_0]$ in $\mathrm{W}[\boldsymbol{k}]$, and so $x = f(x)$ for all $x \in \mathrm{W}[\boldsymbol{k}_0]$, which proves the claim. □

7 AKE in Mixed Characteristic

In this chapter we construe a valued field as a 3-sorted structure

$$\mathcal{K} = (K, \boldsymbol{k}, \Gamma;\ \pi, v)$$

as explained in the beginning of Sect. 5.3. Note that here we do not include an angular component map among the primitives of $\mathcal{K}$. Throughout we fix a prime number p. Given a perfect field $\boldsymbol{k}$ of characteristic p, we are particularly interested in $\mathrm{W}(\boldsymbol{k})$, construed as the valued field $(\mathrm{W}(\boldsymbol{k}), \boldsymbol{k}, \mathbb{Z}; \pi, v)$ where v is the discrete valuation on the field $\mathrm{W}(\boldsymbol{k})$ with valuation ring $\mathrm{W}[\boldsymbol{k}]$ and where $\pi : \mathrm{W}[\boldsymbol{k}] \to \boldsymbol{k}$ is given by $\pi(x) = x_0$.

To state the main results to be established in this chapter, let

$$\mathcal{K} = (K, \boldsymbol{k}, \Gamma; \pi, v), \qquad \mathcal{K}' = (K', \boldsymbol{k}', \Gamma'; \pi', v')$$

be valued fields with $\mathrm{char}(K) = \mathrm{char}(K') = 0$ and $\mathrm{char}(\boldsymbol{k}) = \mathrm{char}(\boldsymbol{k}') = p$. In addition we assume that $\mathcal{K}$ and $\mathcal{K}'$ are *unramified*, that is, $v(p)$ and $v'(p)$ are the smallest positive elements of their respective value groups Γ and Γ', and we assume also that $\boldsymbol{k}$ and $\boldsymbol{k}'$ are *perfect*. Under these assumptions we prove:

Theorem 7.1. *Suppose $\mathcal{K}$ and $\mathcal{K}'$ are henselian. Then:*

$$\mathcal{K} \equiv \mathcal{K}' \iff \boldsymbol{k} \equiv \boldsymbol{k}' \text{ as fields, and } \Gamma \equiv \Gamma' \text{ as ordered abelian groups.}$$

In particular, if $\mathcal{K}$ is henselian and $|\Gamma/n\Gamma| = n$ for all $n \geq 1$, then $\mathcal{K} \equiv \mathrm{W}(\boldsymbol{k})$.

Theorem 7.2. *Suppose $\mathcal{K}$ and $\mathcal{K}'$ are henselian and $\mathcal{K} \subseteq \mathcal{K}'$. Then:*

$$\mathcal{K} \preccurlyeq \mathcal{K}' \iff \boldsymbol{k} \preccurlyeq \boldsymbol{k}' \text{ as fields, and } \Gamma \preccurlyeq \Gamma' \text{ as ordered abelian groups.}$$

Theorem 7.3. *Suppose $\mathcal{K}$ is henselian. Then each subset of $\boldsymbol{k}^n$ which is definable in $\mathcal{K}$ is definable in the field $\boldsymbol{k}$.*

For example, if $\boldsymbol{k}$ is algebraically closed of characteristic p, then the subsets of $\boldsymbol{k}^n$ definable in the valued field $\mathrm{W}(\boldsymbol{k})$ are exactly the finite unions of differences $X \setminus Y$ where X, Y are algebraic subsets of $\boldsymbol{k}^n$.

One can derive these results from a suitable Equivalence Theorem in mixed characteristic, as we did in Chap. 5 in the equicharacteristic zero case. We have opted here for another route, in order to illustrate a technique that is useful in many other situations, namely *coarsening*. This goes as follows. Identify $\mathbb{Z}$ with a convex subgroup of Γ via $k \mapsto kv(p)$, so that $v(p) = 1 \in \mathbb{Z} \subseteq \Gamma$. Make the quotient group $\dot{\Gamma} := \Gamma/\mathbb{Z}$ into an ordered group by $\gamma + \mathbb{Z} > 0 \Leftrightarrow \gamma > \mathbb{Z}$. Coarsen the valuation $v : K^\times \to \Gamma$ to the valuation

$$\dot{v} : K^\times \to \dot{\Gamma} := \Gamma/\mathbb{Z}, \qquad \dot{v}(a) := v(a) + \mathbb{Z}.$$

It is routine to check that K with this coarsened valuation $\dot{v}$ has residue field of characteristic 0, and is henselian if K with the original valuation v is henselian. So far so good. Coarsening does involve, however, a loss of information: the original valuation ring $\mathcal{O}_v$ is not definable in the coarsened valued field $(K, \mathcal{O}_{\dot{v}})$. Fortunately, it is definable (uniformly) in a certain expansion of $(K, \mathcal{O}_{\dot{v}})$, obtained by making its residue field into a valued field. Moreover, if $\mathcal{K}$ is $\aleph_1$-saturated, then the valued residue field of $(K, \mathcal{O}_{\dot{v}})$ is canonically isomorphic to the valued field $\mathrm{W}(\boldsymbol{k})$. This is how our study of Witt vectors will pay off.

Thus coarsening gives a reduction of the mixed characteristic case to the equicharacteristic zero case, but at the cost of requiring extra structure on the residue field (of characteristic zero). It is routine to check that the Equivalence Theorem of Chap. 5 does in fact go through with any extra structure on the residue field. (We can also allow extra structure on the value group.) In the next sections we carry out the program just sketched.

We can relax the conditions in the theorems above that the residue field is perfect and $v(p)$ is the smallest positive element in the value group, but the conclusions are a bit weaker. For example: Suppose (K, A) and (L, B) are henselian valued fields of mixed characteristic $(0, p)$. Then:

$$(K,A) \equiv (L,B) \iff A/p^nA \equiv B/p^nB \text{ for } n = 1,2,3,\ldots, \text{ and}$$
$$\big(\Gamma_A, v_A(p)\big) \equiv \big(\Gamma_B, v_B(p)\big).$$

Here $\big(\Gamma_A, v_A(p)\big)$ is the ordered abelian group Γ_A with $v_A(p)$ as a distinguished element. We leave it to the reader to prove this along the lines sketched above.

7.1 Cross-Sections Revisited

As in Chap. 5 we shall use cross-sections and angular component maps even though these do not show up in the final results as stated above. We improve here Corollary 5.5 on the existence of cross-sections. This is in the form of a digression on abelian groups. Let A, B be (additively written) abelian groups. Call A a *pure subgroup of* B if A is a subgroup of B such that $A \cap nB = nA$ for all $n \geq 1$. If A is a subgroup of B and B/A is torsion-free, then A is a pure subgroup of B. In case B is itself torsion-free, then

$$A \text{ is a pure subgroup of } B \iff B/A \text{ is torsion-free.}$$

In particular, a subgroup Δ of an ordered abelian group Γ is pure in Γ as defined in Sect. 5.1 if and only if Δ is a pure subgroup of Γ as just defined. Also, if A is a direct summand of B (that is, A is a subgroup of B and $B = A \oplus B'$, internally, for some subgroup B' of B), then A is a pure subgroup of B.

Lemma 7.4. *Suppose A is a pure subgroup of B, and the group B/A is finitely generated. Then $B = A \oplus B'$, internally, for some subgroup B' of B.*

Proof. We have $B/A = \mathbb{Z}(b_1 + A) \oplus \cdots \oplus \mathbb{Z}(b_m + A)$ for suitable $b_1, \ldots, b_m \in B$. If $b_i + A$ has finite order $n_i \geq 1$ in B/A, then $n_i b_i \in A$, so $n_i b_i = n_i a_i$ with

$a_i \in A$, and replacing b_i by $b_i - a_i$ we arrange $n_i b_i = 0$. With this adjustment of the b_is one checks easily that $B = A \oplus B'$ where

$$B' = \mathbb{Z}b_1 + \cdots + \mathbb{Z}b_m = \mathbb{Z}b_1 \oplus \cdots \oplus \mathbb{Z}b_m.$$

□

Corollary 7.5. *Suppose A is a pure subgroup of B, $e_{ij} \in \mathbb{Z}$ for $1 \le i \le m$ and $1 \le j \le n$, and $a_1, \dots, a_m \in A$. Suppose the system of equations*

$$\begin{aligned} e_{11}x_1 + \cdots + e_{1n}x_n &= a_1 \\ \dots\dots\dots\dots\dots\dots &= .. \\ e_{m1}x_1 + \cdots + e_{mn}x_n &= a_m \end{aligned}$$

has a solution in B, that is, there are $x_1, \dots, x_n \in B$ for which the above equations hold. Then this system has a solution in A.

Lemma 7.6. *Suppose A is a pure subgroup of B, and $b \in B$. Then there is a pure subgroup A' of B that contains A and b such that A'/A is countable.*

Proof. By the downward Skolem-Löwenheim theorem we can take a subgroup A' of B that contains A and b such that A'/A is countable and $A'/A \preccurlyeq B/A$. It follows easily that A' is a pure subgroup of B. □

Proposition 7.7. *Let $h : A \to U$ be a group morphism into an $\aleph_1$-saturated (additive) abelian group U and suppose A is a pure subgroup of B. Then h extends to a group morphism $B \to U$.*

Proof. By the previous lemma we reduce to the case that B/A is countable. Let $b_0, b_1, b_2, \dots$ generate B over A. If we can find elements $u_0, u_1, u_2, \dots \in U$ such that $\sum_n e_n u_n = h(a)$ whenever $\sum_n e_n b_n = a$ (all $e_n \in \mathbb{Z}$, $e_n = 0$ for all but finitely many n, $a \in A$), then we can extend h as desired by sending b_n to u_n for each n. Note that each finite subset of this countable set of constraints on $(u_0, u_1, u_2, \dots)$ can be satisfied, by Corollary 7.5. The desired result follows. □

Corollary 7.8. *Let U be an $\aleph_1$-saturated (additive) abelian group and a pure subgroup of B. Then U is a direct summand of B.*

Proof. Apply Proposition 7.7 to the identity map $U \to U$. □

These facts on abelian groups are from [10, Chapter V,§5], which treats this material for modules over any ring. For a valued field $\mathcal{K} = (K, \boldsymbol{k}, \Gamma; \pi, v)$ we get:

Lemma 7.9. *If $\mathcal{K}$ is $\aleph_1$-saturated, then it has a cross-section.*

Proof. With U the multiplicative group of units of $\mathcal{O}$, the inclusion $U \to K^\times$ and $v : K^\times \to \Gamma$ yield the exact sequence of abelian groups

$$1 \to U \to K^\times \to \Gamma \to 0.$$

Since Γ is torsion-free, U is a pure subgroup of $K^\times$. If $\mathcal{K}$ is $\aleph_1$-saturated, then so is the group U, and thus the above exact sequence splits by Corollary 7.8. □

We shall also need the following variant:

Lemma 7.10. *Let $\mathcal{K}$ be $\aleph_1$-saturated, let $\mathcal{E} = (E, \boldsymbol{k}_E, \Gamma_E; \dots)$ be a valued subfield of $\mathcal{K}$ such that the abelian group Γ_E is pure in Γ and $\aleph_1$-saturated. Let s_E be a cross-section on $\mathcal{E}$. Then s_E extends to a cross-section on $\mathcal{K}$.*

Proof. By Lemma 7.9 we have a cross-section s on $\mathcal{K}$. By Corollary 7.8 we have an internal direct sum decomposition $\Gamma = \Gamma_E \oplus \Delta$ with Δ a subgroup of Γ. This gives a cross-section on $\mathcal{K}$ that coincides with s_E on Γ_E and with s on Δ. □

7.2 *Enhancing the Equivalence Theorem*

According to the program sketched in the beginning of this chapter we need to extend the Equivalence Theorem of Chap. 5 by allowing extra structure on the residue field and value group. This is what we do in this section.

Let $\mathcal{L}$ be the 3-sorted language of valued fields and $\mathcal{L}(\mathrm{ac})$ the language of ac-valued fields, as introduced in Sect. 5.6. Consider now a language $\mathcal{L}^* \supseteq \mathcal{L}(\mathrm{ac})$ such that every symbol of $\mathcal{L}^* \setminus \mathcal{L}(\mathrm{ac})$ is a relation symbol of positive arity, and of some sort $(\mathrm{r}, \dots, \mathrm{r})$ or $(\mathrm{v}, \dots, \mathrm{v})$. Let $\mathcal{L}^*_{\mathrm{r}}$ be the sublanguage of $\mathcal{L}^*$ involving only the sort r, that is, (a copy of) the language of fields together with the new relation symbols of sort $(\mathrm{r}, \dots, \mathrm{r})$. Also, let $\mathcal{L}^*_{\mathrm{v}}$ be the sublanguage of $\mathcal{L}^*$ involving only the sort v, that is, the language of ordered abelian groups together with the new relation symbols of sort $(\mathrm{v}, \dots, \mathrm{v})$. By a $*$-valued field we mean an $\mathcal{L}^*$-structure whose $\mathcal{L}(\mathrm{ac})$-reduct is an ac-valued field.

Let $\mathcal{K} = (K, \boldsymbol{k}, \Gamma; \cdots)$ be a $*$-valued field. Then we shall view $\boldsymbol{k}$ as an $\mathcal{L}^*_{\mathrm{r}}$-structure, and Γ as an $\mathcal{L}^*_{\mathrm{v}}$-structure, in the obvious way. Any subfield E of K is viewed as a valued subfield of $\mathcal{K}$ with valuation ring $\mathcal{O}_E := \mathcal{O} \cap E$.

A *good substructure* of $\mathcal{K} = (K, \boldsymbol{k}, \Gamma; \cdots)$ is a triple $\mathcal{E} = (E, \boldsymbol{k}_{\mathcal{E}}, \Gamma_{\mathcal{E}})$ such that

(1) E is a subfield of K;
(2) $\boldsymbol{k}_{\mathcal{E}} \subseteq \boldsymbol{k}$ as $\mathcal{L}^*_{\mathrm{r}}$-structures, the underlying ring of $\boldsymbol{k}_{\mathcal{E}}$ is a subfield of $\boldsymbol{k}$, and $\mathrm{ac}(E) \subseteq \boldsymbol{k}_{\mathcal{E}}$;
(3) $\Gamma_{\mathcal{E}} \subseteq \Gamma$ as $\mathcal{L}^*_{\mathrm{v}}$-structures with $v(E^\times) \subseteq \Gamma_{\mathcal{E}}$.

Assume in the rest of this section that $\mathcal{K} = (K, \boldsymbol{k}, \Gamma; \cdots)$ and $\mathcal{K}' = (K', \boldsymbol{k}', \Gamma'; \cdots)$ are $*$-valued fields, and that $\mathcal{E} = (E, \boldsymbol{k}_{\mathcal{E}}, \Gamma_{\mathcal{E}})$, $\mathcal{E}' = (E', \boldsymbol{k}_{\mathcal{E}'}, \Gamma_{\mathcal{E}'})$ are good substructures of $\mathcal{K}$, $\mathcal{K}'$ respectively.

A *good map* $\mathbf{f} : \mathcal{E} \to \mathcal{E}'$ is a triple $\mathbf{f} = (f, f_{\mathrm{r}}, f_{\mathrm{v}})$ consisting of an isomorphism $f : E \to E'$ of fields, an isomorphism $f_{\mathrm{r}} : \boldsymbol{k}_{\mathcal{E}} \to \boldsymbol{k}_{\mathcal{E}'}$ of $\mathcal{L}^*_{\mathrm{r}}$-structures, and an isomorphism $f_{\mathrm{v}} : \Gamma_{\mathcal{E}} \to \Gamma_{\mathcal{E}'}$ of $\mathcal{L}^*_{\mathrm{v}}$-structures, such that

(i) $f_r\big(\mathrm{ac}(a)\big) = \mathrm{ac}'\big(f(a)\big)$ for all $a \in E$, and f_r is elementary as a partial map between the $\mathcal{L}_r^*$-structures $\boldsymbol{k}$ and $\boldsymbol{k}'$;
(ii) $f_v\big(v(a)\big) = v'\big(f(a)\big)$ for all $a \in E^\times$, and f_v is elementary as a partial map between the $\mathcal{L}_v^*$-structures Γ and Γ'.

Theorem 5.21 (the Equivalence Theorem) goes through in this enriched setting, with the same proof except for obvious changes:

Theorem 7.11. *If* $\mathrm{char}(\boldsymbol{k}) = 0$ *and* $\mathcal{K}$, $\mathcal{K}'$ *are henselian, then any good map* $\mathcal{E} \to \mathcal{E}'$ *is a partial elementary map between* $\mathcal{K}$ *and* $\mathcal{K}'$.

The four corollaries of Sect. 5.6 also go through in this enriched setting, with residue fields and value groups replaced by their $\mathcal{L}_r^*$-expansions and $\mathcal{L}_v^*$-expansions, respectively, and with $\mathcal{L}^*$ instead of $\mathcal{L}(\mathrm{ac})$. In defining special r-formulas and special v-formulas the roles of $\mathcal{L}_r$ and $\mathcal{L}_v$ are of course taken over by $\mathcal{L}_r^*$ and $\mathcal{L}_v^*$, respectively. Except for obvious changes the proofs are the same as in Sect. 5.6, using Theorem 7.11 in place of Theorem 5.21.

7.3 Eliminating Angular Components

For use in the mixed characteristic case we need a variant of Theorem 7.11 without angular component maps. Below we derive such a variant at the cost of a purity assumption. So let $\mathcal{K}, \mathcal{K}'$ be as in the previous section except that we do not require angular component maps as part of these structures. We modify the notion of *good substructure* of $\mathcal{K}$ by replacing in clause (2) of its definition the condition $\mathrm{ac}(E) \subseteq \boldsymbol{k}_{\mathcal{E}}$ by $\pi(\mathcal{O}_E) \subseteq \boldsymbol{k}_{\mathcal{E}}$. In defining the notion of a good map $\mathbf{f} = (f, f_v, f_r) : \mathcal{E} \to \mathcal{E}'$ the condition on f_r is to be changed to: $f_r\big(\pi(a)\big) = \pi'\big(f(a)\big)$ for all $a \in \mathcal{O}_E$, and f_r is elementary as a partial map between the $\mathcal{L}_r^*$-structures $\boldsymbol{k}$ and $\boldsymbol{k}'$.

Theorem 7.12. *Suppose that* $\mathrm{char}(\boldsymbol{k}) = 0$ *and* $\mathcal{K}$ *and* $\mathcal{K}'$ *are henselian. Let* $\mathbf{f} : \mathcal{E} \to \mathcal{E}'$ *be a good map between good substructures* $\mathcal{E}$ *of* $\mathcal{K}$ *and* $\mathcal{E}'$ *of* $\mathcal{K}'$, *with* $v(E^\times)$ *pure in* Γ. *Then* $\mathbf{f}$ *is a partial elementary map between* $\mathcal{K}$ *and* $\mathcal{K}'$.

Proof. The desired conclusion holds trivially if $\Gamma = \{0\}$. Let $\Gamma \neq \{0\}$. We first arrange that the valued subfield $\big(E, \pi(\mathcal{O}_E), v(E^\times); \dots\big)$ of $\mathcal{K}$ is $\aleph_1$-saturated by passing to an elementary extension of a suitable many-sorted structure with $\mathcal{K}$, $\mathcal{K}'$, $\mathcal{E}$, $\mathcal{E}'$ and $\mathbf{f}$ as ingredients. Then Lemma 7.9 gives a cross-section $s_E : v(E^\times) \to E^\times$ on this valued subfield of $\mathcal{K}$. As in the beginning of the proof of Theorem 5.21 we arrange next that $\mathcal{K}$ and $\mathcal{K}'$ are κ-saturated, where κ is an uncountable cardinal such that $|\boldsymbol{k}_{\mathcal{E}}|$, $|\Gamma_{\mathcal{E}}| < \kappa$. Then we apply the extension procedure (3) in the proof of Theorem 5.21 to arrange that $\boldsymbol{k}_{\mathcal{E}} = \pi(\mathcal{O}_E)$, without changing $v(E^\times)$. (The valued subfield $(E, \pi(\mathcal{O}_E), v(E^\times); \dots)$ of $\mathcal{K}$ might not be $\aleph_1$-saturated anymore after this extension, but it keeps its cross-section s_E.) To simplify notation we identify $\mathcal{E}$ and $\mathcal{E}'$ via $\mathbf{f}$; we have to show that then $\mathcal{K} \equiv_{\mathcal{E}} \mathcal{K}'$. Since the group $v(E^\times)$ is $\aleph_1$-saturated, Lemma 7.10 yields cross-sections

$$s : \Gamma \to K^\times, \qquad s' : \Gamma' \to (K')^\times$$

such that s and s' extend s_E. These cross-sections induce angular component maps on $\mathcal{E}$, $\mathcal{K}$, and $\mathcal{K}'$, as indicated in Sect. 5.4, and this allows us to apply Theorem 7.11 to obtain the desired conclusion. □

7.4 Coarsening

Let Γ be an ordered abelian group. A *convex* subgroup of Γ is a subgroup Δ of Γ such that for all $\gamma \in \Gamma$ and $\delta \in \Delta$,

$$|\gamma| < |\delta| \implies \gamma \in \Delta.$$

Given any $\gamma \in \Gamma$ there is a smallest convex subgroup of Γ that contains γ, namely

$$\{\delta \in \Gamma : |\delta| \le n|\gamma| \text{ for some } n\},$$

If $\gamma \in \Gamma$ and $\gamma \neq 0$, there is also a largest convex subgroup of Γ that does not contain γ, namely

$$\{\delta \in \Gamma : n|\delta| < |\gamma| \text{ for all } n\}.$$

The subgroups $\{0\}$ and Γ of Γ are convex, and Γ is archimedean iff there are no other convex subgroups of Γ. The set of all convex subgroups of Γ is linearly ordered by inclusion.

Let Δ be a convex subgroup of Γ. Then we make the quotient group Γ/Δ into an ordered abelian group, by defining $\gamma + \Delta > 0 + \Delta$ iff $\gamma > \Delta$. The natural map $\Gamma \to \Gamma/\Delta$ then preserves $\le$: for $\gamma_1, \gamma_2 \in \Gamma$ we have

$$\gamma_1 \le \gamma_2 \implies \gamma_1 + \Delta \le \gamma_2 + \Delta.$$

Next, consider a valuation $v : K^\times \to \Gamma$ on a field K with valuation ring $\mathcal{O}$, maximal ideal $\mathfrak{m}$ of $\mathcal{O}$, and residue field $\boldsymbol{k} = \mathcal{O}/\mathfrak{m}$. Let a convex subgroup Δ of Γ be given. Then we define the *coarsening* $\dot{v}$ of v as follows:

$$\dot{v} : K^\times \to \dot{\Gamma} := \Gamma/\Delta, \qquad \dot{v}(a) := v(a) + \Delta.$$

This coarsening is again a valuation on K, with valuation ring

$$\dot{\mathcal{O}} = \{a \in K : v(a) \ge \delta \text{ for some } \delta \in \Delta\}.$$

The maximal ideal of $\dot{\mathcal{O}}$ is

$$\dot{\mathfrak{m}} = \{a \in K : v(a) > \Delta\}.$$

Thus $\mathcal{O} \subseteq \dot{\mathcal{O}}$ and $\mathfrak{m} \supseteq \dot{\mathfrak{m}}$. For $a \in \dot{\mathcal{O}}$ we set $\dot{a} := a + \dot{\mathfrak{m}} \in \dot{\mathcal{O}}/\dot{\mathfrak{m}}$.

Lemma 7.13. *If the local ring $\mathcal{O}$ is henselian, then so is $\dot{\mathcal{O}}$.*

Proof. Assume $\mathcal{O}$ is henselian. Let $f(x) = 1 + x + a_2x^2 + \cdots + a_nx^n$ with $n \geq 2$, $a_2, \ldots, a_n \in \dot{\mathfrak{m}}$. Then $a_2, \ldots, a_n \in \mathfrak{m}$, so $f(x)$ has a zero $b \in \mathcal{O}$, and so this zero also lies in $\dot{\mathcal{O}}$. Thus $\dot{\mathcal{O}}$ is henselian. □

The information that gets lost by this coarsening can be recovered in the residue field $\dot{K} := \dot{\mathcal{O}}/\dot{\mathfrak{m}} = \{\dot{a} : a \in \dot{\mathcal{O}}\}$ by means of the *induced* valuation

$$v : \dot{K}^{\times} \to \Delta, \qquad v(\dot{a}) := v(a) \text{ for } a \in \dot{\mathcal{O}} \setminus \dot{\mathfrak{m}},$$

as we shall see later in this section. (To keep notations simple we use the same letter v for the initial valuation on K and this induced valuation on $\dot{K}$.) Note that $\dot{\mathfrak{m}}$ is a prime ideal of $\mathcal{O}$, and that the subring $\mathcal{O}/\dot{\mathfrak{m}}$ of $\dot{\mathcal{O}}/\dot{\mathfrak{m}} = \dot{K}$ is the valuation ring of this induced valuation v on $\dot{K}$, with maximal ideal $\mathfrak{m}/\dot{\mathfrak{m}}$.

Thus in some sense the original valuation $v : K^{\times} \to \Gamma$ is decomposed into two simpler valuations, namely $\dot{v} : K^{\times} \to \dot{\Gamma}$ with residue field $\dot{K}$, and $v : \dot{K}^{\times} \to \Delta$. Let us next consider the situation where Γ has a smallest positive element, call it 1. Then $\Delta := \mathbb{Z}1$ is the smallest convex subgroup of Γ that contains 1 and $v : \dot{K} \to \mathbb{Z}1$ is a discrete valuation, and so its valuation ring $\mathcal{O}/\dot{\mathfrak{m}}$ is a DVR. Pick some $t \in \mathcal{O}$ with $v(t) = 1$. It is easy to check that then

$$\dot{\mathcal{O}} = \mathcal{O}[1/t], \qquad \dot{\mathfrak{m}} = \bigcap_{n=0}^{\infty} t^n\mathcal{O}.$$

In this situation and with this $\Delta = \mathbb{Z}1$ we have:

Lemma 7.14. *Suppose the valued field $(K, \mathcal{O})$ is $\aleph_1$-saturated. Then $\mathcal{O}/\dot{\mathfrak{m}}$ is a complete* DVR.

Proof. Let (a_n) be a sequence in $\mathcal{O}$ such that $v(\dot{a}_{n+1} - \dot{a}_n) \geq n \cdot 1$ for all n. Then $v(a_{n+1}-a_n) \geq n{\cdot}1$ for all n, so by saturation we get $a \in \mathcal{O}$ such that $v(a-a_n) \geq n{\cdot}1$ for all n, and thus $\lim_{n\to\infty} \dot{a}_n = \dot{a}$ in $\dot{\mathcal{O}}$. □

In the rest of this section we fix a valued field

$$\mathcal{K} = (K, \boldsymbol{k}, \Gamma; \pi, v)$$

construed as a 3-sorted structure as indicated, with valuation ring $\mathcal{O} := \mathcal{O}_v$, and maximal ideal $\mathfrak{m} := \mathfrak{m}_v$. Let Δ be a convex subgroup of Γ. This gives $\dot{\Gamma} := \Gamma/\Delta$, the coarsening $\dot{v} : K^{\times} \to \dot{\Gamma}$ of v, with valuation ring $\dot{\mathcal{O}} = \mathcal{O}_{\dot{v}}$ and maximal ideal $\dot{\mathfrak{m}}$ of $\dot{\mathcal{O}}$. Let $\dot{K} = \dot{\mathcal{O}}/\dot{\mathfrak{m}}$ be the residue field for $\dot{v}$ (as before), and $\dot{\pi} : \dot{\mathcal{O}} \to \dot{K}$ the canonical map. Then $\dot{\pi}(\mathcal{O}) = \mathcal{O}/\dot{\mathfrak{m}}$ is a valuation ring of $\dot{K}$. This gives a (3-sorted) valued field $(K, \dot{K}, \dot{\Gamma};\ \dot{\pi}, \dot{v})$, as well as the (1-sorted) valued field $\dot{K}(*) := \big(\dot{K}, \dot{\pi}(\mathcal{O})\big)$. We combine these into the 3-sorted structure

$$\mathcal{K}(*) \ :=\ \big(K, \dot{K}(*), \dot{\Gamma};\ \dot{\pi}, \dot{v}\big).$$

As promised, we now recover $\mathcal{O}$ in this enriched coarsening of $\mathcal{K}$: the set $\mathcal{O} \subseteq K$ is definable in the structure $\mathcal{K}(*)$ by a formula independent of $\mathcal{K}$, since

$$\mathcal{O}\ =\ \{a \in \dot{\mathcal{O}} : \dot{\pi}(a) \in \dot{\pi}(\mathcal{O})\}.$$

In this way we reconstruct $\mathcal{K}$ from $\mathcal{K}(*)$. The advantage of working with $\mathcal{K}(*)$ is that it has equicharacteristic 0 if $\mathcal{K}$ has mixed characteristic $(0, p)$ with $v(p) \in \Delta$. Below we also need the surjective ring morphism

$$\pi_\Delta : \mathcal{O}/\dot{\mathfrak{m}} = \dot{\pi}(\mathcal{O}) \to \boldsymbol{k}, \qquad \pi_\Delta(\dot{a}) \ :=\ \pi(a) \text{ for } a \in \mathcal{O}.$$

Assume now that $\operatorname{char} K = 0$, $\operatorname{char} \boldsymbol{k} = p$, and $\boldsymbol{k}$ is perfect, and $\mathcal{K}$ is unramified and $\aleph_1$-saturated. Then $v(p)$ is the smallest positive element of Γ, so $\mathbb{Z} \cdot v(p)$ is a convex subgroup of Γ. We take $\Delta := \mathbb{Z} \cdot v(p)$, so $\operatorname{char} \dot{K} = 0$. From Lemma 7.14 we get that $\dot{\pi}(\mathcal{O})$ is a complete discrete valuation ring of $\dot{K}$. Since $\boldsymbol{k}$ is perfect, Theorem 6.11 gives a unique ring isomorphism

$$\iota : \mathrm{W}[\boldsymbol{k}] \ \cong\ \dot{\pi}(\mathcal{O})$$

such that $\pi_\Delta \circ \iota : \mathrm{W}[\boldsymbol{k}] \to \boldsymbol{k}$ is the projection map $(a_0, a_1, a_2, \dots) \mapsto a_0$. Note that ι extends to a field isomorphism $\mathrm{W}(\boldsymbol{k}) \ \cong\ \dot{K}$. Under these conditions and with these notations we have:

Lemma 7.15. *Let E be a subfield of K and $\boldsymbol{k}_1$ a subfield of $\boldsymbol{k}$ such that the valuation ring $\mathcal{O}_E := \mathcal{O} \cap E$ of E has perfect residue field $\pi(\mathcal{O}_E) \subseteq \boldsymbol{k}_1$. Then the subring $\dot{\pi}(\mathcal{O}_E)$ of $\dot{\pi}(\mathcal{O})$ is a valuation ring contained in $\iota(\mathrm{W}[\boldsymbol{k}_1])$.*

Proof. By assumption, $\boldsymbol{k}_0 := \pi(\mathcal{O}_E)$ is a perfect subfield of $\boldsymbol{k}$. Then Lemma 6.12 gives $\dot{\pi}(\mathcal{O}_E) \subseteq \iota(\mathrm{W}[\boldsymbol{k}_0])$, and the latter is contained in $\iota(\mathrm{W}[\boldsymbol{k}_1])$. □

7.5 Functorial Back-and-Forth

If $\mathcal{M} = (M; \cdots)$ and $\mathcal{N} = (N; \cdots)$ are L-structures with a common subset A of M and N, then $\mathcal{M} \equiv_A \mathcal{N}$ means that $\mathcal{M}$ and $\mathcal{N}$ are elementarily equivalent over A, that is, their natural expansions to L_A-structures are elementarily equivalent.

Lemma 7.16. *Let R and S be rings with a common subring A such that $R \equiv_A S$. Let κ be an uncountable cardinal such that the rings R and S are κ-saturated and $|A| < \kappa$. Then, as rings,*

$$\mathrm{W}[R] \equiv_{\mathrm{W}[A]} \mathrm{W}[S].$$

Proof. We just apply the functor W to a suitable back-and-forth system between R and S. In detail, let B range over the subrings of R with $A \subseteq B$ and $|B| < \kappa$, and let C range over the subrings of S with $A \subseteq C$ and $|C| < \kappa$. Let Φ be the set of all ring isomorphisms $\phi : B \to C$ that are the identity on A and are partial elementary maps between the rings R and S. Note that the identity map on A belongs to Φ, so $\Phi \neq \emptyset$ and Φ is a back-and-forth system between R and S. The functor W yields a back-and-forth system $\mathrm{W}[\Phi]$ between $\mathrm{W}[R]$ and $\mathrm{W}[S]$ consisting of the

$$\mathrm{W}[\phi] : \mathrm{W}[B] \to \mathrm{W}[C]$$

with $\phi : B \to C$ an element of Φ. □

A similar use of functoriality gives:

Lemma 7.17. *Let Γ be an ordered abelian group, and Γ_1 and Γ_2 ordered abelian group extensions of Γ with a common smallest positive element $1 \in \Gamma$. Let κ be an uncountable cardinal such that Γ_1 and Γ_2 are κ-saturated, $|\Gamma| < \kappa$, and $\Gamma_1 \equiv_\Gamma \Gamma_2$. Let Δ be the common convex subgroup $\mathbb{Z} \cdot 1$ of Γ, Γ_1 and Γ_2. Then the ordered quotient groups $\dot{\Gamma}_1 := \Gamma_1/\Delta$ and $\dot{\Gamma}_2 := \Gamma_2/\Delta$ are elementarily equivalent over their common ordered subgroup $\dot{\Gamma} := \Gamma/\Delta$.*

7.6 Equivalence in Mixed Characteristic

In this final section $\mathcal{K} = (K, \boldsymbol{k}, \Gamma;\ \pi, v)$ is an unramified henselian valued field of mixed characteristic $(0, p)$ with perfect residue field $\boldsymbol{k}$. (Note that $\mathcal{K}$ is not equipped here with an angular component map.) A *good substructure* of $\mathcal{K}$ is by definition a triple $\mathcal{E} = (E, \boldsymbol{k}_\mathcal{E}, \Gamma_\mathcal{E})$ consisting of a subfield E of K, a subfield $\boldsymbol{k}_\mathcal{E}$ of $\boldsymbol{k}$ with $\pi(\mathcal{O}_E) \subseteq \boldsymbol{k}_\mathcal{E}$, and an ordered subgroup $\Gamma_\mathcal{E}$ of Γ with $v(E^\times) \subseteq \Gamma_\mathcal{E}$.

For the next theorem, we consider also a second unramified henselian valued field $\mathcal{K}' = (K', \boldsymbol{k}', \Gamma';\ \pi', v')$ of mixed characteristic $(0, p)$ with perfect residue field $\boldsymbol{k}'$, and good substructures $\mathcal{E} = (E, \boldsymbol{k}_\mathcal{E}, \Gamma_\mathcal{E})$ of $\mathcal{K}$ and $\mathcal{E}' = (E', \boldsymbol{k}_{\mathcal{E}'}, \Gamma_{\mathcal{E}'})$ of $\mathcal{K}'$. A *good map* $\mathbf{f} : \mathcal{E} \to \mathcal{E}'$ is then a triple $(f, f_\mathrm{r}, f_\mathrm{v})$ consisting of field isomorphisms $f : E \to E'$ and $f_\mathrm{r} : \boldsymbol{k}_\mathcal{E} \to \boldsymbol{k}_{\mathcal{E}'}$, and an isomorphism $f_\mathrm{v} : \Gamma_\mathcal{E} \to \Gamma_{\mathcal{E}'}$ of ordered abelian groups, such that

(i) $f_\mathrm{r}\big(\pi(a)\big) = \pi'\big(f(a)\big)$ for all $a \in \mathcal{O}_E$, and f_r is elementary as a partial map between the fields $\boldsymbol{k}$ and $\boldsymbol{k}'$;
(ii) $f_\mathrm{v}\big(v(a)\big) = v'\big(f(a)\big)$ for all $a \in E^\times$, and f_v is elementary as a partial map between the ordered abelian groups Γ and Γ'.

Theorem 7.12 has a weak analogue in the present mixed characteristic setting:

Theorem 7.18. *Suppose $\pi(\mathcal{O}_E)$ is perfect, $v(E^\times) = \Gamma_\mathcal{E}$ and $\Gamma_\mathcal{E}$ is pure in Γ. Then any good map $\mathbf{f} : \mathcal{E} \to \mathcal{E}'$ is a partial elementary map between $\mathcal{K}$ and $\mathcal{K}'$.*

Proof. Let $\mathbf{f} : \mathcal{E} \to \mathcal{E}'$ be a good map. To simplify notation we identify $\mathcal{E}$ and $\mathcal{E}'$ via $\mathbf{f}$, so $\mathbf{f}$ becomes the identity on $\mathcal{E}$. We have to show that then $\mathcal{K} \equiv_{\mathcal{E}} \mathcal{K}'$. By passing to suitable elementary extensions we arrange that $\mathcal{K}$ and $\mathcal{K}'$ are κ-saturated, where κ is an uncountable cardinal such that $|\boldsymbol{k}_{\mathcal{E}}|, |\Gamma_{\mathcal{E}}| < \kappa$. With $v(p) = 1$ as the smallest positive element of $\Gamma_{\mathcal{E}}$ and of Γ and Γ' and with $\Delta := \mathbb{Z} \cdot 1$ we have $\dot{\Gamma} \equiv_{\dot{\Gamma}_{\mathcal{E}}} \dot{\Gamma}'$ by Lemma 7.17 and using its notations. From the purity of $v(E^{\times})$ in Γ it follows that $\dot{v}(E^{\times}) = \dot{\Gamma}_{\mathcal{E}}$ is pure in $\dot{\Gamma}$. As explained in Sect. 7.4 we have the (one-sorted) valued fields

$$\dot{K}(*) = \big(\dot{K}, \dot{\pi}(\mathcal{O})\big), \qquad \dot{K}'(*) = \big(\dot{K}', \dot{\pi}'(\mathcal{O}')\big).$$

Now $\mathcal{K}$ and $\mathcal{K}'$ are $\aleph_1$-saturated, so as indicated at the end of Sect. 7.4 we may identify the valuation rings $\dot{\pi}(\mathcal{O})$ of $\dot{K}$ and $\dot{\pi}'(\mathcal{O}')$ of $\dot{K}'$ with $\mathrm{W}[\boldsymbol{k}]$ and $\mathrm{W}[\boldsymbol{k}']$, respectively. By Lemma 7.16 the rings $\mathrm{W}[\boldsymbol{k}]$ and $\mathrm{W}[\boldsymbol{k}']$ are elementarily equivalent over their common subring $\mathrm{W}[\boldsymbol{k}_{\mathcal{E}}]$. Let G be the fraction field of $\mathrm{W}[\boldsymbol{k}_{\mathcal{E}}]$. Taking G as a common subfield of $\dot{K}$ and $\dot{K}'$ gives therefore

$$\dot{K}(*) \equiv_G \dot{K}'(*).$$

Hence the assumptions of Theorem 7.12 are satisfied with

$$\mathcal{K}(*) := \big(K, \dot{K}(*), \dot{\Gamma};\, \dot{\pi}, \dot{v}\big) \text{ and } \mathcal{K}'(*) := \big(K', \dot{K}'(*), \dot{\Gamma}';\, \dot{\pi}', \dot{v}'\big)$$

in the role of $\mathcal{K}$ and $\mathcal{K}'$, and $\mathcal{E}(*) := (E, G, \dot{\Gamma}_{\mathcal{E}})$ in the role of both $\mathcal{E}$ and $\mathcal{E}'$ and with the identity on $\mathcal{E}(*)$ as a good map. (There is a subtlety here: for $(E, G, \dot{\Gamma}_{\mathcal{E}})$ to be a good substructure of $\mathcal{K}(*)$ requires $\dot{\pi}(\dot{\mathcal{O}}_E) \subseteq G$, where $\dot{\mathcal{O}}_E := \dot{\mathcal{O}} \cap E$. To see that this inclusion holds, use that under the above identification of $\dot{\pi}(\mathcal{O})$ with $\mathrm{W}[\boldsymbol{k}]$ we have $\dot{\pi}(\mathcal{O}_E) \subseteq \mathrm{W}[\boldsymbol{k}_{\mathcal{E}}]$ in view of Lemma 7.15. This lemma is applicable since the residue field $\pi(\mathcal{O}_E) \subseteq \boldsymbol{k}$ of $\mathcal{O}_E$ is perfect.)

Applying Theorem 7.12 therefore gives $\mathcal{K}(*) \equiv_{\mathcal{E}(*)} \mathcal{K}'(*)$. This yields $\mathcal{K} \equiv_{\mathcal{E}} \mathcal{K}'$, since by a remark in Sect. 7.4 the valuation rings of $\mathcal{K}$ and $\mathcal{K}'$ can be defined in $\mathcal{K}(*)$ and $\mathcal{K}'(*)$ by the same quantifier-free formula. □

We can now derive the results stated in the introduction to this chapter:

Proof of Theorem 7.1. It is clear that if $\mathcal{K} \equiv \mathcal{K}'$, then $\boldsymbol{k} \equiv \boldsymbol{k}'$ (as fields) and $\Gamma \equiv \Gamma'$ (as ordered abelian groups). For the converse, assume $\boldsymbol{k} \equiv \boldsymbol{k}'$ as fields, and $\Gamma \equiv \Gamma'$ as ordered groups. With the usual identifications, $\mathbb{Q}$ and $\mathbb{F}_p$ are subfields of K and $\boldsymbol{k}$, with $\mathcal{O} \cap \mathbb{Q} = \mathbb{Z}_{(p\mathbb{Z})}$, and $\mathbb{Z}$ is a subgroup of Γ. Thus we have a good substructure $\mathcal{E} := (\mathbb{Q}, \mathbb{F}_p, \mathbb{Z})$ of $\mathcal{K}$, and likewise we have a good substructure $\mathcal{E}' := (\mathbb{Q}, \mathbb{F}_p, \mathbb{Z})$ of $\mathcal{K}'$. It is trivial to verify that the identity is a good map $\mathcal{E} \to \mathcal{E}'$. Now apply Theorem 7.18. □

Theorem 7.2 is just a special case of Theorem 7.18.

Proof of Theorem 7.3. By standard reductions it suffices to prove:

Claim. Suppose $\mathcal{K}$ is $\aleph_1$-saturated, $\mathcal{E} = (E, \boldsymbol{k}_E, \Gamma_E; \dots) \preccurlyeq \mathcal{K}$ is countable, and $r, r' \in \boldsymbol{k}^n$ have the same type over $\boldsymbol{k}_E$, that is, $\text{tp}(r|\boldsymbol{k}_E) = \text{tp}(r'|\boldsymbol{k}_E)$ in the field $\boldsymbol{k}$. Then r and r' have the same type over $(E, \boldsymbol{k}_E, \Gamma_E)$ in $\mathcal{K}$.

To prove this claim, let $\boldsymbol{k}_1 := \boldsymbol{k}_E(r)$ and $\boldsymbol{k}_1' := \boldsymbol{k}_E(r')$ (subfields of $\boldsymbol{k}$). Then $(E, \boldsymbol{k}_1, \Gamma_E)$ and $(E, \boldsymbol{k}_1', \Gamma_E)$ are good substructures of $\mathcal{K}$, and the assumption on types yields a good map $(E, \boldsymbol{k}_1, \Gamma_E) \to (E, \boldsymbol{k}_1', \Gamma_E)$ that is the identity on E, $\boldsymbol{k}_E$ and Γ_E and sends r to r'. Note also that $v(E^\times) = \Gamma_E$ is pure in Γ. It remains to apply Theorem 7.18. □

Final Remarks, and an Erratum. I don't know if we can omit the assumption in Theorem 7.18 that $v(E^\times) = \Gamma_{\mathcal{E}}$. This assumption is tacitly used at the very end of its proof to justify the step from $\mathcal{K}(*) \equiv_E \mathcal{K}'(*)$ to $\mathcal{K} \equiv_E \mathcal{K}'$: without $v(E^\times) = \Gamma_{\mathcal{E}}$ it is not clear how to reconstruct $\Gamma_{\mathcal{E}}$ from $\mathcal{K}(*)$. If we could drop $v(E^\times) = \Gamma_{\mathcal{E}}$ from the hypothesis of the theorem, we could derive a stronger conclusion in Theorem 7.3, namely, that any subset of $\boldsymbol{k}^m \times \Gamma^n$ definable in $\mathcal{K}$ is a finite union of rectangles $Y \times Z$ with $Y \subseteq \boldsymbol{k}^m$ definable in the field $\boldsymbol{k}$ and $Z \subseteq \Gamma^n$ definable in the ordered abelian group Γ.

The paper [7] errs on this issue: it considers valued *difference* fields, so in addition to the valued field structure, $\mathcal{K}$ has also a valuation-preserving automorphism of the underlying field as part of the structure, but otherwise the treatment there in the mixed characteristic case is quite similar to what is done in the present chapter. However, Theorem 8.8 in [7], which corresponds to the above Theorem 7.18, does not have $v(E^\times) = \Gamma_{\mathcal{E}}$ among its assumptions, and so the last sentence of its proof is unjustified. Another error in [7] is that it omits requiring $\pi(\mathcal{O})$ to be perfect in Theorem 8.8, and omits the facts corresponding to the above Lemmas 6.12 and 7.15. Accordingly, Theorem 8.8 in [7] and its Corollary 8.11 should be corrected along the lines of the above Theorem 7.18 and its corollary Theorem 7.3.

8 References with Comments

Before listing books and papers, let me try to give some historical perspective. Valuation theory began with the nineteenth century realization of similarities between function fields in one variable on the one hand, and number fields on the other hand. We should mention in this connection the resemblance between evaluating a function at a point and taking the residue class of an integer modulo a prime number. This extends to the analogy between developing the function into a power series around the point, and considering the integer modulo higher and higher powers of the prime. The subject is now old enough to have a history, with Hensel and Hahn among the pioneers, and Kürschák, Ostrowski, Hasse, Krull, Deuring, Teichmüller, and Witt as other key figures in work done mainly in Germany before the second world war. For the topics of our notes it is still worth consulting

Hahn [26], Ostrowski's monumental [39] (investigating valued fields of rank 1), and Krull [33]. Still in this time period the subject was taken up in America by Mac Lane, Schilling, and Kaplansky, and most prominently by Zariski in his study of local uniformization and resolution of singularities. A nice overview of the early development is in [44]. From the end of this early period Kaplansky's paper [31] foreshadows later AKE-developments. Zariski's use of valuation theory as a tool in algebraic geometry was continued notably by Abhyankar; see for example [1].

Some books in the list of references are mainly within valuation theory: [19, 41, 42, 47]. Others have valuation theory as a substantial chapter in commutative algebra: [9, 38, 53]. I also list some books with a wide variety of connections to valuation theory or local algebra: [11, 12, 24, 25, 28, 43, 48, 52]. Some books are listed for various other reasons: [10, 34, 49, 50].

F.-V. Kuhlmann is preparing an extensive treatise on valuation theory, with much more on the positive characteristic case than in the present notes. Some of the chapters are available on his webpage.

Here are brief comments on several of these books. The first application of model theory to valuation theory is in [43]. The books [9, 38, 53] testify to the vigorous development of local algebra in the period 1930–1960, with the theory of henselian local rings and henselization of particular relevance for the topics of the present notes. The book [28] shows how ideas from stability theory impact valuation theory (and conversely).

Next I comment mainly on articles that appeared after 1960. The original papers by Ax & Kochen are [4–6], and those by Ershov are [20–23]. A counterexample to Artin's conjecture on homogeneous forms is in [51]. Paul Cohen found a constructive way to obtain the AKE-results in [13]. This had its influence on later work such as [18]. The reduction of mixed characteristic to equicharacteristic zero by means of coarsening (for results on elementary equivalence) can be found in [32].

In the Introduction we briefly mentioned Artin's Approximation Theorem [2, 3]. This fundamental theorem in local algebra is quite independent of the AKE results but in a similar spirit.

Macintyre's paper [36] eliminates quantifiers for the field of p-adic numbers in the language of rings augmented for each $n \geq 2$ by a predicate P_n for the set of nth powers. See also [14] for generalizations, and [41] for a highly readable treatment from scratch. A big impact of [36] was its new focus on definable sets, and on the analogy with semialgebraic sets over the field of real numbers. This aspect became very prominent in [15, 18] and [40] by Denef and Pas. The cell decompositions in these papers are similar to cell decomposition in o-minimality, a topic that also got started in the early 1980s. Denef has also shown that eliminating quantifiers for henselian valued fields can be done quite efficiently via resolution of singularities.

These results get extended to p-adic fields equipped with restricted analytic functions in [17]. This yields a p-adic version of the theory of real semianalytic and subanalytic sets due to Łojasiewicz, Gabrielov, and Hironaka. A key new ingredient is Weierstrass division (piecewise uniformly in parameters).

Likewise, algebraically closed valued fields with analytic structure give rise to a theory of rigid subanalytic sets that complements rigid analytic geometry. This has been carried out in a series of papers by Lipshitz and Robinson, of which we mention only [35].

A different kind of extra structure is studied in [45] and [8]. These papers extend the AKE-results to certain classes of valued *differential* fields and of valued *difference* fields, requiring the right notions of "differential-henselian" and "difference-henselian" as well as new tricks with pc-sequences. Thus [8] covers valued fields $W(\boldsymbol{k})$ equipped with its Witt-Frobenius automorphism $\mathbf{F}$, under a condition on the residue field $\boldsymbol{k}$ that is for example satisfied when $\boldsymbol{k}$ is algebraically closed. This condition on the difference residue field is weakened in [7], which also considers definable sets in more detail.

In our Chap. 6 on Witt vectors the shift map $\mathbf{V}$ played a certain role. We note here that $\mathbf{V}$ is definable in the expansion of the valued field $W(\boldsymbol{k})$ by $\mathbf{F}$, since for $a \in W[\boldsymbol{k}]$ we have $\mathbf{V}(a) = p\mathbf{F}^{-1}(a)$.

Scanlon's paper [46] has AKE-type results for valued fields that are equipped simultaneously with restricted analytic functions and with an automorphism such as the Witt-Frobenius. This applies in particular to the $W(\boldsymbol{k})$ above, with substantial consequences for p-adic number theory.

The monograph [52] studies a promising candidate for a model-theoretic analysis along the lines of the present notes, namely the differential field $\mathbb{T}$ of transseries. This differential field is of great intrinsic interest, and has a definable valuation ring with a strong interaction of derivation and valuation, very different from the interaction considered in [45].

Holly's thesis [29] is a first step towards eliminating imaginaries for algebraically closed valued fields. The solution to this difficult problem is in [27].

There is a long series of AKE-related papers by Denef, Loeser, Cluckers, and others on motivic integration. See [16] for a brief survey, and [11, 12] for more.

The next (and last) paper we mention is [30] by Hrushovski and Kazhdan which is rich in new ideas. Their top-down approach to motivic integration contains AKE-theory almost as a byproduct. In this connection one should be aware that model theory has witnessed a tremendous internal development in the last fifty years, quite separate from AKE (which mostly uses only elementary model-theoretic ideas). This development goes under the headings of *stability theory* and *geometric model theory*. In [27, 28, 30] we begin to see the power of this combinatorially and geometrically oriented model theory in dealing with valued fields.

The above is not intended to be a representative overview of AKE-theory and its various evolutions and ramifications. One could mention for example local uniformization, tropical geometry, and Berkovich spaces as topics that are in the process of being related to AKE-theory. There are even signs that the global view of number fields and function fields as fields with a *family* of valuations obeying a product formula is not immune to incorporation into a useful model-theoretic framework (Ben Yaacov, Hrushovski).

References

1. S. Abhyankar, On the valuations centered in a local domain. Am. J. Math. **78**, 321–348 (1956)
2. M. Artin, On the solutions of analytic equations. Invent. Math. **5**, 277–291 (1968)
3. M. Artin, Algebraic approximation of structures over complete local rings. Inst. Hautes Études Sci. Publ. Math. **36**, 23–58 (1969)
4. J. Ax, S. Kochen, Diophantine problems over local fields. I. Am. J. Math. **87**, 605–630 (1965)
5. J. Ax, S. Kochen, Diophantine problems over local fields. II. A complete set of axioms for p-adic number theory. Am. J. Math. **87**, 631–648 (1965)
6. J. Ax, S. Kochen, Diophantine problems over local fields. III. Decidable fields. Ann. Math. **83**, 437–456 (1966)
7. S. Azgin, L. van den Dries, Elementary theory of valued fields with a valuation-preserving automorphism. J. Inst. Math. Jussieu **10**, 1–35 (2011)
8. L. Bélair, A. Macintyre, T. Scanlon, Model theory of the Frobenius on the Witt vectors. Am. J. Math. **129**, 665–721 (2007)
9. N. Bourbaki, *Elements of Mathematics. Commutative Algebra* (Hermann, Paris, 1972). Translated from the French
10. G. Cherlin, *Model Theoretic Algebra—Selected Topics*. Lecture Notes in Mathematics, vol. 521 (Springer, Berlin, 1976)
11. R. Cluckers, J. Nicaise, J. Sebag (eds.) *Motivic Integration and Its Interactions with Model Theory and Non-Archimedean Geometry, Volume I*, vol. 383 of *London Mathematical Society Lecture Note Series* (Cambridge University Press, Cambridge, 2011)
12. R. Cluckers, J. Nicaise, J. Sebag (eds.) *Motivic integration and Its Interactions with Model Theory and Non-Archimedean Geometry. Volume II*, vol. 384 of *London Mathematical Society Lecture Note Series* (Cambridge University Press, Cambridge, 2011)
13. P.J. Cohen, Decision procedures for real and p-adic fields. Comm. Pure Appl. Math. **22**, 131–151 (1969)
14. F. Delon, Quelques propriétés des corps valués en théorie des modeles. PhD thesis, Université Paris VII, 1982
15. J. Denef, The rationality of the Poincaré series associated to the p-adic points on a variety. Invent. Math. **77**, 1–23 (1984)
16. J. Denef, F. Loeser, Motivic integration and the Grothendieck group of pseudo-finite fields. In *Proceedings of the International Congress of Mathematicians, Vol. II (Beijing, 2002)*, pp. 13–23 (Higher Ed. Press, Beijing, 2002)
17. J. Denef, L. van den Dries, p-adic and real subanalytic sets. Ann. Math. **128**, 79–138 (1988)
18. J. Denef, p-adic semi-algebraic sets and cell decomposition. J. Reine Angew. Math. **369**, 154–166 (1986)
19. A.J. Engler, A. Prestel, *Valued Fields*. Springer Monographs in Mathematics (Springer, Berlin, 2005)
20. Ju.L. Ershov, On elementary theories of local fields. Algebra i Logika Sem. **4**, 5–30 (1965)
21. Ju.L. Ershov, On elementary theory of maximal normalized fields. Algebra i Logika Sem. **4**, 31–70 (1965)
22. Ju.L. Ershov, On the elementary theory of maximal normed fields. II. Algebra i Logika Sem. **5**, 5–40 (1966)
23. Ju.L. Ershov, On the elementary theory of maximal normed fields. III. Algebra i Logika Sem. **6**, 31–38 (1967)
24. J. Fresnel, M. van der Put, *Géométrie Analytique Rigide et Applications*, vol. 18 of *Progress in Mathematics* (Birkhäuser Boston, Boston, 1981)
25. M.J. Greenberg, *Lectures on Forms in Many Variables* (W. A. Benjamin, New York-Amsterdam, 1969)
26. H. Hahn, Über die nichtarchimedischen Grössensysteme. S.-B. Akad. Wiss. Wien, Math.-naturw. Kl. Abt. IIa **116**, 601–655 (1907)

27. D. Haskell, E. Hrushovski, D. Macpherson, Definable sets in algebraically closed valued fields: elimination of imaginaries. J. Reine Angew. Math. **597**, 175–236 (2006)
28. D. Haskell, E. Hrushovski, D. Macpherson, *Stable Domination and Independence in Algebraically Closed Valued Fields*, vol. 30 of *Lecture Notes in Logic* (Association for Symbolic Logic, Chicago, 2008)
29. J. Holly, Definable equivalence relations and disc spaces of algebraically closed valued fields. PhD thesis, University of Illinois, 1992
30. E. Hrushovski, D. Kazhdan, Integration in valued fields. In *Algebraic geometry and number theory*, vol. 253 of *Progr. Math.*, pp. 261–405 (Birkhäuser Boston, Boston, 2006)
31. I. Kaplansky, Maximal fields with valuations. Duke Math. J. **9**, 303–321 (1942)
32. S. Kochen, The model theory of local fields. In $1xsy$) *ISILC Logic Conference (Proc. Internat. Summer Inst. and Logic Colloq., Kiel, 1974)*, pp. 384–425. Lecture Notes in Math., vol. 499 (Springer, Berlin, 1975)
33. W. Krull, Allgemeine Bewertungstheorie. J. Reine Angew. Math. **167**, 160–196 (1932)
34. S. Lang, *Algebra*, 2nd edn (Addison-Wesley Publishing Company Advanced Book Program, Reading, 1984)
35. L. Lipshitz, Z. Robinson, Rings of separated power series and quasi-affinoid geometry. Astérisque **264**, vi+171 (2000)
36. A. Macintyre, On definable subsets of p-adic fields. J. Symbolic Logic **41**, 605–610 (1976)
37. A. Macintyre, K. McKenna, L. van den Dries, Elimination of quantifiers in algebraic structures. Adv. Math. **47**, 74–87 (1983)
38. M. Nagata, *Local Rings*. Interscience Tracts in Pure and Applied Mathematics, No. 13 (Interscience Publishers a division of Wiley, New York-London, 1962)
39. A. Ostrowski, Untersuchungen zur arithmetischen Theorie der Körper. Math. Z. **39**, 269–404 (1935)
40. J. Pas, Uniform p-adic cell decomposition and local zeta functions. J. Reine Angew. Math. **399**, 137–172 (1989)
41. A. Prestel, P. Roquette, *Formally p-adic Fields*, vol. 1050 of *Lecture Notes in Mathematics* (Springer, Berlin, 1984)
42. P. Ribenboim, *Théorie des Valuations*, vol. 1964 of *Séminaire de Mathématiques Supérieures, No. 9 (Été). Deuxième édition multigraphiée* (Les Presses de l'Université de Montréal, Montreal, 1968)
43. A. Robinson, *Complete Theories* (North-Holland Publishing, Amsterdam, 1956)
44. P. Roquette, History of valuation theory. I. In *Valuation Theory and Its Applications, Vol. I (Saskatoon, SK, 1999)*, vol. 32 of *Fields Inst. Commun.*, pp. 291–355 (American Mathematical Society, Providence, 2002)
45. T. Scanlon, A model complete theory of valued D-fields. J. Symbolic Logic **65**, 1758–1784 (2000)
46. T. Scanlon, Analytic difference rings. In *International Congress of Mathematicians. Vol. II*, pp. 71–92 (Eur. Math. Soc., Zürich, 2006)
47. O.F.G. Schilling, *The Theory of Valuations*. Mathematical Surveys, No. 4 (American Mathematical Society, New York, 1950)
48. J.-P. Serre, *Corps Locaux*. Publications de l'Institut de Mathématique de l'Université de Nancago, VIII. Actualités Sci. Indust., No. 1296 (Hermann, Paris, 1962)
49. J.-P. Serre, *Cours d'arithmétique*, vol. 2 of *Collection SUP: "Le Mathématicien"* (Presses Universitaires de France, Paris, 1970)
50. J.R. Shoenfield, *Mathematical Logic* (Addison-Wesley Publishing, Reading, London-Don Mills, Ont., 1967)
51. G. Terjanian, Un contre-exemple à une conjecture d'Artin. C. R. Acad. Sci. Paris Sér. A-B **262**, A612 (1966)
52. J. van der Hoeven, *Transseries and Real Differential Algebra*, vol. 1888 of *Lecture Notes in Mathematics* (Springer, Berlin, 2006)
53. O. Zariski, P. Samuel, *Commutative Algebra. Vol. II* The University Series in Higher Mathematics (D. Van Nostrand, Princeton, Toronto-London-New York, 1960)

Undecidability in Number Theory

Jochen Koenigsmann

1 Introduction

These lectures are variations on a theme that is faintly echoed in the following loosely connected counterpointing pairs:

Euclid	versus	Diophantos
geometry	versus	arithmetic
decidability	versus	undecidability
Tarski	versus	Gödel
Hilbert	versus	Matiyasevich

Let me explain how.

With his *Elements* which in the Middle Ages was the most popular 'book' after the Bible, Euclid laid a foundation for modern mathematics already around 300 BC. He introduced the axiomatic method according to which every mathematical statement has to be deduced from (very few) first principles (axioms) that have to be so evident that no further justification is required. The paradigm for this is Euclidean geometry.

It wasn't quite so easy for arithmetic (for good reasons as we know now). In the third century AD, Diophantos of Alexandria, often considered the greatest (if not only) algebraist of antique times, tackled what we call today *diophantine equations*, that is, polynomial equations over the integers, to be solved in integers. Diophantos was the first to use symbols for unknowns, for differences and for powers; in short, he invented the polynomial. He was the first to do arithmetic in its own right, not just embedded into geometry (like, e.g., Pythagorean triples). The goal was to find a systematic method, a procedure, an algorithm by which such diophantine equations

J. Koenigsmann (✉)
Mathematical Institute, Radcliff Observatory Quarter, Oxford OX2 6GG, UK
e-mail: koenigsmann@maths.ox.ac.uk

L. van den Dries et al., *Model Theory in Algebra, Analysis and Arithmetic*, Lecture Notes in Mathematics 2111, DOI 10.1007/978-3-642-54936-6_5,

could be solved (like the well known formulas for quadratic equations). One of the oldest, very efficient such algorithm is the Euclidean (!) algorithm for finding the greatest common divisor $\gcd(a, b)$ for two integers a, b. This *is* a diophantine problem: for any integers a, b, c,

$$c = \gcd(a, b) \Leftrightarrow \begin{cases} \text{the three diophantine equations} \\ cu = a,\ cv = b \text{ and } c = ax + by \\ \text{are solvable} \end{cases}$$

(for even more surprising examples of mathematical problems that are diophantine problems 'in disguise', see Sect. 4.4).

In his 10th problem from the famous list of 23 problems presented to the Congress of Mathematicians in Paris in 1900, David Hilbert, rather than asking for an algorithm to produce solutions to diophantine equations, asked for a more modest algorithm that decides whether or not a given diophantine equation *has* a solution. In modern logic terminology, such an algorithm would mean that the existential 1st-order theory of $\mathbb{Z}$ (in the language of rings, $\mathcal{L}_{ring} := \{+, \times; 0, 1\}$) would be decidable (cf. Sect. 4.1).

It would have been even more challenging—and quite in Hilbert's spirit—to show that the full 1st-order theory of $\mathbb{Z}$, often simply called *arithmetic*, would be decidable. However, in 1931, Gödel showed in the first of his two Incompleteness Theorems that this is not the case: *no algorithm can answer every arithmetic YES/NO-question correctly.* It is called 'Incompleteness Theorem' because it says that every effectively (= algorithmically) producible list of axioms true in $\mathbb{Z}$ is incomplete, i.e., cannot axiomatize the full 1st-order theory of $\mathbb{Z}$ (Sect. 2).

This undecidability result generalises to other number theoretic objects, like all number fields (= finite extensions of $\mathbb{Q}$) and their rings of integers, by showing—following Julia Robinson—that $\mathbb{Z}$ is 1st-order definable in any of these (Sect. 3.3). The key tools are the field $\mathbb{Q}_p$ of p-adic numbers (Sect. 3.1) and the Hasse-Minkowski Local-Global-Principle for quadratic forms (Sect. 3.2).

In contrast, around the same time as Gödel's Incompleteness Theorem, Tarski showed that the other classical mathematical discipline, geometry (at least elementary geometry), is decidable; this holds true not only for Euclidean geometry, but for all of algebraic geometry where, when translated into cartesian coordinates, geometric objects don't necessarily obey linear or quadratic equations, but polynomial equations of arbitrary degree over $\mathbb{R}$ or $\mathbb{C}$: The full 1st-order theory of $\mathbb{R}$ (and hence that of $\mathbb{C}$) is decidable.[1]

[1]Even though Tarski may be better known for his decidability results than his undecidability results, one should point out that his 'Undefinability Theorem' (1936) that *arithmetical truth cannot be defined in arithmetic* is very much in the spirit of Gödel's Incompleteness Theorems (in fact, it was discovered independently by Gödel while proving these). Tarski also proved undecidability of various other first-order theories, like, e.g., abstract projective geometry. This may put our very rough initial picture of the five counterpointing pairs into a more accurate historical perspective.

There is a whole zoo of interesting natural intermediate rings between $\mathbb{Z}$ and $\mathbb{C}$ (or rather between $\mathbb{Z}$ and the field $\tilde{\mathbb{Q}}$ of complex algebraic numbers). To explore the boundaries within this zoo of species that belong to the decidable world (like $\tilde{\mathbb{Q}}$, or the field $\mathbb{Q}_p^{alg} := \mathbb{Q}_p \cap \tilde{\mathbb{Q}}$ of p-adic algebraic numbers or the field of totally real numbers or the ring $\tilde{\mathbb{Z}}$ of all algebraic integers) and those on the undecidable side (like number rings or number fields or the ring of totally real integers) is a fascinating task with more open questions than answers (Sects. 3.4 and 3.5).

With the full 1st-order theory of $\mathbb{Z}$ being undecidable, there still might be an algorithm to solve Hilbert's 10th Problem, i.e., an effective decision procedure for the existential 1st-order theory of $\mathbb{Z}$. That this is also not the case is the celebrated result due to Martin Davis, Hilary Putnam, Julia Robinson and Yuri Matiyasevich (sometimes, for short, referred to as 'Matiyasevich's Theorem' as his contribution in 1970 was the last and perhaps most demanding): *Hilbert's 10th Problem is unsolvable*—no algorithm can decide correctly for all diophantine equations whether or not they have integer solutions (Sect. 4).

Maybe the most prominent open problem in the field is the question whether Hilbert's 10th Problem can be solved over $\mathbb{Q}$, i.e., whether there is an algorithm deciding solvability of diophantine equations with solutions in $\mathbb{Q}$. If we had an existential definition of $\mathbb{Z}$ in $\mathbb{Q}$ (which is still open) the answer would again be no because then Hilbert's original 10th Problem over $\mathbb{Z}$ would be reducible to that over $\mathbb{Q}$, and an algorithm for the latter would give one for the former, contradicting Matiyasevich's Theorem.

Instead, in Sect. 5, we reproduce the author's *universal* definition of $\mathbb{Z}$ in $\mathbb{Q}$ which, at least in terms of logical complexity, comes as close to the desired existential definition as one could get so far (Sect. 5.1) and, modulo either of two conjectures from arithmetic geometry, as one ever possibly gets: assuming Mazur's Conjecture or the Bombieri-Lang Conjecture, there is no existential definition of $\mathbb{Z}$ in $\mathbb{Q}$ (Sect. 5.3).

In Sect. 6 we briefly discuss the question of full/existential decidability for several other important rings, not all arising from number theory.

In these notes we do not aim at an encyclopedic survey of what has been achieved in the area, nor do we provide full detailed proofs of the theorems treated (each proof ought to be followed by an exercise: 'fill in the gaps ...'). We rather try to point to the landmarks in the field and their relative position, to allow glimpses into the colourful variety of beautiful methods developed for getting there. Many (often, but not always long-standing) open problems are mentioned to whet the appetite, while the exercises provide working experience with some of the tools introduced. What makes the topic really attractive, especially for graduate students, is that most results don't use very heavy machinery, are elementary in this sense, though, obviously, people did have very good ideas.

I would like to express my warmest thanks to Dugald Macpherson and Carlo Toffalori for giving me the opportunity to hold these lectures in the superb setting of Cetraro, and to the enthusiastic audience for their immense interest, their encouraging questions and their critical remarks. I am also very grateful to the anonymous referee and the editors for their most valuable suggestions for improving on an earlier version of these notes.

2 Decidability, Turing Machines and Gödel's 1st Incompleteness Theorem

In this lecture we would like to sketch the proof of the first of Gödel's celebrated two Incompleteness Theorems. Denoting by $\mathcal{N} := \langle \mathbb{N}; +, \cdot; 0, 1\rangle$ the natural numbers as $\mathcal{L}_{ring}$-structure, where $\mathcal{L}_{ring} := \{+, \cdot; 0, 1\}$, and by $\mathtt{Th}(\mathcal{N})$ its 1st-order $\mathcal{L}_{ring}$-theory, a weak version of the theorem is the following

Theorem 2.1 ([25]). $\mathtt{Th}(\mathcal{N})$ *is undecidable.*

That is, there is *no* algorithm which, on INPUT any $\mathcal{L}_{ring}$-sentence α, gives

$$\text{OUTPUT} \begin{cases} \textbf{YES} \text{ if } \alpha \in \mathtt{Th}(\mathcal{N}), \text{ i.e., } \mathcal{N} \models \alpha \\ \textbf{NO} \quad \text{otherwise} \end{cases}$$

In order to make this statement precise, we will define the notion of an *algorithm* using Turing machines. There have been many alternative definitions (via register machines, λ-calculus, recursive functions etc.) all of which proved to be equivalent. And, indeed, it is the credo of what has come to be called **Church's Thesis** that, no matter how we pin down an exact (and sensible) notion of algorithm, it is going to be equivalent to the existing ones. Whether or not one should take this as more than an empirical fact about the algorithms checked sofar, is an interesting philosophical question.

2.1 *Turing Machines*

A *Turing machine* **T** over a finite *alphabet* $A = \{a_1, a_2, \ldots, a_n\}$ consists of a *tape*

$\cdots$	a_i	a_j	a_k	$\cdots$

with infinitely many *cells*, each of which is either empty (contains the *empty letter* a_0) or contains exactly one a_i $(1 \leq i \leq n)$.

In each step the *tape head* is on exactly one cell and performs exactly one of the following four *operations*:

$\mathbf{a_i}$ type a_i on the working cell $(0 \leq i \leq n)$
r go to the next cell on the right
l go to the next cell on the left
s stop

The *program* (action table) of **T** is a finite sequence of lines of the shape

$$\mathbf{z}\ a\ \mathbf{b}\ \mathbf{z}'$$

where $a \in A \cup \{a_0\}$, where **b** is one of the above operations, and where **z**, **z**′ are from a finite set $Z = \{z_1, \ldots, z_m\}$ of *states*.

The program determines **T** by asking **T** to interpret

$$\mathbf{z}\ a\ \mathbf{b}\ \mathbf{z}'$$

as '*if* **T** *is in state z and the tape head reads a then do* **b** *and go to state z*′'.

T may stop on a given INPUT after finitely many steps (and then the OUTPUT is what's on the tape then) or it runs forever (with no OUTPUT given). A decision algorithm always stops, by definition.

2.2 Coding 1st-Order $\mathcal{L}_{ring}$-Formulas and Turing Machines

Let $A = \{+, \cdot; 0, 1; \doteq, (,), \neg, \rightarrow, \forall, v, '\}$ be the finite alphabet for 1st-order arithmetic, thinking of the variable v_n as the string $v'^{\cdots\prime}$ of length $1 + n$, and assign to the finitely many elements of the disjoint union

$$A \cup Z \cup \{a_0, \mathbf{r}, \mathbf{l}, \mathbf{s}\}$$

distinct positive integers (their *codes*). Code *formulas* via unique prime decomposition, e.g., if $\neg$, 0, $\doteq$, 1 have codes 2, 4, 1, 3 resp., the formula $\rho = \neg 0 \doteq 1$ has code

$$\ulcorner \rho \urcorner = 2^2 \cdot 3^4 \cdot 5^1 \cdot 7^3 = 555660$$

and can be recovered from it.

Similarly, one can define a unique code $\ulcorner \mathbf{T} \urcorner$ for (the program of) each *Turing machine* **T** by coding the sequence of lines in the program.

2.3 Proof of Gödel's 1st Incompleteness Theorem (Sketch)

Suppose, for the sake of finding a contradiction, that there is a Turing machine **T** which decides for any $\mathcal{L}_{ring}$-sentence α whether $\mathcal{N} \models \alpha$ or $\mathcal{N} \not\models \alpha$. Then there is an arithmetic function $f_{\mathbf{T}} : \mathbb{N} \rightarrow \mathbb{N}$ describing **T** such that for any $\mathcal{L}_{ring}$-sentence α,

$$f(\ulcorner \alpha \urcorner) = \begin{cases} 1 \text{ if } \mathcal{N} \models \alpha \\ 2 \text{ if } \mathcal{N} \not\models \alpha \end{cases}$$

Here we call a function $f : \mathbb{N} \to \mathbb{N}$ *arithmetic* if there is an $\mathcal{L}_{ring}$-formula $\phi(v_0, v_1)$ such that for any $n, m \in \mathbb{N}$,

$$f(n) = m \Leftrightarrow \mathcal{N} \models \phi(n, m).$$

(ϕ is then called a *defining* formula for f.) That there is such a defining formula ϕ_{T} for f_{T} comes from the 1st-order fashion in which Turing programs come along; the logical connectives and quantifiers translate into arithmetic operations.

So for any $\mathcal{L}_{ring}$-sentence α we have

$$\mathcal{N} \models \phi_{\mathrm{T}}(\ulcorner\alpha\urcorner, 1) \Longleftrightarrow \mathcal{N} \models \alpha,$$

which means that we can 'talk' about the truth of a sentence about $\mathcal{N}$ *inside* $\mathcal{N}$, and so we are in a position to simulate the liar's paradox: Define $g : \mathbb{N} \to \mathbb{N}$ such that for all $\mathcal{L}_{ring}$-formulas $\rho(v_0)$

$$g(\ulcorner\rho\urcorner) = \begin{cases} 1 \text{ if } \mathcal{N} \models \neg\rho(\ulcorner\rho\urcorner) \\ 2 \text{ if } \mathcal{N} \models \rho(\ulcorner\rho\urcorner) \end{cases}$$

Then g is arithmetic, say with defining formula ψ, so that

$$\mathcal{N} \models \psi(\ulcorner\rho\urcorner, 1) \Leftrightarrow \mathcal{N} \models \neg\rho(\ulcorner\rho\urcorner).$$

Applied to the formula $\rho_0(v_0) := \psi(v_0, 1)$ this gives

$$\mathcal{N} \models \rho_0(\ulcorner\rho_0\urcorner) \Leftrightarrow \mathcal{N} \models \neg\rho_0(\ulcorner\rho_0\urcorner),$$

the contradiction we were looking after. □

Since there is an effective algorithm (a Turing machine) listing the Peano axioms, and since $\mathtt{Th}(\mathcal{N})$ is complete we get the following immediate

Corollary 2.2. *The Peano axioms don't axiomatise all of* $\mathtt{Th}(\mathcal{N})$.

Nevertheless, a great many theorems about $\mathcal{N}$ do follow from the Peano axioms, and there has been an exciting controversy launched by Angus Macintyre as to whether Fermat's Last Theorem belongs (cf. the Appendix in [35]).

Another immediate consequence is the following

Corollary 2.3. $\mathtt{Th}(\langle\mathbb{Z}; +, \cdot; 0, 1\rangle)$ *is undecidable.*

Proof. Otherwise, as the natural numbers are exactly the sums of four squares of integers, $\mathtt{Th}(\mathcal{N})$ would be decidable. □

3 Undecidability of Number Rings and Fields

3.1 The Field $\mathbb{Q}_p$ of p-adic Numbers

The field of p-adic numbers was discovered, or rather invented, by Kurt Hensel over 100 years ago and has ever since played a crucial role in number theory. It is, like the field $\mathbb{R}$ of real numbers, the completion of $\mathbb{Q}$, though not w.r.t. the ordinary (real) absolute value, but rather w.r.t. a 'p-adic' analogue (for each prime p a different one). This allows to bring new (p-adic) analytic methods into number theory and to reduce some problems about number fields (which are so-called 'global' fields) to the 'local' fields $\mathbb{R}$ and $\mathbb{Q}_p$. This is of particular interest in our context because, as we will see, number fields are undecidable, whereas the fields of real or p-adic numbers are decidable. So whenever a number theoretic problem is reducible to a problem about the local fields $\mathbb{R}$ and $\mathbb{Q}_p$ (one then says that the problem satisfies a *Local-Global-Principle*) the number theoretic problem becomes decidable as well.[2]

Let us fix a rational prime p. The *p-adic valuation* v_p on $\mathbb{Q}$ is defined by the formula

$$v_p(p^r \cdot \frac{m}{n}) = r \text{ for any } r, m, n \in \mathbb{Z} \text{ with } p \not| \, m \cdot n \neq 0,$$

with $v_p(0) := \infty$. It is easy to check that this is a well defined valuation (for background in valuation theory cf. [76] in this volume or [21]). The corresponding valuation ring is

$$\mathbb{Z}_{(p)} = \{q \in \mathbb{Q} \,|\, v_p(q) \geq 0\} = \left\{\frac{a}{b} \,\middle|\, a \in \mathbb{Z}, p \not| \, b \in \mathbb{Z} \setminus \{0\}\right\}$$

with maximal ideal

$$p\mathbb{Z}_{(p)} = \{q \in \mathbb{Q} \,|\, v_p(q) > 0\} = \left\{\frac{a}{b} \,\middle|\, p \mid a \in \mathbb{Z}, p \not| \, b \in \mathbb{Z} \setminus \{0\}\right\}.$$

v_p induces the *p-adic norm* $| \,.\, |_p$ on $\mathbb{Q}$ given by

$$| \, q \, |_p := p^{-v_p(q)} \text{ for } q \neq 0,$$

with $| \, 0 \, |_p := 0$. Observe that the sequence $p, p^2, p^3, \ldots$ converges to 0 w.r.t. $| \,.\, |_p$.

[2] The use of the terms 'local' and 'global' which one more typically encounters in analysis or in algebraic geometry hints at a deep analogy between number theory and algebraic geometry, more specifically between number fields (i.e. finite extensions of $\mathbb{Q}$, the global fields of characteristic 0) and algebraic function fields in one variable over finite fields (i.e., finite extensions of the field $\mathbb{F}_p(t)$ of rational functions over $\mathbb{F}_p$, the global fields of positive characteristic). It is one of the big open problems in model theory whether or not the positive characteristic analogue of $\mathbb{Q}_p$, i.e., the local field $\mathbb{F}_p((t))$ of (formal) Laurent series over $\mathbb{F}_p$, is decidable.

We can now define the *field* $\mathbb{Q}_p$ *of p-adic numbers* as the completion of $\mathbb{Q}$ w.r.t. $| \, . \, |_p$, i.e., the field obtained by taking the quotient of the ring of (p-adic) Cauchy sequences by the maximal ideal of (p-adic) zero sequences (`version 1`). Equivalently (`version 2`), one may define

$$\mathbb{Q}_p := \left\{ \alpha = \sum_{\nu=n}^{\infty} a_\nu p^\nu \,\middle|\, n \in \mathbb{Z}, a_\nu \in \{0, 1, \ldots, p-1\} \right\}$$

as the ring of formal Laurent series in powers of p with coefficients from $0, 1, \ldots, p-1$, where addition is componentwise modulo p starting with the lowest non-zero terms and carrying over whenever $a_\nu + b_\nu \geq p$ (so if this happens for the first time at ν the $(\nu+1)$-th coefficient becomes $1 + a_{\nu+1} + b_{\nu+1}$ modulo p etc.)—like adding decimals, but from the left. Similarly, multiplication is like multiplying polynomials in the 'unknown' p with coefficients modulo p, and again carrying over whenever necessary.

Exercise 3.1. *Show that* $-1 = \sum_{\nu=0}^{\infty}(p-1)p^\nu$ *and* $\frac{1}{1-p} = 1 + p + p^2 + \ldots$.

Given this presentation of p-adic numbers, one defines the p-adic valuation on $\mathbb{Q}_p$, again denoted by v_p, for any non-zero $\alpha \in \mathbb{Q}_p$ as

$$v_p(\alpha) := \min\{\nu \mid a_\nu \neq 0\}.$$

This is a prolongation of the p-adic valuation on $\mathbb{Q}$; its value group is, obviously, still $\mathbb{Z}$, and its valuation ring is

$$\mathbb{Z}_p := \mathcal{O}_{v_p} = \{\alpha \in \mathbb{Q}_p \mid v_p(\alpha) \geq 0\} = \left\{ \sum_{\nu=0}^{\infty} a_\nu p^\nu \,\middle|\, a_\nu \in \{0, 1, \ldots, p-1\} \right\},$$

the *ring of p-adic integers* with maximal ideal

$$p\mathbb{Z}_p = \{\alpha \in \mathbb{Q}_p \mid v_p(\alpha) > 0\} = \left\{ \sum_{\nu=1}^{\infty} a_\nu p^\nu \,\middle|\, a_\nu \in \{0, 1, \ldots, p-1\} \right\},$$

and with residue field

$$\mathbb{Z}_p / p\mathbb{Z}_p \cong \mathbb{Z}_{(p)} / p\mathbb{Z}_{(p)} \cong \mathbb{Z}/p\mathbb{Z} = \mathbb{F}_p.$$

By definition, $\mathbb{Q}$ is dense in $\mathbb{Q}_p$ w.r.t. the p-adic topology induced by (the two) v_p, and $\mathbb{Z}$ is dense in $\mathbb{Z}_p$: in fact, $\mathbb{Z}_p$ is the completion of $\mathbb{Z}$ (w.r.t. the norm induced by $| \, . \, |_p$ on $\mathbb{Z}$).

Yet another way to think about the ring of p-adic integers (`version 3`) is to view it as an inverse limit

$$\mathbb{Z}_p = \lim_{\leftarrow} \mathbb{Z}/p^n\mathbb{Z}$$

of the rings $\mathbb{Z}/p^n\mathbb{Z}$ w.r.t. the canonical projections $\mathbb{Z}/p^n\mathbb{Z} \to \mathbb{Z}/p^m\mathbb{Z}$ for $m \leq n$. Note that, for $n > 1$, the rings $\mathbb{Z}/p^n\mathbb{Z}$ have zero divisors whereas the projective limit $\mathbb{Z}_p$ becomes an integral domain (with $\mathbb{Q}_p$ as field of fractions).

Exercise 3.2. *Prove the equivalence of* `versions 1, 2, 3.`

One of the key facts about $\mathbb{Z}_p$ is *Hensel's Lemma* which uses the analytic tool of Newton approximation to find a precise zero of a polynomial, given an approximate zero. For $\alpha \in \mathbb{Z}_p$, let us denote its image under the canonical residue map $\mathbb{Z}_p \to \mathbb{F}_p$ by $\overline{\alpha}$. Similarly, we will write $\overline{f}$ for the image of the polynomial $f \in \mathbb{Z}_p[X]$ under the coefficient wise extension of the residue map to $\mathbb{Z}_p[X] \to \mathbb{F}_p[X]$.

Lemma 3.3 (Hensel's Lemma). ***Simple zeros lift:*** *Let $f \in \mathbb{Z}_p[X]$ be a monic polynomial and assume $\alpha \in \mathbb{Z}_p$ is such that $\overline{\alpha}$ is a simple zero of $\overline{f}$ (i.e., $\overline{f}(\overline{\alpha}) = 0 \neq \overline{f}'(\overline{\alpha})$). Then there is some $\beta \in \mathbb{Z}_p$ with $f(\beta) = 0$ and $\overline{\beta} = \overline{\alpha}$.*

Exercise 3.4. *The proof is an adaptation of the proof in van den Dries' contribution to this volume, Sect. 3.2, the details being left to the reader as an exercise.*

Example. If $p > 2$ then every 1-unit, that is, every $x \in 1+p\mathbb{Z}_p$ is a square: consider the polynomial $f(X) = X^2 - x$ and let $\alpha = 1$; these satisfy the assumptions of Hensel's Lemma, and so f has a zero β (with $\overline{\beta} = 1$); hence $x = \beta^2$.

Similarly, if one denotes by ζ_n a primitive n-th root of unity, then $\zeta_{p-1} \in \mathbb{Z}_p$: the polynomial $X^{p-1} - 1$ has $(p-1)$ distinct linear factors over $\mathbb{F}_p$, and so, by Hensel's Lemma, the same holds in $\mathbb{Z}_p$.

Thus we can write the multiplicative group of $\mathbb{Q}_p$ as a direct product of three 'natural' subgroups:

$$\mathbb{Q}_p^\times = p^{\mathbb{Z}} \cdot \langle \zeta_{p-1} \rangle \cdot (1 + p\mathbb{Z}_p)$$

From this one immediately reads off that, for $p > 2$, there are precisely four square classes (elements in $\mathbb{Q}_p^\times/(\mathbb{Q}_p^\times)^2$), represented by

$$1,\ p,\ \zeta_{p-1} \text{ and } p\zeta_{p-1}.$$

As a consequence, one obtains the following well-known 1st-order $\mathcal{L}_{ring}$-definition of $\mathbb{Z}_p$ in $\mathbb{Q}_p$:

$$\mathbb{Z}_p = \{x \in \mathbb{Q}_p \mid \exists y \in \mathbb{Q}_p \text{ such that } 1 + px^2 = y^2\}.$$

Exercise 3.5. *Show that, for $p = 2$, a similar definition works with squares replaced by cubes.*

Exercise 3.6. *Show that, for p a prime $\equiv 3 \mod 4$,*

$$\mathbb{Z}_p = \{t \in \mathbb{Q}_p \mid \exists x, y, z \in \mathbb{Q}_p \text{ such that } 2 + pt^2 = x^2 + y^2 - pz^2\},$$

and, if $p \equiv 1 \mod 4$ and $q \in \mathbb{N}$ a quadratic non-residue $\mod p$,

$$\mathbb{Z}_p = \{t \in \mathbb{Q}_p \mid \exists x, y, z \in \mathbb{Q}_p \text{ such that } 2 + pqt^2 = x^2 + qy^2 - pz^2\}.$$

With the valuation ring, also the maximal ideal, the residue field, the group of units, the value group and the valuation map all become interpretable in $\mathcal{L}_{ring}$. Thus, the axiomatization given below can be phrased entirely in $\mathcal{L}_{ring}$-terms. Now here is a milestone in the model theory of $\mathbb{Q}_p$:

Theorem 3.7 (Ax-Kochen/Ershov). $\mathtt{Th}(\mathbb{Q}_p)$ *is decidable. It is effectively axiomatized by the following axioms:*

- *v_p is henselian*
- *the residue field of v_p is $\mathbb{F}_p$*
- *the value group Γ is a $\mathbb{Z}$-group, i.e., $\Gamma \equiv \langle \mathbb{Z}; +; 0; < \rangle$ which can be axiomatized by saying that there is a minimal positive element and that $[\Gamma : n\Gamma] = n$ for all n*
- *$v_p(p)$ is minimal positive*

The proof, again, is similar to the proof of the other Ax-Kochen/Ershov Theorem presented in section 6 of van den Dries' contribution [76] to this volume.

Fields elementarily equivalent to $\mathbb{Q}_p$ are called *p-adically closed.*[3]

Exercise 3.8. *Check that fields which are relatively algebraically closed in a p-adically closed field are again p-adically closed.*

So, for example, the field

$$\mathbb{Q}_p^{alg} := \mathbb{Q}_p \cap \tilde{\mathbb{Q}}$$

of algebraic p-adic numbers is p-adically closed (we use the notation $\tilde{K}$ for the algebraic closure of K). Note that $\mathbb{Q}_p^{alg}$ is countable while $\mathbb{Q}_p$ isn't.

Exercise 3.9. *Show that $K := \mathbb{Q}_p((\mathbb{Q}))$ (in the notation of [76], after Definition 5.3, this is $\mathbb{Q}_p((t^{\mathbb{Q}}))$) is p-adically closed and solve the mystery that, on the one hand, K and $\mathbb{Q}_p$ are elementarily equivalent, on the other, they both have a henselian valuation with the same residue field $\mathbb{Q}_p$, but with non-elementarily equivalent value groups ($\mathbb{Q}$ for K and $\{0\}$ for $\mathbb{Q}_p$).*

[3] Sometimes the term *p-adically closed* refers, more generally, to fields elementarily equivalent to *finite extensions* of $\mathbb{Q}_p$—cf. [59].

3.2 The Local-Global-Principle (LGP) for Quadratic Forms Over $\mathbb{Q}$

Let $q(X_1, \ldots, X_n)$ be a quadratic form over $\mathbb{Q}$, i.e., a homogeneous polynomial of degree 2 in $\mathbb{Q}[X_1, \ldots, X_n]$. An element $a \in \mathbb{Q}$ is said to be *represented by q* if there is $\overline{x} = (x_1, \ldots, x_n) \in \mathbb{Q}^n$ such that $a = q(\overline{x})$.

Theorem 3.10 (Hasse-Minkowski-Theorem). *A rational a is represented by q in $\mathbb{Q}$ if and only if a is represented by q in all $\mathbb{Q}_p$ and in $\mathbb{R}$.*

The proof is trivial for $n = 1$, it uses the so-called *geometry of numbers* for $n = 2$, it requires delicate case distinctions 'modulo 8' for $n = 3$ and $n = 4$, and then follows more easily by general quadratic form tricks for $n > 4$. (cf. [44], for an alternative proof using Hilbert symbols and quadratic reciprocity cf. [69]).

The above LGP-principle is effective: by a simple linear transformation any quadratic form can be brought into diagonal form:

$$q(X_1, \ldots, X_n) = a_1 X_1^2 + \cdots + a_n X_n^2$$

Then the only primes p where representability of a by q in $\mathbb{Q}_p$ need to be checked are $p = 2$, $p = \infty$ (where $\mathbb{Q}_\infty := \mathbb{R}$), those p where $v_p(a) \neq 0$, and those where $v_p(a_i) \neq 0$ for some $i \leq n$. So only these finitely many primes need checking and, since all the $\mathbb{Q}_p$ and $\mathbb{R}$ are decidable, the whole procedure is effective.

For *cubic* forms no such LGP holds: By an example of Selmer [68], 5 is represented by the cubic form

$$3X^3 + 4Y^3$$

in $\mathbb{R}$ and in every $\mathbb{Q}_p$, but not in $\mathbb{Q}$.

3.3 Julia Robinson's Definition for $\mathbb{Z}$ in $\mathbb{Q}$ and in Other Number Fields

Julia Robinson's contribution to questions of decidability in number theory is enormous. Her first big result in this direction is the 1st-order definability of $\mathbb{Z}$ (or $\mathbb{N}$) in $\mathbb{Q}$ from which the undecidability of $\mathtt{Th}(\mathbb{Q})$ immediately follows (1949), given Gödel's 1st Incompleteness Theorem. 10 years later she extended this to arbitrary number fields. Later she became heavily involved in Hilbert's 10th Problem (Sect. 4). The very appealing documentary 'Julia Robinson and Hilbert's Tenth Problem' by George Csicsery came out in 2010 [5].

To give Julia Robinson's explicit definition of $\mathbb{Z}$ in $\mathbb{Q}$, let us introduce the following formulas: for $a, b \in \mathbb{Q}^\times$ and $k \in \mathbb{Q}$, let

$$\phi(a,b,k) := \exists x, y, z(2 + abk^2 + bz^2 = x^2 + ay^2)$$

and let, for $n \in \mathbb{Q}$,

$$\psi(n) := \forall a, b \neq 0\,[\{\phi(a,b,0) \wedge \forall k\langle \phi(a,b,k) \to \phi(a,b,k+1)\rangle\} \to \phi(a,b,n)]$$

Theorem 3.11 ([61]). *For any $n \in \mathbb{Q}$,*

$$\mathbb{Q} \models \psi(n) \Leftrightarrow n \in \mathbb{Z}.$$

Proof. The easy direction '$\Leftarrow$' follows, for $n \in \mathbb{N}$, by the principle of induction, and for $n \in \mathbb{Z}$, from the observation that $\psi(n) \Leftrightarrow \psi(-n)$, because n occurs only squared in ψ.

For the non-trivial direction one first shows, using Exercise 3.6 and Theorem 3.10, that, for a prime $p \equiv 3 \mod 4$ and $k \in \mathbb{Q}$,

$$\phi(1, p, k) \Leftrightarrow v_p(k) \geq 0 \text{ and } v_2(k) \geq 0,$$

and that, for primes p, q with $p \equiv 1 \mod 4$ and q a quadratic non-residue $\mod p$,

$$\phi(q, p, k) \Leftrightarrow v_p(k) \geq 0 \text{ and } v_q(k) \geq 0.$$

So, in either case, the $\{\ldots\}$-bit in ψ is satisfied, and, thus, for $\psi(n)$ to hold we must have $\phi(1, p, n)$ for any prime $p \equiv 3 \mod 4$ resp. $\phi(q, p, n)$ for any pair p, q of primes in the second case. But then, by the equivalences above, $v_p(n) \geq 0$ for any prime p, and so $n \in \mathbb{Z}$. □

Corollary 3.12. $\mathtt{Th}(\mathbb{Q})$ *is undecidable.*

Let us recall, that number fields are finite extensions of $\mathbb{Q}$, and that the *ring of integers in K*, denoted by $\mathcal{O}_K$, is the integral closure of $\mathbb{Z}$ in K, i.e., the set of elements of K satisfying a monic polynomial with coefficients in $\mathbb{Z}$.

Theorem 3.13 ([63]). *For any number field K, $\mathcal{O}_K$ is definable in K and $\mathbb{Z}$ is definable in $\mathcal{O}_K$. In particular, $\mathtt{Th}(\mathcal{O}_K)$ and $\mathtt{Th}(K)$ are undecidable.*

Proof. The definition of $\mathcal{O}_K$ in K proceeds along similar lines as that of $\mathbb{Z}$ in $\mathbb{Q}$, especially as the LGP for quadratic forms holds in arbitrary number fields.

That $\mathbb{N}$ is definable in $\mathcal{O}_K$ uses the fact that for all non-zero $f \in \mathcal{O}_K$ there are only finitely many $a \in \mathcal{O}_K$ such that

$$a + 1 \mid f \wedge \ldots \wedge a + l \mid f,$$

where $l = [K : \mathbb{Q}]$.

Now define, for $a, f, g, h \in \mathcal{O}_K$,

$$\rho(a, f, g, h) := f \neq 0 \wedge (a + 1 \mid f \wedge \ldots \wedge a + l \mid f) \wedge 1 + ag \mid h$$

Then, for any $n \in \mathcal{O}_K$,

$$n \in \mathbb{N} \Leftrightarrow \exists f, g, h \left[\rho(0, f, g, h) \wedge \forall a\{\rho(a, f, g, h) \rightarrow \langle \rho(a+1, f, g, h) \vee a \doteq n \rangle\}\right]$$

To prove the easy direction '$\Leftarrow$', assume n satisfies the right hand side. By the fact above, there are only finitely many a with $\rho(a, f, g, h)$. The inductive form of the definition ensures that $\rho(0, f, g, h)$, $\rho(1, f, g, h), \ldots$ terminating only for $a = n$. Therefore, n must be a natural number.

For the converse direction '$\Rightarrow$', assume $n \in \mathbb{N}$. It suffices to find $f, g, h \in \mathcal{O}_K$ such that

$$\rho(a, f, g, h) \leftrightarrow a = 0 \vee a = 1 \vee \ldots \vee a = n.$$

Put $f := (n + l)!$ and let $S := \{a \in \mathcal{O}_K \mid a + 1 \mid f \wedge \ldots \wedge a + l \mid f\}$. Then, by the fact above, S is finite and we can find some $g \in \mathbb{N}$ large enough so that, for any two distinct $a, b \in S$, $a - b \mid g$ and, for any non-zero $a \in S$, $1 + ag \not| 1$. Then, for any distinct $a, b \in S$, $1 + ag$ and $1 + bg$ are relatively prime (if there is a prime ideal of $\mathcal{O}_K$ containing both $1 + ag$ and $1 + bg$ then it contains $(a - b)g$, hence g^2 and so g, but it cannot contain both g and $1 + ag$).

Now put $h = (1 + g)(1 + 2g) \cdots (1 + ng)$. Then $\rho(a, f, g, h)$ is satisfied for $a = 1, \ldots, n$. If, however, there is some other $a \in S$ then $1 + ag$ is not a unit and is prime to h. Therefore, $1 + ag \not| h$ and $\rho(a, f, g, h)$ does not hold. □

3.4 *Totally Real Numbers*

There are many infinite algebraic extensions of $\mathbb{Q}$ (sometimes misleadingly called 'infinite number fields') which are also known to be undecidable. In fact, most of them are: there are only countably many decision algorithms, but uncountably many non-isomorphic, and hence, in this case, non-elementarily equivalent algebraic extensions of $\mathbb{Q}$. To give an explicit example, let A be an undecidable (= non-recursive, cf. Sect. 4.2) subset of the set of all primes and let

$$K := \mathbb{Q}(\{\sqrt{p} \mid p \in A\}).$$

Then (the 1st-order theory of) K, and hence also $\mathcal{O}_K$, is undecidable: otherwise $A = \{p \in \mathbb{N} \mid p \text{ is prime and } \exists x \in K\ p = x^2\}$ would be decidable.

While this example seems artificial, there is a number of 'natural' infinite algebraic extensions of $\mathbb{Q}$ for which it makes sense to ask about decidability of the field or of its ring of integers. We will treat some of these in this section and the next, and we will list some nice open problems in Sect. 6.3. The field of totally real numbers is special in that its ring of integers is undecidable whereas the field is decidable.

The field of totally real numbers is defined to be the maximal Galois extension T of $\mathbb{Q}$ inside $\mathbb{R}$. T is an infinite algebraic extension of $\mathbb{Q}$, the intersection of all real closures of $\mathbb{Q}$ (inside a fixed algebraic closure $\tilde{\mathbb{Q}}$). T can also be thought of as the compositum of all finite extensions $F/\mathbb{Q}$ for which all embeddings $F \hookrightarrow \mathbb{C}$ are real.

As for finite extensions of $\mathbb{Q}$ one defines $\mathcal{O}_T$, *the ring of integers of* T, as the integral closure of $\mathbb{Z}$ in T.

Theorem 3.14 ([64]). $\mathtt{Th}(\mathcal{O}_T)$ *is undecidable.*

Let us separate the key ingredients of the proof in the two lemmas below.

Lemma 3.15. *Let R be an integral domain with $\mathbb{N} \subseteq R$. Let $\mathcal{F} \subseteq \wp(R)$ be a family of subsets of R which is* arithmetically defined *(or* uniformly parametrised*), say, by an $\mathcal{L}_{ring}$-formula $\phi(x; y_1, \ldots, y_k)$, i.e., for any $F \subseteq R$,*

$$F \in \mathcal{F} \Leftrightarrow \exists \overline{y} \in R^k\ \forall x \in R[x \in F \leftrightarrow \phi(x; \overline{y})].$$

Assume that all $F \in \mathcal{F}$ are finite and each initial segment $\{0, 1, \ldots, n\}$ of $\mathbb{N}$ is in $\mathcal{F}$. Then $\mathbb{N}$ is definable in R.

Proof. For any $n \in R$,

$$n \in \mathbb{N} \Leftrightarrow \exists \overline{y} \in R^k\ [\phi(0, \overline{y}) \wedge \forall x\{\phi(x, \overline{y}) \to (\phi(x+1, \overline{y}) \vee x = n)\}].$$

□

For $a, b \in T$, we use the notation '$a \ll b$' to indicate that $a < b$ for any ordering $<$ on T. Note that this is expressible by an $\mathcal{L}_{ring}$-formula:

$$a \ll b \Longleftrightarrow a \neq b \wedge \exists x\ b = a + x^2$$

(totally positive elements are always sums of squares, and, in T, every sum of squares is a square).

Lemma 3.16.

$$\min\{M \in \mathbb{R} \mid \exists\infty\text{-}ly\ many\ t \in \mathcal{O}_T\ s.t.\ 0 \ll t \ll M\} = 4$$

Proof. That there are infinitely many $t \in \mathcal{O}_T$ with $0 \ll t \ll 4$ is easy: for any $n > 1$ and any n-th root ζ_n of unity, $t_n := 2 + \zeta_n + \zeta_n^{-1}$ has this property.

Conversely, any t with this property is one of these t_n: This follows from Kronecker's 1857 Theorem [32] that *algebraic integers all of whose conjugates have absolute value* ≤ 1 *are roots of unity*: if $\alpha = \alpha_1, \alpha_2, \ldots, \alpha_n$ are all the conjugates of such an algebraic integer α, then, for any k, the coefficients $a_{k,s}$ of the polynomial

$$f_k(X) := (X - \alpha_1^k) \cdots (X - \alpha_n^k) = X^n + a_{k,n-1}X^{n-1} + \ldots + a_{k,0}$$

satisfy $| a_{k,s} | \leq \binom{n}{s}$; being integers as well, there can only be finitely many such $a_{k,s}$, hence only finitely many such f_k, and so $\alpha^k = \alpha^l$ for some $k < l$, making α an $(l-k)$-th root of unity.

From this one obtains that *totally real integers all of whose conjugates have absolute value* ≤ 2 *are of the shape* $\alpha + \alpha^{-1}$ *for some root of unity* α*:* let $\beta \in \mathcal{O}_T$ be such an element with conjugates $\beta = \beta_1, \ldots, \beta_n$ and let

$$\alpha = \frac{\beta}{2} + \sqrt{\frac{\beta^2}{4} - 1};$$

then α is an algebraic integer with $\alpha^2 - \beta\alpha + 1 = 0$ and any conjugate α' of α (over $\mathbb{Q}$) satisfies $\alpha'^2 - \beta_i\alpha' + 1 = 0$ for some i; as $| \beta_i | \leq 2$ (and, in fact, w.l.o.g., < 2) the two roots of this equation are the two complex conjugates of α', hence

$$| \alpha' |^2 = \frac{\beta_i^2}{4} + 1 - \frac{\beta_i^2}{4} = 1,$$

so, by Kronecker's Theorem, α is a root of unity and $\beta = \alpha + \alpha^{-1}$ is of the indicated shape.

Now the Lemma follows easily. □

Proof of Theorem 3.14. The family $\mathcal{F} \subseteq \wp(\mathcal{O}_T)$ defined by

$$\phi(x; p, q) \Leftrightarrow 0 \ll qx \ll p \wedge p \ll 4q$$

contains, by Lemma 3.16, only finite sets, but arbitrarily large ones. Hence, by a variant of Lemma 3.15, $\mathbb{N}$ is definable in $\mathcal{O}_T$. □

3.5 Large Algebraic Extensions of $\mathbb{Q}$ and Geometric LGP's

As in the previous section, we denote the field of totally real numbers by T. In the literature, it is also often denoted by $\mathbb{Q}^{tot-r}$ or $\mathbb{Q}^{t.r.}$.

Theorem 3.17. *T is* pseudo-real-closed *('PRC'), i.e. T satisfies the following geometric LGP: for each (affine) algebraic variety V/T,*

$$V(T) \neq \emptyset \Leftrightarrow V(R) \neq \emptyset \textit{ for all real closures } R \textit{ of } T.$$

The theorem was first proved by Moret-Bailly [43] using heavy machinery from algebraic geometry, with a more elementary proof given later by Green et al. [26].

By explicitly describing the structure of the absolute Galois group G_T of T, that is, the Galois group of the algebraic closure $\tilde{T} = \tilde{\mathbb{Q}}$ of T over T, Fried et al. showed in [23], using the above theorem:

Theorem 3.18. $\mathtt{Th}(T)$ *is decidable.*

The axiomatization expresses the PRC property in elementary terms (it is not at all obvious how to do this, but it had long been established, e.g., in [58]) as well as the fact that G_T is the free (profinite) product of all G_R, where R runs through a set of representatives of the conjugacy classes of all real closures of T (so each $G_R \cong \mathbb{Z}/2\mathbb{Z}$). The latter can be 'axiomatized' via so-called embedding problems in a similar fashion as, by a famous Theorem of Iwasawa, free profinite groups of infinite countable rank can be characterized.

An interesting immediate consequence of this tension between T being decidable and $\mathcal{O}_T$ not, is the following:

Corollary 3.19. $\mathcal{O}_T$ *is not definable in* T.

The decidability of (the 1st-order theory of) T implies that the field $T(\sqrt{-1})$ is decidable as well. By Theorem 10.7 of [6], even $\mathcal{O}_{T(\sqrt{-1})}$ is decidable.

A p-adic analogue of Theorem 3.17 and 3.18 was given by Pop in [57]: One defines the field $\mathbb{Q}^{tot-p}$, the *field of totally p-adic numbers*, as the maximal Galois extension of $\mathbb{Q}$ inside $\mathbb{Q}_p$, that is, the intersection of all conjugates of $\mathbb{Q}_p^{alg}$ over $\mathbb{Q}$. Pop showed that $\mathbb{Q}^{tot-p}$ is *pseudo-p-adically-closed*, i.e. it satisfies an analogous LGP (with real closures replaced by p-adic closures), and that the absolute Galois group is similarly well behaved. As a consequence, $\mathbb{Q}^{tot-p}$ is decidable. And, again by Theorem 10.7 of [6], $\mathcal{O}_{\mathbb{Q}^{tot-p}}$ is also decidable.

Let us close this section by mentioning another celebrated LGP, that is, *Rumely's Local-Global-Principle* [67] which concerns the ring $\tilde{\mathbb{Z}}$ of all algebraic integers, i.e., the integral closure of $\mathbb{Z}$ in the algebraic closure $\tilde{\mathbb{Q}}$ of $\mathbb{Q}$:

Theorem 3.20. *Let V be an affine variety defined over $\tilde{\mathbb{Z}}$. Then*

$$V(\tilde{\mathbb{Z}}) \neq \emptyset \Leftrightarrow V(\mathcal{O}) \neq \emptyset \text{ for all valuation rings } \mathcal{O} \text{ of } \tilde{\mathbb{Q}}.$$

Using this, van den Dries [75] showed the following Theorem via some quantifier elimination, Prestel and Schmid [60] showed it via an explicit axiomatization:

Theorem 3.21. $\mathtt{Th}(\tilde{\mathbb{Z}})$ *is decidable.*

Note that $\tilde{\mathbb{Z}}$ is not definable in $\tilde{\mathbb{Q}}$ (by quantifier elimination in ACF_0, every definable subset of $\tilde{\mathbb{Q}}$ is finite or cofinite), so there is no cheap way of proving the above Theorem.

For a survey on geometric LGP's and many more results in this direction cf. [6].

4 Hilbert's 10th Problem and the DPRM-Theorem

4.1 The Original Problem and First Generalisations

In 1900, at the Conference of Mathematicians in Paris, Hilbert presented his celebrated and influential list of 23 mathematical problems [27]. One of them is

Hilbert's 10th Problem ('H10'). *Find an algorithm which gives on INPUT any* $f(X_1, \ldots, X_n) \in \mathbb{Z}[X_1, \ldots, X_n]$

$$\textit{OUTPUT}\begin{cases} \textbf{YES} \textit{ if } \exists \overline{x} \in \mathbb{Z}^n \textit{ such that } f(\overline{x}) = 0 \\ \textbf{NO} \textit{ else} \end{cases}$$

Hilbert did not ask to prove that there is such an algorithm. He was convinced that there should be one, and that it was all a question of producing it—one of those instances of Hilbert's optimism reflected in his famous slogan '*wir müssen wissen, wir werden wissen*' ('*we must know, we will know*'). As it happens, Hilbert was too optimistic: after previous work since the 1950s by Martin Davis, Hilary Putnam and Julia Robinson, in 1970, Yuri Matiyasevich showed that there is no such algorithm (Corollary 4.7).

The original formulation of Hilbert's 10th problem was weaker than the standard version we have given above in that he rather asked *'Given a polynomial f, find an algorithm ...'*. So maybe one could have different algorithms depending on the number of variables and the degree. However, it is even possible to find a single polynomial for which no such algorithm exists (Corollary 4.14)—this is essentially because there are universal Turing Machines.

One should, however, mention that, in the special case of $n = 1$, that is, for polynomials in one variable, there is an easy algorithm: if, for some $x \in \mathbb{Z}$, $f(x) = 0$ then $x \mid f(0)$; hence one only has to check the finitely many divisors of $f(0)$. Similarly, by the effective version of the Hasse-Minkowski-LGP (Theorem 3.10) and some extra integrality considerations, one also has an algorithm for polynomials in an arbitrary number of variables, but of total degree ≤ 2. And, even if there is no general algorithm, it is one of the major projects of computational arithmetic geometry to exhibit other families of polynomials for which such algorithms exist.

To conclude these introductory remarks let us point in a different direction of generalizing Hilbert's 10th Problem, namely, generalizing it to rings other than $\mathbb{Z}$: If R is an integral domain, there are two natural ways of generalizing **H10**:

$$\mathbf{H10}/R = \mathbf{H10} \text{ with the 2nd occurrence of } \mathbb{Z} \text{ replaced by } R$$

$$\mathbf{H10}^+/R = \mathbf{H10} \text{ with both occurrences of } \mathbb{Z} \text{ replaced by } R$$

Observation 4.1. *Let R be an integral domain whose field of fractions does not contain the algebraic closure of the prime field ($\mathbb{F}_p$ resp. $\mathbb{Q}$). Then*

$$\mathbf{H10}/R \textit{ is solvable} \Leftrightarrow \mathtt{Th}_{\exists^+}(R) \textit{ is decidable}$$
$$\mathbf{H10}^+/R \textit{ is solvable} \Leftrightarrow \mathtt{Th}_{\exists^+}(\langle R; r \mid r \in R\rangle) \textit{ is decidable,}$$

where $\mathtt{Th}_{\exists^+}$ denotes the positive existential theory consisting of existential sentences where the quantifier-free part is a conjunction of disjunctions of polynomial equations (no inequalities).

Note that the language on the right hand side of the 2nd line contains a constant symbol for each $r \in R$.

Proof. '$\Leftarrow$' is obvious in both cases. For '$\Rightarrow$' one has to see that a disjunction of two polynomial equations is equivalent to (another) single equation, and, likewise, for conjunctions: By our assumption we can find some monic $g \in \mathbb{Z}[X]$ of degree > 1 which is irreducible over R. Then, for any polynomials f_1, f_2 over $\mathbb{Z}$ resp. R and for any tuple $\overline{x}$ over R,

$$f_1(\overline{x}) = 0 \vee f_2(\overline{x}) = 0 \Longleftrightarrow f_1(\overline{x}) \cdot f_2(\overline{x}) = 0$$

$$f_1(\overline{x}) = 0 \wedge f_2(\overline{x}) = 0 \Longleftrightarrow g\left(\frac{f_1(\overline{x})}{f_2(\overline{x})}\right) \cdot f_2(\overline{x})^{\deg g} = 0$$

□

Since in fields, inequalities can be expressed by a positive existential formula $(f(\overline{x}) \neq 0 \leftrightarrow \exists y\ f(\overline{x}) \cdot y = 1)$, we immediately obtain the following:

Corollary 4.2. *Let K be a field not containing the algebraic closure of the prime field. Then*

$$\mathbf{H10}/K \textit{ is solvable } \Leftrightarrow \mathtt{Th}_{\exists}(K) \textit{ is decidable.}$$

In fact, the same is true for $\mathcal{O}_K$, the ring of integers of a number field K:

Exercise 4.3. *Show that, if K is a number field,*

$$\mathcal{O}_K \models \forall x[x \neq 0 \leftrightarrow \exists y\ x \mid (2y-1)(3y-1)].$$

Deduce that $\mathtt{Th}_{\exists}(\mathcal{O}_K) = \mathtt{Th}_{\exists^+}(\mathcal{O}_K)$.

One of the biggest open questions in the area is

Question 4.4. *Is* $\mathbf{H10}/\mathbb{Q}$ *solvable?*

4.2 Listable, Recursive and Diophantine Sets

A subset $A \subseteq \mathbb{Z}$ is called

- *diophantine* if there is some $m \in \mathbb{N}$ and some polynomial $p \in \mathbb{Z}[T; X_1, \ldots, X_m]$ such that

$$A = \{a \in \mathbb{Z} \mid \exists \overline{x} \in \mathbb{Z}^m \text{ with } p(a; \overline{x}) = 0\}$$

  e.g., $\mathbb{N}$ is diophantine in $\mathbb{Z}$: take $p = T - X_1^2 - X_2^2 - X_3^2 - X_4^2$

- *listable* (= *recursively enumerable*) if there is an algorithm (= a Turing machine) printing out the elements of A (and only those), e.g., the set $\mathbb{P}$ of primes or $S := \{a^3 + b^3 + c^3 \mid a, b, c \in \mathbb{Z}\}$
- *recursive* (= *decidable*) if there is an algorithm deciding membership in A, e.g., $\mathbb{N}$ and $\mathbb{P}$ are recursive, about S it is not known.

It is clear that every diophantine set is listable and that every recursive set is listable. That, conversely, every listable set is diophantine is the content of the 'DPRM-Theorem' (next section).

That not every listable set is recursive follows from the following

Proposition 4.5 (The Halting Problem of Computer Science is Undecidable). *There is no algorithm to decide whether a program (with code) p halts on INPUT x.*

Proof. Otherwise define a new program H by:

$$H \text{ halts on input } x \Leftrightarrow x \text{ does not halt on input } x$$

(we identify x with $\ulcorner x \urcorner$). For $x = H$ we are in trouble then. □

Using this, we find a listable, but non-decidable set:

$$A = \{2^p 3^x \mid p \text{ halts on input } x\}$$

It is non-decidable by the proposition, but we can list it: for $x, p \leq N$ print $2^p 3^x$ if p halts on input x in $\leq N$ steps.

4.3 *The Davis-Putnam-Robinson-Matiyasevich (= DPRM)–Theorem*

..., often for short referred to as Matiyasevich's Theorem, is the following remarkable

Theorem 4.6 ([37], Conjectured by Davis 1953, Building on Work of Davis, Putnam and Robinson). *Every listable subset of $\mathbb{Z}$ is diophantine.*

Corollary 4.7. *Hilbert's 10th problem is unsolvable.*

Proof. The set A at the end of the previous section is listable, hence, by the Theorem, diophantine. So there is some $m \in \mathbb{N}$ and some polynomial $p \in \mathbb{Z}[T; X_1, \ldots, X_m]$ such that $A = \{a \in \mathbb{Z} \mid \exists \overline{x} \in \mathbb{Z}^m \text{ with } p(a; \overline{x}) = 0\}$. By construction, however, it is not decidable. Hence there is no algorithm which decides, on input $t \in \mathbb{Z}$, whether or not $p(t; \overline{x}) = 0$ has a solution $\overline{x} \in \mathbb{Z}^m$. □

We will only give a brief history and sketch of the proof of the Theorem. For a full account of the history see the excellent survey article [39], and, for a full self-contained proof (not always following the historic path), see [8].

Whether, in **H10**, we ask for solutions in $\mathbb{Z}^n$ or in $\mathbb{N}^n$ doesn't make a difference: as every integer is a difference of two natural numbers and as every natural number is the sum of 4 squares of integers we can easily transform a polynomial equation into another such that the former has integer solutions if and only if the latter has solutions in natural numbers, and we find a similar transformation for the other way round. Following history (and because $\mathbb{N}$ has the advantage of having a least element) we will stick to finding solutions in $\mathbb{N}$.

Theorem 4.8 ([7]). *If $A \subseteq \mathbb{N}$ is listable then A is* almost diophantine, *i.e., there is a polynomial $g \in \mathbb{Z}[T; \overline{X}; Y, Z]$ such that for all $a \in \mathbb{N}$*

$$a \in A \Leftrightarrow \exists z \forall y \leq z\, \exists \overline{x}\; g(a; \overline{x}; y, z) = 0.$$

Note that every A with such a presentation, later called 'Davis normal form', *is* listable.

Exercise 4.9. *Observe that $2^{\mathbb{N}}$ is listable. Find an almost diophantine presentation.*

A big challenge at the time was to find a diophantine presentation for $2^{\mathbb{N}}$, or, more generally, for exponentiation. Julia Robinson showed that, in order to achieve this, it suffices to find a diophantine relation 'of exponential growth' (what then went under the name 'Julia Robinson-predicate'):

Theorem 4.10 ([62]). *There is a polynomial $q \in \mathbb{Z}[A, B, C; \overline{X}]$ such that for all $a, b, c \in \mathbb{N}$*

$$a = b^c \Leftrightarrow \exists \overline{x}\; q(a, b, c; \overline{x}) = 0,$$

provided *there is a diophantine relation $J(u, v)$* of exponential growth, *i.e., for all $u, v \in \mathbb{N}$*

- *$J(u, v) \Rightarrow v < u^u$*
- *$\forall k \in \mathbb{N}\; \exists u, v$ with $J(u, v)$ and $v > u^k$.*

That it is, indeed, enough to show that exponentiation is diophantine, was then proved in the joint paper of Davis, Putnam and Julia Robinson.[4] They used the term *exponential polynomial* to refer to expressions obtained by applying the usual operations of addition, multiplication and exponentiation to integer coefficients and the variables:

[4]That we write the first name only for the lady is neither gallantry nor sexism: the reason is that there are other Robinsons in the same area: Abraham Robinson, one of the founders of model theory, and the logician and number theorist Raphael Robinson, also Julia's husband.

Theorem 4.11 ([9]). *If $A \subseteq \mathbb{N}$ is listable then there are exponential polynomials E_L and E_R such that, for all $a \in \mathbb{N}$,*

$$a \in A \Leftrightarrow \exists \overline{x}\ E_L(a; \overline{x}) = E_R(a; \overline{x}).$$

As predicted by Martin Davis, it then needed a young Russian mathematician, Yuri Matiyasevich from St Petersborough, to fill the gap:

Theorem 4.12 ([37]). *There is a diophantine relation $J(u, v)$ of exponential growth.*

The original proof used Fibonacci numbers. In [8], Davis gave a simpler proof using the so called *Pell equation*

$$x^2 - dy^2 = 1,$$

where $d = a^2 - 1 \in \mathbb{N}$ is a non-square. It is not hard to verify that, for the number field $K := \mathbb{Q}(\sqrt{d})$,

$$\begin{aligned} \mathcal{O}_K^\times &\supseteq \{x + y\sqrt{d} \mid x, y \in \mathbb{Z} \text{ with } x^2 - dy^2 = 1\} \\ &= \{\pm 1\} \cdot (a + \sqrt{a^2 - 1})^{\mathbb{Z}} \end{aligned}$$

It then requires a series of elementary computations to actually check that, on $\mathbb{N} \times \mathbb{N}$, the relation

$$J(x, y) \Leftrightarrow x^2 - (a^2 - 1)y^2 = 1 \Leftrightarrow x + y\sqrt{d} = (a + \sqrt{a^2 - 1})^m \text{ for some } m \in \mathbb{N}$$

is of exponential growth, or, at least, close to it.

DPRM-Theorem 4.6 = Theorem 4.10 + 4.11 + 4.12

4.4 Consequences of the DPRM-Theorem

One of the reasons why Davis' conjecture (the later DPRM-Theorem) was considered dubious was the fact that it implies the existence of a *prime producing polynomial* ($\mathbb{P}$ denotes the set of prime numbers):

Corollary 4.13. *There is some $n \in \mathbb{N}$ and a polynomial $f \in \mathbb{Z}[X_1, \ldots, X_n]$ such that $\mathbb{P} = f(\mathbb{Z}^n) \cap \mathbb{N}_{>0}$.*

Today, even an explicit polynomial (with $n = 10$) is known [38].

Proof. Obviously, $\mathbb{P}$ is listable. Hence, by the DPRM-Theorem 4.6, there is a polynomial $p \in \mathbb{Z}[T; \overline{X}]$ such that

$$\mathbb{P} = \{a \in \mathbb{Z} \mid \exists \overline{x}\ p(a; \overline{x}) = 0\}.$$

But then the polynomial

$$f(T_1, \ldots, T_4; \overline{X}) := [1 - p(T_1^2 + \ldots + T_4^2; \overline{X})](T_1^2 + \ldots + T_4^2)$$

does the job. □

As indicated at the beginning of this section, Hilbert's 10th problem has a negative solution even if one asks it for a single polynomial, or, to put it less misleadingly, for polynomials of a fixed shape, in particular of a fixed number of variables and a fixed degree:

Corollary 4.14. *There is a polynomial $U \in \mathbb{Z}[T; \overline{X}]$ and an algorithm producing, for each algorithm $\mathcal{A}$, some $t_{\mathcal{A}}$ (a counterexample) such that $\mathcal{A}$ fails to answer correctly whether there is some $\overline{x} \in \mathbb{Z}^n$ with $U(t_{\mathcal{A}}; \overline{x}) = 0$.*

The proof relies on the fact that there are *universal* recursive functions/Turing machines which, in addition to an INPUT-tuple $\overline{x}$, take an INPUT-code $t_{\mathcal{A}}$, and give as OUTPUT the OUTPUT of the algorithm $\mathcal{A}$ on input $\overline{x}$.

Let us close this section by mentioning that many famous mathematical problems can be translated into diophantine problems. For example, Goldbach's Conjecture that every even number > 2 is the sum of two prime numbers: Clearly, the set of counterexamples to this conjecture is listable, hence, by DPRM, diophantine, i.e., we can find a polynomial $g \in \mathbb{Z}[T; \overline{X}]$ such that, for any $t \in \mathbb{N}$,

$$g(t; \overline{x}) = 0$$

has a solution if and only if t spoils the conjecture. So Goldbach's Conjecture is equivalent to the statement that $g(t; \overline{x}) = 0$ has no solution at all.

Similar translations can be found for Fermat's Last Theorem, the Four Colour Theorem and the Riemann Hypothesis. This may serve as an 'explanation' why Hilbert's 10th problem had to be unsolvable: such difficult and diverse mathematical problems cannot be expected to be solvable by just one universal process.

5 Defining $\mathbb{Z}$ in $\mathbb{Q}$

Hilbert's 10th problem over $\mathbb{Q}$, i.e., the question whether $\mathtt{Th}_{\exists}(\mathbb{Q})$ is decidable, is still open.

If one had an *existential* (= *diophantine*) definition of $\mathbb{Z}$ in $\mathbb{Q}$ (i.e., a definition by an existential 1st-order $\mathcal{L}_{ring}$-formula) then $\mathtt{Th}_{\exists}(\mathbb{Z})$ would be interpretable in $\mathtt{Th}_{\exists}(\mathbb{Q})$, and the answer would, by (for short) Matiyasevich's Theorem, again be no. But it is still open whether $\mathbb{Z}$ is existentially definable in $\mathbb{Q}$.

We have seen the earliest 1st-order definition of $\mathbb{Z}$ in $\mathbb{Q}$, due to Julia Robinson [61], in Sect. 3.3. It can be expressed by an $\forall\exists\forall$-formula of the shape

$$\phi(t) : \ \forall x_1 \forall x_2 \exists y_1 \ldots \exists y_7 \forall z_1 \ldots \forall z_6 \ f(t; x_1, x_2; y_1, \ldots, y_7; z_1, \ldots, z_6) = 0$$

for some $f \in \mathbb{Z}[t; x_1, x_2; y_1, \ldots, y_7; z_1, \ldots, z_6]$, i.e., for any $t \in \mathbb{Q}$,

$$t \in \mathbb{Z} \text{ iff } \phi(t) \text{ holds in } \mathbb{Q}.$$

In 2009, Bjorn Poonen [55] managed to find an $\forall\exists$-definition with two universal and seven existential quantifiers (earlier, in [3], an $\forall\exists$-definition with just one universal quantifier was proved modulo an open conjecture on elliptic curves). In this section we present our $\forall$-definition of $\mathbb{Z}$ in $\mathbb{Q}$:

Theorem 5.1 ([30]). *There is a polynomial $g \in \mathbb{Z}[T; X_1, \ldots, X_{418}]$ such that, for all $t \in \mathbb{Q}$,*

$$t \in \mathbb{Z} \textit{ iff } \forall \overline{x} \in \mathbb{Q}^{418}\ g(t; \overline{x}) \neq 0.$$

If one measures logical complexity in terms of the number of changes of quantifiers then this is the simplest definition of $\mathbb{Z}$ in $\mathbb{Q}$, and, in fact, it is the simplest possible:

Exercise 5.2. *Show that there is no quantifier-free definition of $\mathbb{Z}$ in $\mathbb{Q}$.*

Corollary 5.3. *$\mathbb{Q} \setminus \mathbb{Z}$ is diophantine in $\mathbb{Q}$.*

Corollary 5.4. *$Th_{\forall\exists}(\mathbb{Q})$ is undecidable.*

Theorem 5.1 came somewhat unexpected because it does not give what one would like to have, namely an existential definition of $\mathbb{Z}$ in $\mathbb{Q}$. However, if one had the latter the former would follow:

Observation 5.5. *If there is an existential definition of $\mathbb{Z}$ in $\mathbb{Q}$ then there is also a universal one.*

Proof. If $\mathbb{Z}$ is diophantine in $\mathbb{Q}$ then so is

$$\mathbb{Q} \setminus \mathbb{Z} = \{x \in \mathbb{Q} \mid \exists m, n, a, b \in \mathbb{Z} \text{ with } n \neq 0, \pm 1,\ am + bn = 1 \text{ and } m = xn\}$$

□

In fact, we will indicate in Sect. 5.3 why we do not expect there to be an existential definition of $\mathbb{Z}$ in $\mathbb{Q}$.

Using heavier machinery from number theory, Jennifer Park has recently generalised Theorem 5.1 to number fields:

Theorem 5.6 ([45]). *For any number field K, the ring of integers $\mathcal{O}_K$ is universally definable in K.*

5.1 Key Steps in the Proof of Theorem 5.1

Like all previous definitions of $\mathbb{Z}$ in $\mathbb{Q}$, we use the Hasse-Minkowski Local-Global-Principle for quadratic forms (Theorem 3.10). What is new in our approach is the use of the Quadratic Reciprocity Law and, inspired by the model theory of local fields, the transformation of some existential formulas into universal formulas.

5.1.1 *Step 1:* Poonen's Diophantine Definition of Quaternionic Semi-local Rings

The first step essentially copies Poonen's proof [55]. We adopt his terminology:

Definition 5.7. For $a, b \in \mathbb{Q}^\times$, let

- $H_{a,b} := \mathbb{Q}\cdot 1 \oplus \mathbb{Q}\cdot\alpha \oplus \mathbb{Q}\cdot\beta \oplus \mathbb{Q}\cdot\alpha\beta$ be the quaternion algebra over $\mathbb{Q}$ with multiplication defined by $\alpha^2 = a$, $\beta^2 = b$ and $\alpha\beta = -\beta\alpha$,
- $\Delta_{a,b} := \{l \in \mathbb{P}\cup\{\infty\} \mid H_{a,b}\otimes\mathbb{Q}_l \not\cong M_2(\mathbb{Q}_l)\}$ the set of primes (including ∞) where $H_{a,b}$ does not split locally ($\mathbb{Q}_\infty := \mathbb{R}$)—$\Delta_{a,b}$ is always finite, and $\Delta_{a,b} = \emptyset$ iff $a \in N(b)$, i.e., a is in the image of the norm map $\mathbb{Q}(\sqrt{b}) \to \mathbb{Q}$,
- $S_{a,b} := \{2x_1 \in \mathbb{Q} \mid \exists x_2, x_3, x_4 \in \mathbb{Q} : x_1^2 - ax_2^2 - bx_3^2 + abx_4^2 = 1\}$ the set of traces of norm-1 elements of $H_{a,b}$, and
- $T_{a,b} := S_{a,b} + S_{a,b}$—note that $T_{a,b}$ is an existentially defined subset of $\mathbb{Q}$.

Lemma 5.8. *$T_{a,b} = \bigcap_{l\in\Delta_{a,b}} \mathbb{Z}_{(l)}$, where, for $l \in \mathbb{P}$, $\mathbb{Z}_{(l)} = \mathbb{Z}_l \cap \mathbb{Q}$ is $\mathbb{Z}$ localised at l, and $\mathbb{Z}_{(\infty)} := \{x \in \mathbb{Q} \mid -4 \le x \le 4\}$. ($T_{a,b} = \mathbb{Q}$ if $\Delta_{a,b} = \emptyset$.)*

The proof follows essentially that of [55], Lemma 2.5, using Hensel's Lemma, the Hasse bound for the number of rational points on genus-1 curves over finite fields, and the local-global principle for quadratic forms. Poonen then obtains his $\forall\exists$-definition of $\mathbb{Z}$ in $\mathbb{Q}$ from the fact that

$$\mathbb{Z} = \bigcap_{l\in\mathbb{P}} \mathbb{Z}_{(l)} = \bigcap_{a,b>0} T_{a,b}.$$

Note that $\infty \notin \Delta_{a,b}$ iff $a > 0$ or $b > 0$.

5.1.2 *Step 2:* Towards a Uniform Diophantine Definition of All $\mathbb{Z}_{(p)}$'s in $\mathbb{Q}$

We will present a diophantine definition for the local rings $\mathbb{Z}_{(p)} = \mathbb{Z}_p \cap \mathbb{Q}$ (i.e., $\mathbb{Z}$ localized at $p\mathbb{Z}$) depending on the congruence of the prime p modulo 8, and involving p (and if $p \equiv 1 \mod 8$ an auxiliary prime q) as a parameter. However, since in any first-order definition of a subset of $\mathbb{Q}$ we can only quantify over the elements of $\mathbb{Q}$, and not, e.g., over all primes, we will allow arbitrary (non-zero) rationals p and q as parameters in the following definition.

Definition 5.9. For $p, q \in \mathbb{Q}^\times$ let

- $R_p^{[3]} := T_{-p,-p} + T_{2p,-p}$
- $R_p^{[5]} := T_{-2p,-p} + T_{2p,-p}$

- $R_p^{[7]} := T_{-p,-p} + T_{2p,p}$
- $R_{p,q}^{[1]} := T_{2pq,q} + T_{-2pq,q}$

The R's are all existentially defined subrings of $\mathbb{Q}$ containing $\mathbb{Z}$, since for any $a,b,c,d \in \mathbb{Q}^\times$

$$T_{a,b} + T_{c,d} = \bigcap_{l \in \Delta_{a,b} \cap \Delta_{c,d}} \mathbb{Z}_{(l)},$$

and since in each case at least one of a,b,c,d is > 0, so $\infty \notin \Delta_{a,b} \cap \Delta_{c,d}$.

Definition 5.10. (a) $\mathbb{P}^{[k]} := \{l \in \mathbb{P} \mid l \equiv k \mod 8\}$, where $k = 1,3,5$ or 7
(b) For $p \in \mathbb{Q}^\times$, define

- $\mathbb{P}(p) := \{l \in \mathbb{P} \mid v_l(p) \text{ is odd}\}$, where v_l denotes the l-adic valuation on $\mathbb{Q}$
- $\mathbb{P}^{[k]}(p) := \mathbb{P}(p) \cap \mathbb{P}^{[k]}$, where $k = 1,3,5$ or 7
- $p \equiv_2 k \mod 8$ iff $p \in k + 8\mathbb{Z}_{(2)}$, where $k \in \{0,1,2,\ldots,7\}$
- for l a prime, the *generalized Legendre symbol* $\left(\!\left(\frac{p}{l}\right)\!\right) = \pm 1$ to indicate whether or not the l-adic unit $pl^{-v_l(p)}$ is a square modulo l.

Lemma 5.11. *(a)* $\mathbb{Z}_{(2)} = T_{3,3} + T_{2,5}$
(b) For $p \in \mathbb{Q}^\times$ *and* $k = 3,5$ *or* 7, *if* $p \equiv_2 k \mod 8$ *then*

$$R_p^{[k]} = \begin{cases} \bigcap_{l \in \mathbb{P}^{[k]}(p)} \mathbb{Z}_{(l)} & \text{if } \mathbb{P}^{[k]}(p) \neq \emptyset \\ \mathbb{Q} & \text{if } \mathbb{P}^{[k]}(p) = \emptyset \end{cases}$$

In particular, if p *is a prime* ($\equiv k \mod 8$) *then* $\mathbb{Z}_{(p)} = R_p^{[k]}$.
(c) For $p,q \in \mathbb{Q}^\times$ *with* $p \equiv_2 1 \mod 8$ *and* $q \equiv_2 3 \mod 8$,

$$R_{p,q}^{[1]} = \begin{cases} \bigcap_{l \in \mathbb{P}(p,q)} \mathbb{Z}_{(l)} & \text{if } \mathbb{P}(p,q) \neq \emptyset \\ \mathbb{Q} & \text{if } \mathbb{P}(p,q) = \emptyset \end{cases}$$

where

$$l \in \mathbb{P}(p,q) :\Leftrightarrow l \in \begin{cases} \mathbb{P}(p) \setminus \mathbb{P}(q) \text{ with } \left(\!\left(\frac{q}{l}\right)\!\right) = -1, \text{ or} \\ \mathbb{P}(q) \setminus \mathbb{P}(p) \text{ with } \left(\!\left(\frac{2p}{l}\right)\!\right) = \left(\!\left(\frac{-2p}{l}\right)\!\right) = -1, \text{ or} \\ \mathbb{P}(p) \cap \mathbb{P}(q) \text{ with } \left(\!\left(\frac{2pq}{l}\right)\!\right) = \left(\!\left(\frac{-2pq}{l}\right)\!\right) = -1 \end{cases}$$

In particular, if p is a prime $\equiv 1 \mod 8$ *and q is a prime* $\equiv 3 \mod 8$ *with* $\left(\frac{q}{p}\right) = -1$ *then* $\mathbb{Z}_{(p)} = R^{[1]}_{p,q}$.

Corollary 5.12.

$$\mathbb{Z} = \mathbb{Z}_{(2)} \cap \bigcap_{p,q \in \mathbb{Q}^\times} (R^{[3]}_p \cap R^{[5]}_p \cap R^{[7]}_p \cap R^{[1]}_{p,q})$$

The proof of the Lemma uses explicit norm computations for quadratic extensions of $\mathbb{Q}_2$, the Quadratic Reciprocity Law and the following

Observation 5.13. *For $a, b \in \mathbb{Q}^\times$ and for an odd prime l,*

$$l \in \Delta_{a,b} \Leftrightarrow \begin{cases} v_l(a) \text{ is odd, } v_l(b) \text{ is even, and } \left(\left(\frac{b}{l}\right)\right) = -1, \text{ or} \\ v_l(a) \text{ is even, } v_l(b) \text{ is odd, and } \left(\left(\frac{a}{l}\right)\right) = -1, \text{ or} \\ v_l(a) \text{ is odd, } v_l(b) \text{ is odd, and } \left(\left(\frac{-ab}{l}\right)\right) = -1 \end{cases}$$

Corollary 5.14. *The following properties are diophantine properties for any $p \in \mathbb{Q}^\times$:*

- $p \equiv_2 k \mod 8$ *for* $k \in \{0, 1, 2, \ldots, 7\}$
- $\mathbb{P}(p) \subseteq \mathbb{P}^{[1]} \cup \mathbb{P}^{[k]}$ *for* $k = 3, 5$ *or* 7
- $\mathbb{P}(p) \subseteq \mathbb{P}^{[1]}$

5.1.3 *Step 3:* From Existential to Universal

In Step 3, we try to find universal definitions for the R's occurring in Corollary 5.12 imitating the local situation: First one observes that for $R = \mathbb{Z}_{(2)}$, or for $R = R^{[k]}_p$ with $k = 3, 5$ or 7, or for $R = R^{[1]}_{p,q}$, the Jacobson radical $J(R)$ (which is defined as the intersection of all maximal ideals of R) can be defined by an existential formula using Observation 5.13. Now let

$$\tilde{R} := \{x \in \mathbb{Q} \mid \neg\exists y \in J(R) \text{ with } x \cdot y = 1\}.$$

Proposition 5.15. *(a) $\tilde{R}$ is defined by a* universal *formula in $\mathbb{Q}$.*
(b) If $R = \bigcap_{l \in \mathbb{P} \setminus R^\times} \mathbb{Z}_{(l)}$ then $\tilde{R} = \bigcup_{l \in \mathbb{P} \setminus R^\times} \mathbb{Z}_{(l)}$, provided $\mathbb{P} \setminus R^\times \neq \emptyset$, i.e., provided $R \neq \mathbb{Q}$.
(c) In particular, if $R = \mathbb{Z}_{(l)}$ then $\tilde{R} = R$.

The proviso in (b), however, can be guaranteed by diophantinely definable conditions on the parameters p, q:

Lemma 5.16. *(a) Define for* $k = 1, 3, 5$ *and* 7,

$$\begin{aligned}\Phi_k &:= \left\{p \in \mathbb{Q}^\times \,\middle|\, p \equiv_2 k \mod 8 \text{ and } \mathbb{P}(p) \subseteq \mathbb{P}^{[1]} \cup \mathbb{P}^{[k]}\right\}\\ \Psi &:= \left\{(p,q) \in \Phi_1 \times \Phi_3 \,\middle|\, p \in 2\cdot(\mathbb{Q}^\times)^2\cdot(1 + J(R_q^{[3]}))\right\}.\end{aligned}$$

Then Φ_k *and* Ψ *are diophantine in* $\mathbb{Q}$.

(b) Assume that

- $R = R_p^{[k]}$ *for* $k = 3, 5$ *or* 7, *where* $p \in \Phi_k$, *or*
- $R = R_{p,q}^{[1]}$ *where* $(p,q) \in \Psi$.

Then $R \neq \mathbb{Q}$.

The proof of this lemma is somewhat involved, though purely combinatorial, playing with the Quadratic Reciprocity Law and Observation 5.13.

The universal definition of $\mathbb{Z}$ in $\mathbb{Q}$ can now be read off the equation

$$\mathbb{Z} = \widetilde{\mathbb{Z}_{(2)}} \cap \Big(\bigcap_{k=3,5,7}\ \bigcap_{p\in\Phi_k} \widetilde{R_p^{[k]}}\Big) \cap \bigcap_{(p,q)\in\Psi} \widetilde{R_{p,q}^{[1]}},$$

where Φ_k and Ψ are the diophantine sets defined in Lemma 5.16.

The equation is valid by Lemma 5.11, Proposition 5.15 (b), (c) and Lemma 5.16 (b). The definition is universal as one can see by spelling out the equation and applying Lemma 5.15 (a) and Corollary 5.14: for any $t \in \mathbb{Q}$,

$$\begin{aligned} t \in \mathbb{Z} \Leftrightarrow t \in \widetilde{\mathbb{Z}_{(2)}} \wedge &\\ &\forall p \textstyle\bigwedge_{k=3,5,7}(t \in \widetilde{R_p^{[k]}} \vee p \notin \Phi_k) \wedge\\ &\forall p, q(t \in \widetilde{R_{p,q}^{[1]}} \vee (p,q) \notin \Psi)\end{aligned}$$

Theorem 5.1 is now obtained by diophantine routine arguments and counting quantifiers.

5.2 More Diophantine Predicates in $\mathbb{Q}$

From the results and techniques of Sect. 5.1, one obtains new diophantine predicates in $\mathbb{Q}$. Among them are

- $x \notin \mathbb{Q}^2$
- $x \notin N(y)$, where $N(y)$ is the image of the norm $\mathbb{Q}(\sqrt{y}) \to \mathbb{Q}$

The first was also obtained in [56], using a deep result of Colliot-Thélène et al. on Châtelet surfaces—our techniques are purely elementary.

5.3 Why $\mathbb{Z}$ Should Not be Diophantine in $\mathbb{Q}$

There are two conjectures in arithmetic geometry that imply that $\mathbb{Z}$ is not diophantine in $\mathbb{Q}$, Mazur's Conjecture and, what one may call the Bombieri-Lang Conjecture.

Mazur's Conjecture ([41]). *For any affine variety V over $\mathbb{Q}$ the (real) topological closure of $V(\mathbb{Q})$ in $V(\mathbb{R})$ has only a finite number of connected components.*

It is clear that, under this conjecture, $\mathbb{Z}$ cannot be diophantine in $\mathbb{Q}$, as the latter would mean that $\mathbb{Z}$ is the projection of $V(\mathbb{Q})$ for some affine (not necessarily irreducible) variety V over $\mathbb{Q}$, but then, passing to the topological closure in $\mathbb{R}$, $V(\mathbb{R})$ would have finitely many connected components whereas the projection (which is still the closed subset $\mathbb{Z}$ of $\mathbb{R}$) has infinitely many: contradiction.

The next conjecture, though never explicitly formulated by Lang and Bombieri in this form, may (arguably) be called 'Bombieri-Lang Conjecture' (following [28]). In order to state it we define, given a projective algebraic variety V over $\mathbb{Q}$, the *special set* $\mathtt{Sp}(V)$ to be the Zariski closure of the union of all $\phi(A)$, where $\phi : A \to V$ runs through all non-constant morphisms from abelian varieties A over $\mathbb{Q}$ to V.

Bombieri-Lang Conjecture. *If V is a projective variety over $\mathbb{Q}$ then $V(\mathbb{Q}) \setminus (\mathtt{Sp}(V)(\mathbb{Q}))$ is finite.*

We shall use the following consequence of the conjecture:

Lemma 5.17. *Assume the Bombieri-Lang Conjecture. Let $f \in \mathbb{Q}[x_1, \ldots, x_{n+1}] \setminus \mathbb{Q}[x_1, \ldots, x_n]$ be absolutely irreducible and let $V = V(f) \subseteq \mathbb{A}^{n+1}$ be the affine hypersurface defined by f over $\mathbb{Q}$. Assume that $V(\mathbb{Q})$ is Zariski dense in V. Let $\pi : \mathbb{A}^{n+1} \to \mathbb{A}^1$ be the projection on the 1st coordinate. Then $V(\mathbb{Q}) \cap \pi^{-1}(\mathbb{Q} \setminus \mathbb{Z})$ is also Zariski dense in V.*

The proof uses a highly non-trivial finiteness result on integral points on abelian varieties by Faltings [22].

Theorem 5.18 ([30]). *Assume the Bombieri-Lang Conjecture as stated above. Then there is no infinite subset of $\mathbb{Z}$ existentially definable in $\mathbb{Q}$. In particular, $\mathbb{Z}$ is not diophantine in $\mathbb{Q}$.*

Proof. Suppose $A \subseteq \mathbb{Z}$ is infinite and definable in $\mathbb{Q}$ by an existential formula $\phi_A(x)$ in the language of rings. Replacing, if necessary, A by $-A$, we may assume that $A \cap \mathbb{N}$ is infinite.

Choose a countable proper elementary extension $\mathbb{Q}^\star$ of $\mathbb{Q}$ realizing the type $\{\phi_A(x) \wedge x > a \mid a \in A\}$ and let $A^\star = \{x \in \mathbb{Q}^\star \mid \phi_A(x)\}$. Then $A^\star$ contains some nonstandard natural number $x \in \mathbb{N}^\star \setminus \mathbb{N}$. The map $\begin{cases} \mathbb{N} \to \mathbb{N} \\ n \mapsto 2^n \end{cases}$ is definable in $\mathbb{N}$ and hence in $\mathbb{Q}$, so $2^x \in \mathbb{N}^\star$. As 2^x is greater than any element algebraic over $\mathbb{Q}(x)$, the elements $x, 2^x, 2^{2^x}, \ldots$ are algebraically independent over $\mathbb{Q}$. We therefore find an infinite countable transcendence base $\xi_1, \xi_2, \ldots$ of $\mathbb{Q}^\star$ over $\mathbb{Q}$ with $\xi_1 \in A^\star$.

Let $K = \mathbb{Q}(\xi_1, \xi_2, \ldots)$. As $\mathbb{Q}^\star$ is countable we find $\alpha_i \in \mathbb{Q}^\star$ ($i \in \mathbb{N}$) such that

$$K(\alpha_1) \subseteq K(\alpha_2) \subseteq \cdots \text{ with } \bigcup_{i=1}^{\infty} K(\alpha_i) = \mathbb{Q}^\star,$$

where we may in addition assume that, for each $i \in \mathbb{N}$, the minimal polynomial $f_i \in K[Z]$ of α_i over K has coefficients in $\mathbb{Q}[\xi_1, \ldots, \xi_i]$. As $\mathbb{Q}$ is relatively algebraically closed in $\mathbb{Q}^\star$, all the $f_i \in \mathbb{Q}[X_1, \ldots, X_i, Z]$ are absolutely irreducible over $\mathbb{Q}$.

Now consider the following set of formulas in the free variables $x_1, x_2, \ldots$:

$$\begin{aligned} p = p(x_1, x_2, \ldots) := &\{g(x_1, \ldots, x_i) \neq 0 \mid i \in \mathbb{N}, g \in \mathbb{Q}[x_1, \ldots, x_i] \setminus \{0\}\} \\ &\cup \{\exists z\, f_i(x_1, \ldots, x_i, z) = 0 \mid i \in \mathbb{N}\} \\ &\cup \{x_1 \text{ is not an integer}\} \end{aligned}$$

Then p is finitely realizable in $\mathbb{Q}$: Let $p_0 \subseteq p$ be finite and let j be the highest index occurring in p_0 among the formulas from line 2. Since the $K(\alpha_j)$ are linearly ordered by inclusion all formulas from line 2 with index $< j$ follow from the one with index j. Hence one only has to check that $V(f_j)$ has $\mathbb{Q}$-Zariski dense many $\mathbb{Q}$-rational points $(x_1, \ldots, x_j, z) \in \mathbb{A}^{j+1}$ with $x_1 \notin \mathbb{Z}$. But this is, assuming the Bombieri-Lang Conjecture, exactly the conclusion of the above Lemma. Note that $V(f_j)(\mathbb{Q})$ is $\mathbb{Q}$-Zariski dense in $V(f_j)$ because there is a point $(\xi_1, \ldots, \xi_j, \alpha_j) \in V(f_j)(\mathbb{Q}^\star)$ with $\xi_1, \ldots, \xi_i$ algebraically independent over $\mathbb{Q}$.

Hence p is a type that we can realize in some elementary extension $\mathbb{Q}^{\star\star}$ of $\mathbb{Q}$. Calling the realizing ω-tuple in $\mathbb{Q}^{\star\star}$ again $\xi_1, \xi_2, \ldots$ our construction yields that we may view $\mathbb{Q}^\star$ as a subfield of $\mathbb{Q}^{\star\star}$.

But now $\xi_1 \in A^\star \subseteq \mathbb{Z}^\star$ and $\xi_1 \notin \mathbb{Z}^{\star\star}$, hence $\xi_1 \notin A^{\star\star}$. This implies that there is after all no existential definition for A in $\mathbb{Q}$. □

Let us conclude with a collection of closure properties for pairs of models of Th($\mathbb{Q}$) (in the ring language), one a substructure of the other, which might have a bearing on the final (unconditional) answer to the question whether or not $\mathbb{Z}$ is diophantine in $\mathbb{Q}$.

Proposition 5.19. *Let $\mathbb{Q}^\star, \mathbb{Q}^{\star\star}$ be models of Th($\mathbb{Q}$) (i.e., elementary extensions of $\mathbb{Q}$) with $\mathbb{Q}^\star \subseteq \mathbb{Q}^{\star\star}$, and let $\mathbb{Z}^\star$ and $\mathbb{Z}^{\star\star}$ be their rings of integers. Then*

(a) $\mathbb{Z}^{\star\star} \cap \mathbb{Q}^\star \subseteq \mathbb{Z}^\star$.
(b) $\mathbb{Z}^{\star\star} \cap \mathbb{Q}^\star$ *is integrally closed in* $\mathbb{Q}^\star$.
(c) $(\mathbb{Q}^{\star\star})^2 \cap \mathbb{Q}^\star = (\mathbb{Q}^\star)^2$, *i.e.* $\mathbb{Q}^\star$ *is quadratically closed in* $\mathbb{Q}^{\star\star}$.

(d) If $\mathbb{Z}$ is diophantine in $\mathbb{Q}$ then $\mathbb{Z}^{\star\star} \cap \mathbb{Q}^{\star} = \mathbb{Z}^{\star}$ and $\mathbb{Q}^{\star}$ is algebraically closed in $\mathbb{Q}^{\star\star}$.

(e) $\mathbb{Q}$ is not model complete, i.e., there are $\mathbb{Q}^{\star}$ and $\mathbb{Q}^{\star\star}$ such that $\mathbb{Q}^{\star}$ is not existentially closed in $\mathbb{Q}^{\star\star}$.

Proof. (a) is an immediate consequence of our universal definition of $\mathbb{Z}$ in $\mathbb{Q}$. The very same definition holds for $\mathbb{Z}^{\star}$ in $\mathbb{Q}^{\star}$ and for $\mathbb{Z}^{\star\star}$ in $\mathbb{Q}^{\star\star}$ (it is part of Th($\mathbb{Q}$) that all definitions of $\mathbb{Z}$ in $\mathbb{Q}$ are equivalent). So if this universal formula holds for $x \in \mathbb{Z}^{\star\star} \cap \mathbb{Q}^{\star}$ in $\mathbb{Q}^{\star\star}$ it also holds in $\mathbb{Q}^{\star}$, i.e., $x \in \mathbb{Z}^{\star}$.

(b) is true because $\mathbb{Z}^{\star\star}$ is integrally closed in $\mathbb{Q}^{\star\star}$.

(c) follows from the fact that both being a square and, by Sect. 5.2, not being a square are diophantine in $\mathbb{Q}$.

(d) If $\mathbb{Z}$ is diophantine in $\mathbb{Q}$ then $\mathbb{Z}^{\star\star} \cap \mathbb{Q}^{\star} \supseteq \mathbb{Z}^{\star}$ and hence equality holds, by (a).

To show that then also $\mathbb{Q}^{\star}$ is algebraically closed in $\mathbb{Q}^{\star\star}$, let us observe that, for each $n \in \mathbb{N}$,

$$A_n := \{(a_0, \ldots, a_{n-1}) \in \mathbb{Z}^n \mid \exists x \in \mathbb{Z} \text{ with } x^n + a_{n-1}x^{n-1} + \ldots + a_0 = 0\}$$

is decidable: zeros of polynomials in one variable are bounded in terms of their coefficients, so one only has to check finitely many $x \in \mathbb{Z}$. In particular, by (for short) Matiyasevich's Theorem, there is an $\exists$-formula $\phi(t_0, \ldots, t_{n-1})$ such that

$$\mathbb{Z} \models \forall t_0 \ldots t_{n-1} \left(\{\forall x[x^n + t_{n-1}x^{n-1} + \ldots + t_0 \neq 0]\} \leftrightarrow \phi(t_0, \ldots, t_{n-1})\right).$$

Since both A_n and its complement in $\mathbb{Z}^n$ are diophantine in $\mathbb{Z}$, the same holds in $\mathbb{Q}$, by our assumption of $\mathbb{Z}$ being diophantine in $\mathbb{Q}$, i.e., $A_n^{\star\star} \cap (\mathbb{Q}^{\star})^n = A_n^{\star}$. As any finite extension of $\mathbb{Q}^{\star}$ is generated by an integral primitive element this implies that $\mathbb{Q}^{\star}$ is relatively algebraically closed in $\mathbb{Q}^{\star\star}$.

(e) Choose a recursively enumerable subset $A \subseteq \mathbb{Z}$ which is not decidable. Then $B := \mathbb{Z} \setminus A$ is definable in $\mathbb{Z}$, and hence in $\mathbb{Q}$. If B were diophantine in $\mathbb{Q}$ it would be recursively enumerable. But then A would be decidable: contradiction.

So not every definable subset of $\mathbb{Q}$ is diophantine in $\mathbb{Q}$, and hence $\mathbb{Q}$ is not model complete. Or, in other words, there are models $\mathbb{Q}^{\star}, \mathbb{Q}^{\star\star}$ of Th($\mathbb{Q}$) with $\mathbb{Q}^{\star} \subseteq \mathbb{Q}^{\star\star}$ where $\mathbb{Q}^{\star}$ is not existentially closed in $\mathbb{Q}^{\star\star}$. □

We are confident that with similar methods as used in this paper one can show for an arbitrary prime p that the unary predicate '$x \notin \mathbb{Q}^p$' is also diophantine. This would imply that, in the setting of the Proposition, $\mathbb{Q}^{\star}$ is always radically closed in $\mathbb{Q}^{\star\star}$. However, we have no bias towards an answer (let alone an answer) to the following (unconditional)

Question 5.20. *For $\mathbb{Q}^{\star} \equiv \mathbb{Q}^{\star\star} \equiv \mathbb{Q}$ with $\mathbb{Q}^{\star} \subseteq \mathbb{Q}^{\star\star}$, is $\mathbb{Q}^{\star}$ always algebraically closed in $\mathbb{Q}^{\star\star}$?*

6 Decidability and Hilbert's 10th Problem Over Other Rings

In this section we only report on major achievements under this heading and on a small choice of big open problems. There is a multitude of surveys on the subject, each with its own emphasis. For the interested reader, let us mention at least some of them: [17, 40, 49, 51, 53, 54, 65, 71, 72].

6.1 Number Rings

For number rings and number fields, the question of decidability has been answered in the negative by Julia Robinson (Theorem 3.13). The question whether Hilbert's 10th Problem is solvable is much harder. Given that we don't know the answer over $\mathbb{Q}$ (though almost everyone working in the field believes it to be no) there is even less hope that we find the answer for arbitrary number fields in the near future. For number *rings* the situation is much better.

Let K be a number field with ring of integers $\mathcal{O}_K$. Then Hilbert's 10th Problem could be shown to be unsolvable over $\mathcal{O}_K$ in the following cases:

- if K is totally real (i.e., $K \subseteq T$) or a quadratic extension of a totally real number field [11, 13, 15]
- if $[K : \mathbb{Q}] \geq 3$ and $c_K = 2$ [47].[5]
- if $K/\mathbb{Q}$ is abelian [73].

In each of the proofs the authors managed to find an existential definition of $\mathbb{Z}$ in $\mathcal{O}_K$ using Pell-equations, the Hasse-Minkowski Local-Global Principle (which holds in all number fields) and ad hoc methods that are very specific to each of these special cases.

The hope for a uniform proof of the existential undecidability of all number rings only emerged when elliptic curves were brought into the game:

Theorem 6.1 ([52]). *Let K be a number field. Assume*[6] *there is an elliptic curve E over $\mathbb{Q}$ with $\mathtt{rk}(E(\mathbb{Q})) = \mathtt{rk}(E(K)) = 1$. Then $\mathbb{Z}$ is existentially definable in $\mathcal{O}_K$ and so Hilbert's 10th Problem over $\mathcal{O}_K$ is unsolvable.*

In his proof, Poonen uses divisibility relations for denominators of x-coordinates of $n \cdot P$, where $P \in E(K) \setminus E_{tor}(K)$ and $n \cdot P \in E(\mathbb{Q})$ (for a similar approach cf. [4]).

---

[5] c_K denotes the *class number* of K, that is, the size of the ideal class group of K. It measures how far $\mathcal{O}_K$ is from being a PID: $c_K = 1$ iff $\mathcal{O}_K$ is a PID, so $c_K = 2$ is 'the next best'. It is not known whether there are infinitely many number fields with $c_K = 1$.

[6] The set $E(K)$ of K-rational points of E is a finitely generated abelian group isomorphic to the direct product of its torsion subgroup $E_{tor}(K)$ and a free abelian group of rank '$\mathtt{rk}(E(K))$'.

The assumption made in the theorem turns out to hold modulo a generally believed conjecture, the so called *Tate-Shafarevich Conjecture*. For an elliptic curve E over a number field K, it refers to the *Tate-Shafarevich group* (or *Shafarevich-Tate group*) $Ш_{E/K}$, an abelian group defined via cohomology groups. It measures the deviation from a local-global principle for rational points on E.

Tate-Shafarevich Conjecture. $Ш_{E/K}$ *is finite.*

Weak Tate-Shafarevich Conjecture. $\dim_{\mathbb{F}_2} Ш_{E/K}/2$ *is even*

The latter follows from the former due to the Cassels pairing (Theorem 4.14 in [74] which is an excellent reference on elliptic curves).

Theorem 6.2 ([42]). *Let K be a number field. Assume the weak Tate-Shafarevich Conjecture for all elliptic curves E/K. Then there is an elliptic curve $E/\mathbb{Q}$ with* $\mathtt{rk}(E(\mathbb{Q})) = \mathtt{rk}(E(K)) = 1$.

Taking those two theorems together one obtains immediately the following

Corollary 6.3. *Let K be a number field. Assume the weak Tate-Shafarevich Conjecture for all elliptic curves E/K. Then Hilbert's 10th Problem is unsolvable over $\mathcal{O}_K$.*

6.2 Function Fields

It is natural to ask decidability questions not only over number fields, but also over global fields of positive characteristic, i.e., algebraic function fields in one variable over finite fields, and also, more generally, for function fields.

Hilbert's 10th Problem (with t resp. t_1, t_2 in the language) has been shown to be unsolvable for the following function fields:

- $\mathbb{R}(t)$ [12]
- $\mathbb{C}(t_1, t_2)$ [31]
- $\mathbb{F}_q(t)$ [48,78]
- finite extensions of $\mathbb{F}_q(t)$ [19,70]

The first two cases were achieved by existentially defining $\mathbb{Z}$ in the field, and then applying Matiyasevich's Theorem. This is, clearly, not possible in the last two cases. Instead of existentially *defining* $\mathbb{Z}$ the authors existentially *interpret* $\mathbb{Z}$ via elliptic curves: the multiplication by n-map on an elliptic curve E/K where $E(K)$ contains non-torsion points easily gives a diophantine interpretation of the additive group $\langle \mathbb{Z}; + \rangle$. The difficulty is to find an elliptic curve E/K such that there is also an existential definition for multiplication on that additive group.

For the ring of *polynomials* $\mathbb{F}_q[t]$, Demeyer has even shown the analogue of the DPRM-Theorem: listible subsets are diophantine [10].

Generalizing earlier results [2, 18, 50], it is shown in [20], that the *full* first-order theory of any function field of characteristic > 2 is undecidable.

For analogues of Hilbert's 10th Problem for fields of meromorphic or analytic functions cf., e.g., [46, 66, 77].

6.3 Open Problems

Hilbert's 10th Problem is open for

- $\mathbb{Q}$ and all number fields
- the ring of totally real integers $\mathcal{O}_T$ (cf. Sects. 3.4 and 3.5)

Hilbert's 10th Problem and full 1st-order decidability are open for

- $\mathbb{C}(t)$: this may well be considered the most annoying piece of our ignorance in the area. On the other hand, $\mathbb{C}[t]$ and, in fact, $R[t]$ for any integral domain R is known to be existentially undecidable in $\mathcal{L}_{ring} \cup \{t\}$ [12, 14].
- $\mathbb{F}_p((t))$ and $\mathbb{F}_p[[t]]$—in this case the answer to either question will be the same for the field and the ring: in his recent thesis [1], Will Anscombe found a parameter-free existential definition of $\mathbb{F}_p[[t]]$ in $\mathbb{F}_p((t))$. In [16], Jan Denef and Hans Schoutens show that Hilbert's 10th Problem *is* solvable, if one assumes resolution of singularities in characteristic p.
- the field Ω of constructible numbers (= the maximal pro-2 Galois extension of $\mathbb{Q}$). What is known here is that $\mathcal{O}_\Omega$ is definable in Ω [79], and, more generally [80], that for any prime p and any pro-p Galois extension F of a number field, $\mathcal{O}_F$ is definable in F. As a consequence, the field of real numbers constructible with ruler and scale, i.e., the maximal totally real Galois subextension $\Omega \cap T$ of Ω is undecidable.
- the maximal abelian extension $\mathbb{Q}^{ab}$ of $\mathbb{Q}$ and its ring of integers $\mathbb{Z}^{ab}$—recall that, by the Kronecker-Weber Theorem, $\mathbb{Q}^{ab}$ is the maximal cyclotomic extension of $\mathbb{Q}$, obtained from $\mathbb{Q}$ by adjoining all roots of unity. Here the answer may be related to the famous Shafarevich Conjecture that the absolute Galois group of $\mathbb{Q}^{ab}$ is a free profinite group (if this is true decidability becomes more likely, cf. the remarks following Theorem 3.18).

  Let us remark that the ring of integers $\mathbb{Z}^{rab}$ of the field $\mathbb{Q}^{rab} := \mathbb{Q}^{ab} \cap \mathbb{R}$ of real abelian algebraic numbers is undecidable, by the identical proof as Theorem 3.14. Note that $\mathbb{Q}^{ab} = \mathbb{Q}^{rab}(i)$ and that $\mathbb{Q}^{rab}$ is the fixed field of $\mathbb{Q}^{ab}$ under complex conjugation. We do not know whether $\mathbb{Q}^{rab}$ is definable in $\mathbb{Q}^{ab}$, but we conjecture that $\mathbb{Z}^{rab}$ is definable in $\mathbb{Z}^{ab}$ which would result in undecidability of $\mathbb{Z}^{ab}$.
- the maximal solvable extension $\mathbb{Q}^{solv}$ of $\mathbb{Q}$ and its ring of integers $\mathbb{Z}^{solv}$—here the answer may be related to the longstanding open question whether $\mathbb{Q}^{solv}$ is pseudo-algebraically-closed (PAC) (Problem 10.16 in [24], or Problem 11.5.9 (a) in the 3rd edition; a field K is PAC if every absolutely irreducible algebraic variety defined over K has a K-rational point): if the answer to this question is yes and if Shafarevich's Conjecture holds then $\mathbb{Q}^{solv}$ *is* decidable, axiomatized by

being PAC, by the algebraic part and 'by its absolute Galois group', the minimal normal subgroup of a free group with prosolvable quotient.

All known 'number theoretic' examples either have both the full theory and the existential theory decidable or both undecidable. We have no answer to the following

Question 6.4. *Is there a 'naturally occurring' subring R of $\tilde{\mathbb{Q}}$ with* $\mathrm{Th}_\exists(R)$ *decidable, but* $\mathrm{Th}(R)$ *not?*

A positive answer to an analogue of this question is Lipshitz's result that addition and divisibility over $\mathbb{Z}$ is $\exists$-decidable, but $\forall\exists$-undecidable [34]. On the other hand, if one doesn't restrict to subrings of $\tilde{\mathbb{Q}}$, a natural example would be $\mathbb{R}(t)$: it is easy to check that $\mathrm{Th}_\exists(\mathbb{R}(t)) = \mathrm{Th}_\exists(\mathbb{R})$ (by specialising t adequately), so $\mathrm{Th}_\exists(\mathbb{R}(t))$ is decidable, whereas the full theory of $\mathbb{R}(t)$ is not. [In the light of Denef's negative solution of Hilbert's 10th Problem for $\mathbb{R}(t)$ with t in the language, this nicely illustrates the subtle difference between **H10**$/R$ and **H10**$^+/R$, cf. Sect. 4.1.]

Let us conclude these notes by mentioning decidability questions for fields that come up in other parts of this volume:

- Let k be a field of characteristic 0 and let Γ be an ordered abelian group. Then, by the Ax-Kochen/Ershov principle (see [76]),

$$k((\Gamma)) \text{ is decidable} \quad \Leftrightarrow \quad k \text{ and } \Gamma \text{ are decidable}$$
$$k((\Gamma)) \text{ is } \exists\text{-decidable} \quad \Leftrightarrow \quad k \text{ and } \Gamma \text{ are } \exists\text{-decidable}$$

- Let $\mathbb{R}_{exp}$ be the real exponential field and let SC+ be the 'souped up' version of Schanuel's Conjecture as in [36]. Then

$$\mathbb{R}_{exp} \text{ is decidable} \quad \Leftrightarrow \quad \text{SC+ holds} \quad \Leftrightarrow \quad \mathbb{R}_{exp} \text{ is } \exists\text{-decidable}$$

- The complex exponential field $\mathbb{C}_{exp}$ is, clearly, undecidable, as $\mathbb{Z}$ is definable via the kernel of exponentiation (this was already known to Tarski). In fact, $\mathbb{Z}$ is even existentially definable in $\mathbb{C}_{exp}$: In the 1980s, Angus Macintyre observed that, for any $x \in \mathbb{C}$,

$$x \in \mathbb{Q} \Leftrightarrow \exists t, v, u\, [(v-u)t = 1 \wedge e^v = e^u = 1 \wedge vx = u],$$

and in 2002 ([33]), Miklós Laczkovich found, that for any $x \in \mathbb{C}$,

$$x \in \mathbb{Z} \Leftrightarrow x \in \mathbb{Q} \wedge \exists z\, (e^z = 2 \wedge e^{zx} \in \mathbb{Q}).$$

Hence, $\mathbb{C}_{exp}$ is even existentially undecidable (cf. Sect. 2.2 in [29]).

References

1. W. Anscombe, Definability in Henselian fields. Ph.D. thesis, Oxford, 2012
2. G.L. Cherlin, Undecidability of rational function fields in nonzero characteristic. Stud. Log. Found. Math. **112**, 85–95 (1984)
3. G. Cornelissen, K. Zahidi, Elliptic divisibility sequences and undecidable problems about rational points. J. Reine Angew. Math. **613**, 1–33 (2007)
4. G. Cornelissen, T. Pheidas, K. Zahidi, Division-ample sets and the Diophantine problem for rings of integers. J. Theor. Nombres Bord. **17**, 727–735 (2005)
5. G. Csicsery, Julia Robinson and Hilbert's Tenth Problem. Documentary film by Zala Films (2010)
6. L. Darnière, Decidability and local-global principles, in *Hilbert's Tenth Problem: Relations with Arithmetic and Algebraic Geometry*. Contemporary Mathematics, vol. 270 (American Mathematical Society, Providence, 2000), pp. 139–167
7. M. Davis, Arithmetical problems and recursively enumerable predicates. J. Symb. Log. **18(1)**, 33–41 (1953)
8. M. Davis, Hilbert's tenth problem is unsolvable. Am. Math. Monthly **80**(3), 233–269 (1973)
9. M. Davis, H. Putnam, J. Robinson, The decision problem for exponential Diophantine equations. Ann. Math. (2) **74**, 425–436 (1961)
10. J. Demeyer, Recursively enumerable sets of polynomials over a finite field are Diophantine. Invent. Math. **170(3)**, 655–670 (2007)
11. J. Denef, Hilbert's 10th Problem for quadratic rings. Proc. AMS **48**, 214–220 (1975)
12. J. Denef, The diophantine problem for polynomial rings and fields of rational functions. Trans. AMS **242**, 391–399 (1978)
13. J. Denef, Diophantine sets over algebraic integer rings II. Trans. AMS **257**, 227–236 (1980)
14. J. Denef, The diophantine problem for polynomial rings of positive characteristic, in *Logic Colloquium 78* (North Holland, Amsterdam, 1984), pp. 131–145
15. J. Denef, L. Lipshitz, Diophantine sets over some rings of algebraic integers. J. Lond. Math. Soc. **18**, 385–391 (1978)
16. J. Denef, H. Schoutens, On the decidability of the existential theory of $\mathbb{F}_p[[T]]$, in *Valuation Theory and Its Applications 2*. Fields Institute Communications, vol. 33, ed. by F.-V. Kuhlmann et al. (2003), pp. 43–60, AMS
17. J. Denef, L. Lipshitz, T. Pheidas, J. Van Geel, *Hilbert's Tenth Problem: Relations with Arithmetic and Algebraic Geometry*. Contemporary Mathematics, vol. 270 (American Mathematical Society, Providence, 2000)
18. J.-L. Duret, Sur la théorie élémentaire des corps de fonctions. J. Symb. Log. **51**(4), 948–956
19. K. Eisenträger, Hilbert's 10th Problem for algebraic function fields of characteristic 2. Pac. J. Math. **210**(2), 261–281 (2003)
20. K. Eisenträger, A. Shlapentokh, Undecidability in function fields of positive characteristic. Int. Math. Res. Not. **2009**, 4051–4086 (2009)
21. A.J. Engler, A. Prestel, *Valued Fields* (Springer, Berlin, 2005)
22. G. Faltings, Diophantine approximation on abelian varieties. Ann. Math. **133**, 549–576 (1991)
23. M. Fried, D. Haran, H. Völklein, Real hilbertianity and the field of totally real numbers. Contemp. Math. **74**, 1–34 (1994)
24. M. Fried, M. Jarden, *Field Arithmetic* (Springer, Berlin, 1986) [3rd edn., 2008]
25. K. Gödel, Über formal unentscheidbare Sätze der Principia Mathematica und verwandter Systeme. Monatsh. Math. Phys. **38**, 173–198 (1931)
26. B. Green, F. Pop, P. Roquette, On Rumely's local-global principle. Jber. Dt. Math.-Verein. **97**, 43–74 (1995)
27. D. Hilbert, Mathematische Probleme. Vortrag, gehalten auf dem internationalen Mathematiker Kongress zu Paris 1900. Nachr. K. Ges. Wiss. Göttingen, Math.-Phys. Kl. **3**, 253–297 (1900)
28. M. Hindry, J.H. Silverman, *Diophantine Geometry*. Graduate Texts in Mathematics 201 (Springer, Berlin, 2000)

29. J. Kirby, A. Macintyre, A. Onshuus, The algebraic numbers definable in various exponential fields. J. Inst. Math. Jussieu **11**(04), 825–834 (2012)
30. J. Koenigsmann, Defining $\mathbb{Z}$ in $\mathbb{Q}$. Ann. Math. arXiv:1011.3424v1 (2010, to appear)
31. K.H. Kim, F.W. Roush, Diophantine undecidability of $\mathbb{C}(t_1, t_2)$. J. Algebra **150**(1), 35–44
32. L. Kronecker, Zwei Sätze über Gleichungen mit ganzzahligen Coefficienten. J. Reine Angew. Math. **53**, 173–175 (1857)
33. M. Laczkovich, The removal of π from some undecidable problems involving elementary functions. Proc. AMS **131(7)**, 2235–2240 (2002)
34. L. Lipshitz, The diophantine problem for addition and divisibility. Trans. AMS **235**, 271–283 (1978)
35. A. Macintyre, The impact of Gödel's incompleteness theorems on mathematics, in *Kurt Gödel*, ed. by M. Baaz et al. (Cambridge University Press, Cambridge, 2011), pp. 3–25
36. A. Macintyre, A. Wilkie, On the decidability of the real exponential field, in *Kreiseliana*, ed. by P. Odifreddi (A.K. Peters, Wellesley, 1996), pp. 441–467
37. Y.V. Matiyasevich, Diofantovost' perechislimykh mnozhestv. Dokl. AN SSSR **191**(2), 278–282 (1970) (Translated in: Soviet Math. Doklady **11**(2), 354–358, 1970)
38. Y.V. Matiyasevich, Prostye chisla perechislyayutsya polinomom ot 10peremennykh. J. Sov. Math. **15**(19), 33–44 (1981) (translated)
39. Y.V. Matiyasevich, Hilbert's 10th Problem: what was done and what is to be done, in*Hilbert's Tenth Problem: Relations with Arithmetic and Algebraic Geometry*. Contemporary Mathematics, vol. 270 (American Mathematical Society, Providence, 2000), pp. 1–47
40. B. Mazur, Questions of decidability and undecidability in number theory. J. Symb. Log. **59**(2), 353–371 (1994)
41. B. Mazur, Open problems regarding rational points on curves and varieties, in *Galois Representations in Arithmetic Algebraic Geometry*, ed. by A.J. Scholl, R.L. Taylor (Cambridge University Press, Cambridge, 1998), pp. 239–266
42. B. Mazur, K. Rubin Ranks of twists of elliptic curves and Hilbert's tenth problem. Invent. Math. **181(3)**, 541–575 (2010)
43. L. Moret-Bailly, Groupes de Picard et problèmes de Skolem I and II. Ann. Scien. Ec. Norm. Sup. **22**(2), 161–179; 181–194 (1989)
44. O.T. O'Meara, *Introduction to Quadratic Forms* (Springer, Berlin, 1973)
45. J. Park, A universal first order formula defining the ring of integers in a number field (2012). arXiv:1202.6371v1
46. H. Pasten, Powerful values of polynomials and a conjecture of Vojta. J. Number Theory **133**(9), 2843–3206 (2013)
47. T. Pheidas, Hilbert's tenth problem for a class of rings of algebraic integers. Proc. AMS **104**, 611–620 (1988)
48. T. Pheidas, Hilbert's 10th Problem for fields of rational functions over finite fields. Invent. Math. **103**, 1–8 (1991)
49. T. Pheidas, Extensions of Hilbert's 10th Problem. J. Symb. Log. **59**(2), 372–397 (1994)
50. T. Pheidas, Endomorphisms of elliptic curves and undecidability in function fields of positive characteristic. J. Algebra **273**(1), 395–411 (2004)
51. T. Pheidas, K. Zahidi, Undecidability of existential theories of rings and fields: a survey, in *Hilbert's Tenth Problem: Relations with Arithmetic and Algebraic Geometry*. Contemporary Mathematics, vol. 270 (American Mathematical Society, Providence, 2000), pp. 49–105
52. B. Poonen, Using elliptic curves of rank one towards the undecidability of Hilbert's 10th Problem over rings of algebraic integers, in *Algorithmic Number Theory*, ed. by C. Fieker, D.R. Kohel (Springer, Berlin, 2002), pp. 33–42
53. B. Poonen, Hilbert's 10th problem over rings of number-theoretic interest (2003). www-math.mit.edu/ poonen/papers/aws2003.pdf
54. B. Poonen, Undecidability in number theory. Not. AMS **55**(3), 344–350 (2008)
55. B. Poonen, Characterizing integers among rational numbers with a universal-existential formula. Am. J. Math. **131**(3), 675–682 (2009)

56. B. Poonen, The set of nonsquares in a number field is diophantine. Math. Res. Lett. **16**(1), 165–170 (2009)
57. F. Pop, Embedding problems over large fields. Ann. Math. (2) **144**(1), 1–34 (1996)
58. A. Prestel, Pseudo real closed fields, in *Set Theory and Model Theory*, ed. by R.B. Jensen, A. Prestel. Springer Lecture Notes, vol. 872 (1981), pp. 127–156, Springer, Berlin
59. A. Prestel, P. Roquette, *Formally p-Adic Fields*. Springer Lecture Notes, vol. 1050 (1984), Springer, Berlin
60. A. Prestel, J. Schmid, Existentially closed domains with radical relations. J. Reine Angew. Math. **407**, 178–201 (1990)
61. J. Robinson, Definability and decision problems in arithmetic. J. Symb. Log. **14**(2), 98–114 (1949)
62. J. Robinson, Existential definability in arithmetic. Trans. AMS **72**, 437–449 (1952)
63. J. Robinson, The undecidability of algebraic rings and fields. Proc. AMS **10**, 950–957 (1959)
64. J. Robinson, On the decision problem for algebraic rings, in *Studies in Mathematical Analysis and Related Topics*, ed. by G. Szegö (Stanford University Press, Stanford, 1962), pp. 297–304
65. R.M. Robinson, Undecidable rings. Trans. AMS **70**(1), 137–159 (1951)
66. L.A. Rubel, An essay on diophantine equations for analytic functions. Exp. Math **13**, 81–92 (1995)
67. R. Rumely, Arithmetic over the ring of all algebraic integers. J. Reine Angew. Math. **368**(5), 127–133 (1986)
68. E.S. Selmer, The diophantine equation $ax^3 + by^3 + cz^3 = 0$. Acta Math. **85**, 203–362 (1951) and **92**, 191–197 (1954)
69. J.-P. Serre, *A Course in Arithmetic* (Springer, Berlin, 1973)
70. A. Shlapentokh, Hilbert's 10th Problem for rings of algebraic functions in one variable over fields of constants of positive characteristic. Trans. AMS **333**, 275–298 (1992)
71. A. Shlapentokh, Hilbert's 10th problem over number fields, a survey, in *Hilbert's Tenth Problem: Relations with Arithmetic and Algebraic Geometry*. Contemporary Mathematics, vol. 270 (American Mathematical Society, Providence, 2000), pp. 107–137
72. A. Shlapentokh, *Hilbert's Tenth Problem, Diophantine Classes and Extensions to Global Fields* (Cambridge University Press, Cambridge, 2007)
73. A. Shlapentokh, H.N. Shapiro, Diophantine relationships between algebraic number fields. Comm. Pure Appl. Math. **42**, 1113–1122 (1989)
74. J.H. Silverman, *The Arithmetic of Elliptic Curves* (Springer, Berlin, 1986)
75. L. van den Dries, Elimination theory for the ring of algebraic integers. J. Reine Angew. Math. **388**, 189–205 (1988)
76. L. van den Dries, *Lectures on the Model Theory of Valued Fields* (Springer, Heidelberg, 2014) (this volume)
77. X. Vidaux, An analogue of Hilbert's 10th problem for fields of meropmorphic functions over non-Archimedean valued fields. J. Number Theory **101**, 48–73 (2003)
78. C. Videla, Hilbert's 10th Problem for rational function fields in characteristic 2. Proc. AMS **120**(1), 249–253 (1994)
79. C. Videla, On the constructible numbers. Proc. AMS **127**(3), 851–860 (1999)
80. C. Videla, Definability of the ring of integers in pro-p Galois extensions of number fields. Isr. J. Math. **118**(1), 1–14 (2000)

LECTURE NOTES IN MATHEMATICS

Springer

Edited by J.-M. Morel, B. Teissier; P.K. Maini

Editorial Policy (for Multi-Author Publications: Summer Schools / Intensive Courses)

1. Lecture Notes aim to report new developments in all areas of mathematics and their applications - quickly, informally and at a high level. Mathematical texts analysing new developments in modelling and numerical simulation are welcome. Manuscripts should be reasonably selfcontained and rounded off. Thus they may, and often will, present not only results of the author but also related work by other people. They should provide sufficient motivation, examples and applications. There should also be an introduction making the text comprehensible to a wider audience. This clearly distinguishes Lecture Notes from journal articles or technical reports which normally are very concise. Articles intended for a journal but too long to be accepted by most journals, usually do not have this "lecture notes" character.

2. In general SUMMER SCHOOLS and other similar INTENSIVE COURSES are held to present mathematical topics that are close to the frontiers of recent research to an audience at the beginning or intermediate graduate level, who may want to continue with this area of work, for a thesis or later. This makes demands on the didactic aspects of the presentation. Because the subjects of such schools are advanced, there often exists no textbook, and so ideally, the publication resulting from such a school could be a first approximation to such a textbook. Usually several authors are involved in the writing, so it is not always simple to obtain a unified approach to the presentation.
   For prospective publication in LNM, the resulting manuscript should not be just a collection of course notes, each of which has been developed by an individual author with little or no coordination with the others, and with little or no common concept. The subject matter should dictate the structure of the book, and the authorship of each part or chapter should take secondary importance. Of course the choice of authors is crucial to the quality of the material at the school and in the book, and the intention here is not to belittle their impact, but simply to say that the book should be planned to be written by these authors jointly, and not just assembled as a result of what these authors happen to submit.
   This represents considerable preparatory work (as it is imperative to ensure that the authors know these criteria before they invest work on a manuscript), and also considerable editing work afterwards, to get the book into final shape. Still it is the form that holds the most promise of a successful book that will be used by its intended audience, rather than yet another volume of proceedings for the library shelf.

3. Manuscripts should be submitted either online at www.editorialmanager.com/lnm/ to Springer's mathematics editorial, or to one of the series editors. Volume editors are expected to arrange for the refereeing, to the usual scientific standards, of the individual contributions. If the resulting reports can be forwarded to us (series editors or Springer) this is very helpful. If no reports are forwarded or if other questions remain unclear in respect of homogeneity etc, the series editors may wish to consult external referees for an overall evaluation of the volume. A final decision to publish can be made only on the basis of the complete manuscript; however a preliminary decision can be based on a pre-final or incomplete manuscript. The strict minimum amount of material that will be considered should include a detailed outline describing the planned contents of each chapter.
   Volume editors and authors should be aware that incomplete or insufficiently close to final manuscripts almost always result in longer evaluation times. They should also be aware that parallel submission of their manuscript to another publisher while under consideration for LNM will in general lead to immediate rejection.

4. Manuscripts should in general be submitted in English. Final manuscripts should contain at least 100 pages of mathematical text and should always include
   - a general table of contents;
   - an informative introduction, with adequate motivation and perhaps some historical remarks: it should be accessible to a reader not intimately familiar with the topic treated;
   - a global subject index: as a rule this is genuinely helpful for the reader.

   Lecture Notes volumes are, as a rule, printed digitally from the authors' files. We strongly recommend that all contributions in a volume be written in the same LaTeX version, preferably LaTeX2e. To ensure best results, authors are asked to use the LaTeX2e style files available from Springer's web-server at
   ftp://ftp.springer.de/pub/tex/latex/svmonot1/ (for monographs) and
   ftp://ftp.springer.de/pub/tex/latex/svmultt1/ (for summer schools/tutorials).
   Additional technical instructions, if necessary, are available on request from:
   lnm@springer.com.

5. Careful preparation of the manuscripts will help keep production time short besides ensuring satisfactory appearance of the finished book in print and online. After acceptance of the manuscript authors will be asked to prepare the final LaTeX source files and also the corresponding dvi-, pdf- or zipped ps-file. The LaTeX source files are essential for producing the full-text online version of the book. For the existing online volumes of LNM see:
   http://www.springerlink.com/openurl.asp?genre=journal&issn=0075-8434.
   The actual production of a Lecture Notes volume takes approximately 12 weeks.

6. Volume editors receive a total of 50 free copies of their volume to be shared with the authors, but no royalties. They and the authors are entitled to a discount of 33.3 % on the price of Springer books purchased for their personal use, if ordering directly from Springer.

7. Commitment to publish is made by letter of intent rather than by signing a formal contract. Springer-Verlag secures the copyright for each volume. Authors are free to reuse material contained in their LNM volumes in later publications: a brief written (or e-mail) request for formal permission is sufficient.

Addresses:

Professor J.-M. Morel, CMLA,
École Normale Supérieure de Cachan,
61 Avenue du Président Wilson, 94235 Cachan Cedex, France
E-mail: morel@cmla.ens-cachan.fr

Professor B. Teissier, Institut Mathématique de Jussieu,
UMR 7586 du CNRS, Équipe "Géométrie et Dynamique",
175 rue du Chevaleret,
75013 Paris, France
E-mail: teissier@math.jussieu.fr

For the "Mathematical Biosciences Subseries" of LNM:

Professor P. K. Maini, Center for Mathematical Biology,
Mathematical Institute, 24-29 St Giles,
Oxford OX1 3LP, UK
E-mail: maini@maths.ox.ac.uk

Springer, Mathematics Editorial I,
Tiergartenstr. 17,
69121 Heidelberg, Germany,
Tel.: +49 (6221) 4876-8259
Fax: +49 (6221) 4876-8259
E-mail: lnm@springer.com

Printed in the United States
By Bookmasters